T0335468

the Reed Warblers

Diversity in a uniform bird family

See for the appendix: **www.knnvuitgeverij.nl/EN/appendix-the-reedwarblers**

the Reed Warblers

Bernd Leisler & Karl Schulze-Hagen

Diversity
in a **uniform**
bird family

Artwork by **David Quinn**

Max Planck Institute
for Ornithology

KNNV Publishing

KNNV
vereniging
voor veldbiologie

Preface

About 200 years ago, the great naturalist Alexander von Humboldt wanted 'to understand nature through its inner power and lively whole.' At that time, such integrative organismal biology was just a mental framework, wishful – and yet pioneering – thinking. Roughly 150 years later, at a time when such detailed, painstaking, methodical, and often tedious studies were still very rare, two young scientists embarked on a similar life voyage, to understand a group of organisms in its lively whole – 'little brown job' Reed warblers.

The result we have in hand now, summing up 50 years of their own research and that of excellent research groups around the world, is a mind blowing science treasure that will instantly become a classic. Hallmarks of Bernd Leisler's and Karl Schulze-Hagen's approach are their never-ending, deep love for their study subjects, the clarity of their writing and the intrinsic beauty of their pictures and illustrations.

As readers we are fortunate that Bernd and Karl took their time to finish the book because we now have an integrative view of this fascinating group of organisms and their evolution. Science often goes in fashions, favoring a specific topic and interpretation for a decade or so. Four decades ago, the fashion was habitat selection and its influence on behavior and morphology, then genetically-informed systematics, host-parasite interactions and the latest interest, sexual selection. Every few decades, these topics are brought up again and revisited. This has the major advantage that the original information is checked for its explanatory power in the light of all new knowledge, and only the data and hypotheses that weathered time remain.

Such solid knowledge is what is presented in this book, giving rise to many ideas about what to study in the future. The coming decades may see a lot of studies in the areas of movement and migration, perhaps allowing us to understand the interconnectedness of Polish marshes with the wetlands in the Mediterranean and Africa, or between Russia and Southeast Asia. We need this information, not least to better protect the future habitat and migration routes of our beloved warblers. Given the general interconnectedness of life on earth, perhaps knowledge about and protection of the little brown jobs will turn out to be more important to us than we can yet envisage.

Professor Martin Wikelski
Director Max Planck Institute for Ornithology, Radolfzell

Introducing the reed warblers

A reed warbler in its world (Eurasian Reed Warbler).

Reed warblers are remarkably rewarding birds to study.

In the time that has passed since our early work in the late 1960s, we have to confess that we have become rather infatuated with reed warblers and related species – the Acrocephalidae family, which form the subject matter of this book. In all the years of field and aviary work we have come to know their charm. Indeed, we have got to know some individuals really well as they clamber around in the reeds, just a metre or two in front of us, looking us boldly in the eye, or bravely sitting tight on the nest when we wish to check on their brood, scolding us angrily if the young are thought to be in danger. One hand-reared female Aquatic Warbler became almost a personal companion, patiently waiting for her breakfast of wax moths or flies every morning for nine years.

The Acrocephalidae are now recognized as a family in their own right (Alström et al. 2006, Johansson et al. 2008, Fregin et al. 2009) and currently encompass 6 genera and around 50 species in total. These are the reed warblers sensu stricto (Acrocephalus, with several subgenera), the 'tree' and 'scrub' warblers of the genera Hippolais and Iduna, the monotypic genera Phragamaticola (Thick-billed Warblers) and Calamonastides (Papyrus Yellow Warblers), and the Nesillas 'brush' warblers.

They are one of several apparently uniform families in the large group of the Old World warblers. Together with the 'grass' and 'bush' warblers (Locustellidae, Cettidae) they epitomize the 'little brown jobs' which tease birdwatchers all around the world. At first sight, they may seem unspectacular, rather colourless, and difficult to identify, but for these very reasons they have long been valued for the inviting challenge they present to both birders and ringers (Kennerley & Pearson 2010). The 50 or so species that are the subject of this book represent an entity whose members all show great similarity in body shape (Leisler et al. 1997, Ottosson et al. 2005, Bairlein 2006), which makes it all the more fascinating to compare them. Some so resemble each other as to be separable only in the hand by using a variety of scores and indices (chapter 2). Conversely, there is a striking variation in body weight, from the 8 g of Booted Warblers to Tahiti Warblers which are more than five times heavier.

In contrast to others, reed warblers have long cast off their image of dullness to take their place among the most studied bird groups of all. Since they are widely distributed, have developed a great variety of lifestyles, occur in large numbers, and are easy to find and study, they are very attractive for the numerous teams of researchers who have devoted themselves to these remarkable birds. The

number of investigations carried out on this group has grown enormously in recent decades, impressively reflecting just how diverse the habitats and life strategies of an apparently uniform group of birds really are. The wealth of data that has come from a multitude of correlative and experimental field and laboratory studies, as well as from ringing and conservation programmes, makes a comparison of the life strategies of the reed warblers and their allies especially appealing.

The large number and wide spectrum of these studies provide an excellent opportunity to compare the biology and lifestyles of reed warblers, and to review and evaluate in one publication the many stimuli and contributions to ornithological science that have resulted. It might appear surprising, but there are few avian families which lend themselves so well to comparative studies that are not only comprehensive, but also touch on so many different aspects of their biology. Indeed, the ease with which they can be observed and their plentiful occurrence were for us the very reasons why we selected reed warblers in the first place, as they are almost perfectly suited for fieldwork.

Forty-five years ago, it occurred to us that the ideal subjects for field study would perhaps be the Moustached Warblers in the huge reedbeds of the Neusiedler See on the Austrian-Hungarian border, and the Marsh Warblers in the stands of stinging nettles in the rich arable landscapes of the German Rhineland, though for different reasons in each case. In the former, it was the opportunity to clarify the systematic position of Moustached Warblers, since their habits could be compared with those of the four other acrocephalids occurring in the same habitat (Leisler 1970). In the latter, because Marsh Warblers were so common in that location, it would be possible to

Photos 1.1 and 1.2 Reed warblers are among the species characteristic of the extensive reedswamps of the Camargue in southern France. Here Moustached Warblers (right) are frequently seen and heard, usually in the layer of broken-down *Phragmites* stems. On their return from their winter quarters Great Reed Warblers (left) sing continuously from sunrise to sunset. When a partner is found, the song, which is then principally directed at other males, is much reduced in both frequency and length (photos Brendan Doe, Roland Mayr).

determine how far the breeding success of a songbird is dependent on the structure of its habitat (Schulze-Hagen 1975, 1984).

One great advantage in studying reed warblers is the simple structure of their habitats, which are varied enough to require differing adaptations, yet uniform enough to be characterized using the same measuring methods. In such habitats, nests can easily be found and the birds themselves trapped in great numbers; indeed in many ringing projects they are the species most often trapped (e.g. Berthold *et al.* 1991, Bairlein 1998). Furthermore, it is possible to recreate the habitat of some species in a laboratory aviary, and/or to manipulate it in order to carry out experiments. Reed warblers can be hand-raised and

kept in captivity without difficulty. Long-term databases exist that have been running for over 30 years with breeding data for some species (Schaefer *et al.* 2006, Halupka *et al.* 2008) and, more significantly, with phenotypic, genetic, and fitness data for all individuals – still a rare exception in open-nesting songbirds as opposed to cavity-breeders. Such long-term data on natural populations are increasingly being used to run longitudinal comparisons of individuals and their descendants and to test theoretical models (Komdeur 2003a,b, Hansson *et al.* 2004b, Akesson *et al.* 2008, Brouwer *et al.* 2010). Over one-third of acrocephalid species have settled oceanic islands, while others occupy 'habitat islands', allowing comparison of their biology with that of species inhabiting the mainland or with an uninterrupted distribution.

What is more, they have colonized different habitats in almost all climatic zones, thus enabling us to compare resident tropical species with long-distance migrants of the temperate latitudes.

It is therefore easy to see why reed warbler studies have contributed so much and in so many ways, not only to ornithology, but to biology in general. Sometimes it was pioneering work that served to trigger similar studies. We could mention, for example, the comparison by Jilka & Leisler (1974) of three reed warbler species, which was one of the first to show that the structure of birdsong appears to be adapted to the acoustic properties of the environment; the demonstration by Bibby & Green (1981) that Sedge Warblers and Eurasian Reed Warblers reach their similar winter quarters using completely different migration strategies; the first key factor analysis of survival rates of British Sedge Warblers that recognized the significance of conditions in the wintering grounds (Peach *et al.* 1991), and the evidence of active female choice of mate which was gathered when Bensch & Hasselquist (1992) followed radio-tagged free-living Great Reed Warblers in Sweden.

Reed warbler studies have repeatedly set new standards in modern biological disciplines, especially within behavioural and molecular ecology. They have often become textbook examples, particularly in the fields of birdsong and sexual selection, or the ecology and evolution of mating systems (e.g. Krebs & Davies 1993, Gill 2007, Catchpole & Slater 2008, Dugatkin 2009) or citation classics, such as the study on song complexity as a signal of male quality by Hasselquist and colleagues (1996). The molecular work on reed warblers is also at the cutting edge, whether it is studies on avian dispersal (Hansson *et al.* 2003a, b), the colonization history of oceanic islands (Cibois *et al.* 2007, 2008, 2011b) and the inbreeding or immune system of island dwellers (Richardson *et al.* 2004, Beadell *et al.* 2007), or what we can learn about speciation from secondary contact zones (Reullier *et al.* 2006, Secondi *et al.* 2006).

Other studies in which reed warblers have played an important role in answering wider biological questions can be found in review articles, such as those addressing how competition among individuals of dominant and subordinate species contributes to the partitioning of

Fig. 1.1 (Map 1) Map of the world showing the distribution of the members of the Acrocephalidae family. The arrows indicate some of the extreme island populations. The scale shows species density, with its centre in western central Asia. The family does not occur in the New World.

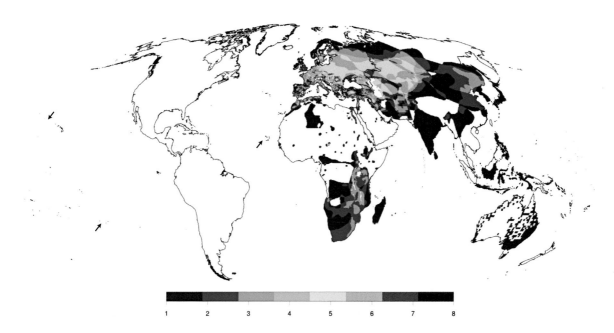

their niches, or how natal experience affects later habitat selection behaviour (Orians 2000, Davis & Stamps 2004). In research into the co-evolution of brood parasites and their hosts, reed warblers have even become almost the field experiment equivalent of 'laboratory white mice' (Davies 2000, Krüger 2007, Stokke *et al.* 2007, 2008). The discovery by Komdeur and co-workers (1997) that female Seychelles Warblers are capable of adaptively adjusting the sex ratio in their clutches, according to environmental conditions, was of far-reaching importance, as was the phenomenon of 'grandmotherly' help in some broods of this species, as revealed by Richardson *et al.* (2007). Some reed warbler studies are classic examples of ecomorphological methodology, examining how the relationship between morphology and ecology is mediated by behaviour (Leisler *et al.* 1989), and demonstrating how climate change can influence the breeding biology of the inhabitants of a variety of habitats in entirely different ways (Both *et al.* 2006). Reed warblers have even managed to make it into the headlines as record-breakers, with their extraordinary abilities in the field of vocal mimicry (Dowsett-Lemaire 1979a), for instance, or for the longest copulations in the avian world (Schulze-Hagen *et al.* 1995).

This very successful family is distributed across approximately 260 degrees of longitude, from the Cape Verde Islands to eastern Polynesia, and is absent only from the New World (fig. 1). It is only because the member species have diversified in their biologies that such a wide range is possible, and we shall be turning our attention to these strategies in the later chapters of this book. For the time being, we would like to invite you to take an imaginary journey around the world, to eight regions, from the Camargue in France to Henderson Island in Polynesia, which will give an impression of just how varied the habitats and lifestyles of this family are, and how great the differences between its species can be.

Photos 1.3 and 1.4 During the spring floods, the wetlands and meadows of the Hortobágy in Hungary form a transition zone of short-grass and long-grass meadows, where Aquatic (below) and Sedge Warblers (top) occur side by side as close neighbours (Hungary; photos Grzegorz & Tomasz Klosowski, Oldrich Mikulica).

Photos 1.5 to 1.7 Two of the inhabitants of the papyrus-dominated reedswamp on Lake Victoria are Greater Swamp Warblers (top) and Papyrus Yellow Warblers (bottom). Their large feet are an adaptation to the papyrus vegetation (photos Nick Borrow). Not too far away, in the montane forests of the 4200 m Mt. Elgon, Mountain Yellow Warblers live in bamboo thickets (photo Hans Winkler).

Photo 1.8 The Seychelles Warblers in the jungle-like forest on Cousin Island build their nests in trees. Both parents participate in raising the single young, often assisted by related helpers (photo Dany Ellinger).

An imaginary journey round the world's reed warblers

Camargue, France

The wetlands of the Camargue are shaped by a mixture of river and sea, freshwater and saltwater. If we struggle through the dense Mediterranean reedbeds we can obtain fascinating glimpses into the lives of three *Acrocephalus* species. The water is not yet very deep, barely 30 cm, and among the 2 m-high reeds are conspicuous clumps of the prickly sea rush. Together with broken reed stems, the stalks of rush create a metre-high tangled and dense undergrowth. From a perch in this matted layer of broken stems a Moustached Warbler sings not far from his nest. Further into the reedbed the water becomes half a metre deep, and in a mixed stand of reed and reedmace we find a second Moustached Warbler nest. Beneath a roof-like structure of old, bent-over *Typha* leaves it is, like the first nest, partly supported from below, partly fixed to the stalks. It is now the end of April, and the three young are already almost fledged. They are cared for by both parents, who bring bundles of invertebrates in their bills. The icy mistral of the last few days has hardly hindered the adults in their search for food, which they pick from the dense undergrowth at the water surface.

In the surrounding deeper water, where higher, more robust reeds form pure stands, some Eurasian and Great Reed Warblers are singing again, having been silenced by the wind and cold. Now there is bustling activity as the Eurasian Reed Warbler males attempt to attract females. One male, who has already found a mate, follows her wherever she goes. Elsewhere we observe how a Great Reed Warbler, fluttering and hopping, with his crown feathers raised, successfully drives a Eurasian Reed Warbler away.

Hortobágy, Hungary

Although the greater part of the Hungarian puszta, the steppe-like plains of eastern Hungary, is today farmland, it has been designated a UNESCO World Heritage Site because of the traditional way of life of its shepherds. In this completely flat landscape, with its endless horizons, it quickly becomes clear that the slightest change in altitude will have far-reaching ecological consequences. For instance, the lowest-lying areas of Hortobágy become inundation sinks during the spring flooding of the River Tisza, and consequently harbour extensive marshlands. Bordering these depressions, wetlands and meadows form a transition to the slightly higher short-grass pastures dotted with the traditional sweep wells. Depending on the salinity and wetness of the soil, an interlocking mosaic of large habitat patches are formed, with certain types of saltmarshes (grey club-rush) and long-grass meadows (slough grass and foxtail). In both saltmarshes and meadows, sea club-rush is common and forms an undergrowth. These damp depressions are favourite grazing grounds of the grey Hungarian Steppe Cattle, as well as the habitat of two striped *Acrocephalus* species. Sedge Warblers share these swampy areas with Aquatic Warblers, though the former are no longer found where the vegetation becomes too short and dry and lacks rigid stems.

In early June, we can only hear a few Sedge Warbler males still singing and launching themselves into their acrobatic song-flights from the tops of high thistles. In most of their territories in these hollows, both males and females are already busy raising their young. By contrast, the song activity of the male Aquatic Warblers continues uninterrupted and they utter their rasping song throughout the day from low vegetation; later they will all participate in an 'evening chorus'.

Since the spring floods were especially high this year, and exceptional heat in May brought a rapid fall in the water level, the shallow depressions are alive with the larvae of great silver water beetles and *Dytiscus* diving beetles, dragonfly larvae, and aquatic spiders and snails. The air is filled with flies (Tabanidae, Tipulidae, Chironomidae, Culicidae), mayflies, and dragonflies, while masses of cicadas and small grasshoppers and crickets swarm in the vegetation.

The Aquatic Warbler nests are found in the slightly raised wet meadows, 10 cm above the water in last year's tussocks of bent grass, usually close to the old stalks of purple loosestrife. We can watch the female from a hide; she feeds her five young, with no help from the male, but seems to have no real difficulty in finding enough items for her brood. When she creeps or runs to the nest with a billful of prey every few minutes, she looks more like a mouse than a bird.

Lake Victoria and Mount Elgon, Uganda

As we look out from a fishing boat drifting in a bay on Lake Victoria, we can recognize that the lush green fringe of the shore vegetation is much more varied than would appear at first sight. Uniform stands of swaying papyrus alternate with beds of eulalia reeds and great fen sedge,

interspersed with stretches of open water, on which float pads of plants, such as waterlilies and rosettes of water lettuce. Elsewhere water-hyacinth creates thick, almost lawn-like mats on the surface, and rafts of vegetation have entangled to form islands which appear to float.

A luxuriously rich birdlife indicates the high productivity of this habitat and presents itself in diverse forms: the cries of an African Fish Eagle fill the air, and fish-eating birds of all sizes abound, from little Malachite Kingfishers to the extraordinary Shoebills, all accompanied by several duck species, those floating vegetation specialists that are the jacanas and rails, and, of course, innumerable songbirds. Hopping from stem to stem, just above the water surface, a Greater Swamp Warbler calls. The bird reminds us of a Great Reed Warbler, but darker, with grey legs and large feet, well adapted to grasping the thick papyrus stalks. Now and again he utters his short song, a boisterous outburst of chuckling notes.

To the east, the Ugandan border with Kenya runs over the 4200 m Mount Elgon, an ancient eroded volcano. On its slopes the altitudinal succession of Afromontane forest types is easily discernible: a belt of Elgon 'teak' and 'cedar' is followed by mighty yellowwood trees, with their dark green foliage, many with twisted trunks. Higher still we find montane bamboo, which finally gives way to Afroalpine moorland. On a moisture-laden slope with bamboo, a few rosewood trees, and lush herbaceous undergrowth, the view opens up as a valley cuts through the jungle, where the sun is just dispersing the last of the morning mist. Here a sweet melodious song attracts our attention. It comes from a Mountain Yellow Warbler, that flits about in the shrubbery like a large olive-yellow leaf warbler. It is an Afrotropical representative of the genus *Iduna* (formerly *Chloropeta)*, which only has very few species.

Cousin Island, Seychelles

One important measure in saving the endangered Seychelles Warblers on the 29-ha Cousin Island was to uproot the coconut palms that had been planted in the 1920s and to re-establish the endemic island vegetation. Today, large-leaved mapou trees and Indian mulberry (noni fruit) are growing there again, forming a closed, shady jungle together with the aerial-rooted screwpine. The granite rocks seem almost in the embrace of the mass of snake-like roots of the huge strangler figs.

Photo 1.9 and 1.10 In the bushy scrub of the dry mountain regions along the Gulf coast of Iran, Upcher's Warblers are a characteristic species. The coast, only a few kilometres distant, is characterized by mangroves in which the ubiquitous Clamorous Reed Warblers sing (photos Mike Pope, Peter Castell).

In the centre of the territory of a group of Seychelles Warblers, we find a single nest. It is in a small evergreen tree, about 2.5 m above the ground, suspended in an upright fork just below the dense crown. It seems to be constructed of coconut fibres, brown strips of bast, and grass. The three colour-ringed birds – the dominant pair and their daughter as a helper – are working full-time to find enough food to satisfy the lone chick in the nest. We can observe them at our leisure as they forage, acrobatically climbing around in the topmost twigs and long-stemmed *Pisonia* leaves, searching for edible morsels from the underside of the light-green leaves. Every so often they dart forward, or stretch up on the alert, with extended legs and fluttering wings, to glean arthropods from beneath the leaves, without having to leave their perch. In this endeavour they are well served by their long, straight, pointed bill, spindle-shaped body, and long graduated tail. One bird energetically repulses a Seychelles Fody that has approached the nest too closely.

Bandar Abbas and adjacent coast, southern Iran

Some 30 km northwest of Bandar Abbas, Kuh-i Geno is an isolated mountain, 2370 m high, which rises above the coastal plain on The Gulf. As we leave behind its foothills, covered with sparse *Acacia* scrub, we follow the steep hairpin road upwards. Halfway up the mountain we stop in a rocky valley with rugged boulder-strewn slopes, overgrown with patches of wild almond and pistachio bushes. Here sings an Upcher's Warbler, a representative of the *Hippolais* genus, having recently returned from its hot winter quarters in east Africa. The song of this medium-sized, round-headed warbler, with grey upper- and dull white underparts, is melodious, unsteady in tempo, and highly repetitive. While singing, the bird frequently changes place and perches on twigs, rocks, and boulders, in a fairly upright stance, cocking and waving its tail.

Just 30 km from Bandar Abbas we are now exploring a completely different habitat, the island of Gheshm. On a rising tide we move slowly in a small motorboat through the mangrove nature reserve of Hara, where inlets, sandbanks, mudflats, small bays, and creeks form a confusing labyrinth. Twisted grey mangroves and their asparagus-like aerial roots border the navigation channel. Western Reef Herons, Terek Sandpipers, and a small troop of Crab Plovers are flushed by the boat. The mangrove crowns are

Photos 1.11 and 1.11a Where southern taiga and forest steppe come together. Booted Warblers and Blyth's Reed Warblers are common here in the willow-birch thickets, while Icterine Warblers (left) can be found in clearings and at the edge of the tall birch forests (photos Peter Castell, Markus Varesvuo).

full of Clamorous Reed Warblers (a large *Acrocephalus* species), conspicuous by their harsh, raucous, powerful, almost frog-like song, but only fleetingly visible in the dense foliage. They resemble Great Reed Warblers, or perhaps rather Greater Swamp Warblers, because of their dark plumage. On our return journey their habitat is hardly recognizable as the tide rises, since now only the tops of the mangroves can be seen above the muddy water.

Novosibirsk, western Siberia, Russia

Starting from Novosibirsk, in the southeastern part of the west Siberian plain, we can get to know three acrocephalids in the transition zone between the southern taiga and the forest steppe. It is a region of numberless shallow lakes, appropriately described by the local people as 'blue eyes'. Here, in a varying landscape mosaic, three common species have colonized vegetation of different heights. Icterine Warblers (another *Hippolais* species) can be found in clearings and at the edge of the tall birch forest, or in open mixed forest with a rich tall undergrowth. In contrast to central Europe, the birds have to make do with a small number of tree species here: birch, pine, or aspen. At the end of May individual males sing in the treetops,

often in concert with Golden Orioles. Almost without interrupting their song, they fly and hop here and there in the open foliage, looking for food, snapping on all sides at small invertebrates.

By contrast, Blyth's Reed Warblers prefer lower trees or high bushes such as willow, with a scattering of open places covered by herbaceous vegetation such as nettles. We find a high density of the species in one area, where overgrown gardens at the edge of a village merge with bushy meadowland and woodland borders. A singing male can be distinguished from other species by his slower tempo and repeated phrases. In contrast to the restless Icterine Warblers, Blyth's Reed Warbler males tend to perch motionless in the open, especially at twilight.

Booted Warblers (another *Iduna* species) are the smallest species of the three and are found in the most open landscape with low vegetation: low willow-birch thickets with old grass and tall perennials. This very lively species has a face pattern reminiscent of reed warblers, but a jizz more reminiscent of a *Phylloscopus* leaf warbler as it slips through the undergrowth, which it is reluctant to leave, flying up only briefly before hiding quickly again in cover. The short song-phrases are an almost ecstatic chatter, with a rippling babble of unique spiralling quality that the Russians call 'murmuring'.

Khao Sam Roi Yot, Thailand

Khao Sam Roi Yot, or 'the Mountain of 300 peaks', is a small national park in Thailand, to the south of Pran Buri and over 300 km southwest of Bangkok. It is famous for its sandy beaches, spectacular caves, and superb mountain viewpoints. The ringing activities of ornithologists are not aimed at the waterfowl, herons, and egrets, for which the area is renowned, but at various *Acrocephalus* species that overwinter and moult here. The mist nets have been erected in a freshwater marsh that creates one part of an important southeast-Asian wetland, along with mangrove swamps. The marshes lie at the foot of picturesque steep grey limestone cliffs that rise from the Gulf of Thailand. On one side they grade into rice paddies and stands of *Borassus* palms.

Our attention is focussed on an extensive belt of mature reed, which is the preferred habitat of four *Acrocephalus* warblers: Oriental, Black-browed, Manchurian Reed Warblers, and Blunt-winged Warblers (Round & Rumsey 2003). The large Oriental Reed Warblers can only be mistaken for the Thick-billed Warblers, but the latter are very rarely trapped here. However, in worn plumage, Black-browed and Manchurian Reed Warblers are so similar as to be easily confused. Black-browed show a distinctive long, broad, buffy-white supercilium, bordered above by a prominent blackish lateral crown-stripe, whereas in Manchurian Reed Warblers, the blackish line on the crown side is narrower and contrasts less with the darker crown centre. Blunt-winged Warblers, another species on the smaller side, are plain coloured and lack the distinctive head markings.

While all Oriental Reed Warblers that we have ever examined arrive with already moulted flight feathers, the Manchurian Reed Warblers only begin to renew their flight feathers in their winter quarters. When monitoring our nets we occasionally see Manchurian Reed Warblers foraging; one striking aspect of their behaviour is that they climb to the tops of the tall reed stems with their tail cocked. We are reminded that we have often noted similar behaviour in the Moustached Warblers of the Camargue.

Henderson Island, Polynesia

Henderson Island in the Pitcairn group is one of the remotest islands in the South Pacific. It is another World Heritage Site, an uninhabited raised reef with an ecology virtually unaltered by humans. From the sea the island appears to be an entirely forested level plateau. During our brief landing we are only able to make observations in the coastal woodland, as it would be impossible to climb the 25 m or so cliffs and penetrate the interior, covered as it is with relatively low tangled pristine forest. Nevertheless, the 6 m tall beach-front forest of screwpines, portia trees, and beach cordia is in itself highly productive.

Henderson Island Warblers abound. We watch a family group, a pair with subadult young, robust brownish warblers with beige underparts. They are not at all timid, rather bold in fact, as they search for food in all vegetation strata, from the ferny herb layer right up to the tops of trees and bushes. The brownish upperparts of the adults exhibit considerable individual variation, often mottled with white, though the young birds are a uniform olive-brown. Two birds, probably a pair, allow themselves to be observed for a longish period. The slightly larger male is strongly mottled with white on the breast and his dark brown primaries contrast with the almost pure white underparts. The female is busy at the foot of a palm trunk, pecking in crevices, then climbing vertically upwards a little way, using her tail as a support. Higher up, the male examines a hanging palm frond. Even over the noise of the surf, we can hear the high-pitched chirping notes and, on one occasion, a thinner, longer call.

2 ▪ A factual journey round the book's chapters

These observations from our imaginary journey around the world have revealed that the supposedly 'unprepossessing' reed warblers have in fact colonized an astonishing variety of habitats and have evolved many very different lifestyles. These simple statements cover such a broad spectrum that we must initially ask ourselves four questions, which all revolve around matters of phylogeny and environment: (1) Do these superficially similar warblers form a monophyletic group at all? (2) Who is most closely related to whom? (3) Which character changes (or adaptations) have species or subgroups developed during the diversification of the family Acrocephalidae? (4)

In this process, what role was played by the various environments and/or the respective phylogenetic starting points?

In Chapter 2 we attempt an answer to questions 1 and 2, which relate to phylogeny. Are these superficially similar, small to medium-sized warblers, with their thin, mid-length bills and more or less uniform grey-yellow-brownish plumage, really a monophyletic group and which species are most closely related? For instance, do Manchurian and Black-browed Reed Warblers belong together, as suggested by the similarity of their head markings? Are the four large species (Great Reed Warblers, Greater Swamp Warblers, Clamorous Reed Warblers, Oriental Reed Warblers) really as closely related to each other as their similar size and appearance suggest? Why are Mountain Yellow Warblers and Booted Warblers now placed together in the same genus, when they differ completely in colouration and distribution?

A reliable answer to these questions has only been possible in the last few years with the development of molecular character analysis, using DNA sequences, that are largely independent of environmental selection, to establish an objective basis (genetic distances) for the relatedness of taxa (Leisler *et al.* 1997, Helbig & Seibold 1999, Cibois *et al.* 2007, 2008, Fregin *et al.* 2009). We shall present the latest molecular phylogeny and compare it with traditional systematics, whereby some unexpected new species groupings arise, which have necessitated dramatic changes in classification. The latest identification of cryptic species, or the occurrence of hybrids and secondary contact zones, provide us with concrete material for the discussion of general problems of speciation. Finally in Chapter 2 we shall discuss where the Acrocephalidae family might have originated, and which other family might be their sister taxon within the Sylvioidea. The rapid progress being made in molecular phylogenetics and new findings in palaeogeography have been of enormous benefit in this area.

The resolution of the true relationships, and the construction of a reliable phylogenetic tree in Chapter 2 are the prerequisites and background for tackling all the aspects relating to evolutionary changes raised in questions 3 and 4 above, aspects that return again and again to the key terms of 'environment' and 'phylogeny', which run like a thread through all the subsequent chapters of the book. Answers to these two questions are not straightforward, since they demand a complex methodological approach. We shall attempt to analyse the historical diversification of the group by using broad interspecific comparisons of species which are related to each other to varying degrees. The degree of relatedness of the species being compared is important; because they have a longer shared history behind them, closely related species are more likely to have traits in common than those that are more distantly related (Cunningham *et al.* 1998).

From time to time we may reverse the traditional approach of deducing relationships from the similarity of phenotypic characters, such as song or morphology, which can produce totally different 'trees' or clusters that deviate from the molecular phylogenetic tree. Such a comparison between different species in the group can illustrate to what extent their phenotypic characters agree or disagree with their genetic relationship. In addition to the possibility of ascertaining the degree of correspondence between phenotype and genotype among species – in other words, investigating the lability of traits – the resolved molecular phylogeny allows us to investigate further evolutionary aspects, such as the reconstruction of the ancestral state of a character in the group. To tackle these questions, we now have at our disposal a variety of sophisticated methods developed over the past 25 years and we shall make use of them in the relevant examples.

If it should emerge that distantly related species nevertheless possess striking similarities in certain phenotypic characters, then we may be looking at the phenomenon of parallel, convergent characters that have evolved independently of each other as adaptations to the same environmental conditions. Such a situation is by no means unusual in birds (Sibley & Ahlquist 1990) and would be frequently expected in the evolution of such a uniform grouping as the reed warblers. In fact, in the course of the book, we shall come across several such parallelisms, which result in species resembling each other, despite the fact that they are only distantly related. These convergences have long led systematists astray and have contributed to a tendency to increase the uniformity of the group.

Given that the habitats of many reed warblers differ from each other only slightly and gradually, (as illustrated by the example of Aquatic and Sedge Warblers in the Hun-

garian puszta on the second stage of our imaginary journey above), then most adaptations to the environment
can be revealed only by the closest of study. Because the
environment provides the parameters – sets the stage for
the actors so to speak – in Chapter 3 we look closely at
habitats, their characteristics and productivity, and the
variation of resources between temperate and tropical
regions. Since no members of our warbler group occupy
complex habitats, such as forest interiors, but only simple, less thermally 'buffered' ones, we find among them
many migratory species, whose environment extends
through many degrees of latitude at different times of the
year. Other topics of investigation in this chapter are how
important resources such as nest sites and food are distributed in habitats, whether they can be economically
defended, and which structural features of habitats offer
the varying foraging opportunities.

Chapter 4 concentrates on foraging and diet, and how
differences in food selection and habitat exploitation contribute to niche differentiation. How, when, and where do
reed warblers catch their prey? Here we shall see how diet
composition differs in the six well-studied species that

Photo 1.12 Southeast Asia is the overwintering region of many
of the Acrocephalidae. Rare Manchurian Reed Warblers
have been recorded regularly in the coastal marshes of
the Khao Sam Roi Yot National Park south of Bangkok
(photo Phil Round).

can be found together in a central European wetland
(Great Reed Warblers, Eurasian Reed Warblers, Moustached Warblers, Marsh Warblers, Sedge Warblers, and
Aquatic Warblers). Some species are able to take advantage of a rich food supply, while others have to be content
with nutrient-poor niches. What techniques do the birds
employ in order to capture large fleeing prey animals, in
contrast to small hidden ones? We shall also discuss what
role the development of the young (ontogeny) and early
experience play in the learning of specialized feeding
skills. Does foraging in specialist and generalist species
become more diverse in the same way when they have
young to provision? For reed warblers, the means of
exploiting their habitat in reedbeds are limited, though
still very varied, and, if we widen our horizon to look at the
songbirds which share their habitat, then we shall see that
the spectrum of possibilities can actually be extended.

Chapter 5 is concerned with ecomorphology, the analysis
of the relationship between morphology, behaviour, and
ecology. Using again the example of the six central-Euro-
pean wetland species, we show that the connection
between morphology and ecology (habitat structures) is
made through an intervening variable, namely, forms of
behaviour. In the specific case of the reed warblers, this
variable is clinging ability. In turn, this leads to the wider
question of how fixed or plastic the species-specific climbing behaviour of a reed warbler actually is. Is it largely
genetically programmed or can it be altered through experience as the young birds develop? Furthermore, which
adaptations are necessary for migratory birds to carry out
their extreme flight performances, and do habitat use and
migration constrain each other, as has been shown for
other bird groups? Which selection forces form wing
length and shape in individual Eurasian Reed Warbler
populations? By comparing overall morphology in the
various lineages, the basic morphological adaptations of
the clades to their respective preferred habitats can be
discerned, as can the specialist adaptations that have
been developed by each individual species.

In Chapter 6 we look at competition and coexistence
between species. Territoriality and spacing play an important role in this regard, and many questions are raised in
this context. What is actually being defended – a mating
site, or resources, or perhaps both? The undisturbed and
rival-free space in which to breed and forage? Why do
Great Reed Warblers repulse not only conspecifics, but

also Eurasian Reed Warblers? What are the advantages of such behaviour and why is interspecific aggression so widespread among marshland warblers compared with the inhabitants of other habitats? How do the different reactions towards other species arise, and what are the consequences for the subordinate species? We give examples of interference and apparent competition, in which the presence of one species can decrease the fitness of another, through the increased number of shared enemies such as predators or parasites. In conclusion we discuss how a reedbed guild is structured.

In all these interactions song plays a vital part, although it can be less or more pronounced in different species, and in Chapter 7 we turn to these issues, including the function of song, which can have both sexual and territorial significance. This dual function, and the question of which benefits females can expect from selecting a male with a complex song, have been particularly intensively studied in reed warblers, partly because of their very varied mating systems. How can we test whether complex songs are reliable signals of male quality and if they are costly for the singer? How and why do the songs of migratory species differ so much from those of tropical residents? Female choice appears to exert a stronger influence on the temporally limited singing of long-distance migrants than is the case in tropical species. Apart from sexual selection, other pressures, such as body mass and the acoustic properties of the environment, also shape song. Which traits of the very diverse reed warbler songs are more strongly subject to environmental influences and which show a stronger phylogenetic component?

Chapter 8 covers a broad range of aspects of breeding biology and life history, starting with the highly developed acrocephalid nest-building behaviour. Not all nests are equally elaborate and nest type and site vary considerably between species. Does nest type reflect the phylogenetic relatedness of species? What ecological factors determine the timing of the breeding season of a tropical and of a northern temperate species? Field experiments on tropical Seychelles Warblers have supplied important new insights here. How does it come about that the reproductive strategies of tropical residents and northern migrants differ so markedly? Which stages of the breeding cycle – incubation, nestling, or fledgling period – are compressed in northern migrants compared with breeders in southern latitudes, and which factors in turn shape the varying nestling growth rates in the six northern wetland model species? Given the high predation rates on marsh-nesting birds, we discuss the anti-predator behaviour of both adults and nestlings. Finally we compare annual reproductive output, a crucial contributing factor to recruitment and age structure in a population.

Chapter 9 is devoted to a single, though very important, cause of breeding failure, namely brood parasitism by Common Cuckoos. While some passerine species have successfully out-manoeuvred parasitization by Cuckoos through evolutionary adaptation, and now possess (temporarily) effective countermeasures, several reed warbler species are among Cuckoos' most important hosts. Many of their populations are currently engaged in an evolutionary arms race with this brood parasite. Because their nests can easily be found by researchers, and because several species that are potential Cuckoo hosts are often present in the same habitat, reed warblers represent an ideal model group for studying the details and dynamics of the interactions between hosts and parasite using comparative methods. How and why do they differ in their strategies when coping with brood parasitism?

Chapter 10 describes the breeding systems or social and mating systems of the acrocephalids, which are often quite different and strongly associated with ecological conditions. For songbirds, breeding systems in the acrocephalids show an astonishing diversity of brood-rearing patterns, ranging from biparental to purely maternal care, and from monogamy and cooperative systems to polygyny and promiscuity. This very diversity raises an interesting series of problems. How does it come about that, in one species, Seychelles Warblers, three adults devote themselves to raising a single nestling, while in another, Aquatic Warblers, five or even six young can be reared by a 'single mother'? What was the ancestral mating system of the group? Is the male of a socially monogamous species, Moustached Warblers, which we watch feeding the brood in the nest, also the genetic father? We discuss which direct and indirect (genetic) benefits females gain through their choice of a particular male as a social or extra-pair mate, and which male traits they can depend on.

Migration, dispersal, and moult are dealt with in Chapter 11. These are processes that have to be well coordinated with each other, and, above all, with the breeding season.

We compare migration routes and wintering grounds, which can vary greatly between different acrocephalid species, and pose various intriguing questions.

For example, how do different populations of Eurasian Reed Warblers find their destination on autumn migration? Do they also have separate overwintering regions in Africa? This taxon has been the subject of ground-breaking studies into how migratory birds efficiently divide their migration into phases of flight and stopovers to replenish their fat reserves. An answer has also been found to the second question: Eurasian Reed Warbler populations are markedly site-faithful to both their population-specific breeding quarters and wintering grounds. High breeding- and winter-site fidelity is also known in other acrocephalids. Studies of Eurasian Reed Warblers have also shown how young birds prepare themselves for their first such journey.

Other questions arise, such as how large are the winter quarters of Aquatic Warblers, whose whereabouts were completely unknown until very recently? Have individuals in isolated populations reduced their dispersal, and do those on oceanic islands disperse at all? Finally, one central question in the life of any bird: when does its annual cycle allow it to moult its plumage and how rigidly or flexibly programmed should the moult be? What role is played here by the length of the migration routes and the location of the winter quarters?

Many acrocephalids occupy habitat islands. In addition, more than 40% of the species in the genus *Acrocephalus* occur only as endemics on small oceanic islands in the Atlantic, Indian, and Pacific Oceans. Chapter 12 is devoted to these island dwellers and their special adaptations. Which islands have been settled by reed warblers? We examine the complex picture of repeated colonizations, which is only now emerging from molecular studies, and which is turning many traditional notions upside-down. Have all island dwellers evolved the same adaptations to island life, or has each species developed in its own way? For example, do they follow the 'island rule', which predicts that small species should become larger and evolve longer bills on oceanic islands? We not only answer these two questions, but will also demonstrate some even more spectacular cases of morphological convergences. Furthermore, we shall present some surprising examples of niche expansion and unexpected feeding innovations in

these island residents. Impressive 'insular trends' can also be demonstrated in breeding systems and song. We conclude by discussing the predispositions of these 'supertramps' for their colonization success, and also the reasons for the extinction of island species and the dangers faced by the extant members of the Acrocephalidae family on oceanic islands.

In Chapter 13 we examine the themes of population and range sizes, as well as the threats to which many acrocephalids are now exposed in a rapidly changing world. Why do Sedge Warblers have such a large distribution range, while other striped relatives such as Streaked Reed Warblers and Aquatic Warblers have such tiny ones? Why are some species strikingly common while others are so rare? Such considerations involve a discussion of long-term population trends and environmental changes, in particular those caused by human activity, such as habitat fragmentation and global warming. While some species can cope perfectly well with anthropogenic changes and can increase their range or improve their breeding success, others react much more sensitively. Aquatic Warblers serve as a worrying example when we consider how their populations have declined so rapidly in recent times and whether the conservation measures in place can reverse this process. Meanwhile, the populations of a number of threatened acrocephalids are so small that a major loss of genetic diversity would seem inevitable. Without the enthusiastic commitment of nature conservationists, some species would now be long extinct. We are thinking of some reed warbler species that have disappeared only recently from the Marianas, Society Islands, and Aldabra.

In Chapter 14 we turn our attention to the large blank space on the map of acrocephalid distribution that is the New World. Have other bird groups there occupied the ecological niche of reed warblers? If so, which groups are they? We can then ask if convergence has taken place, if these unrelated taxa have evolved characters similar to reed warblers in morphology and behaviour as independent adaptations to the same habitat? We shall see that the ecological equivalents of reed warblers in the marshlands of North and South America – wrens and icterids in the former, ovenbirds and tyrannids in the latter – show impressive convergences in plumage pattern, shape, and mating systems. Using comparative methods we will separate those characters attributable to the different phylogenies from those which are convergent adaptations, that is, the same 'solutions' to the same ecological problems.

With this, we have come full circle back to our enquiries during our imaginary journey around the world, concerning the evolution of adaptations in the diversification of the group. In order to answer such questions, we must focus on the precise nature of the selection pressures in the environment that have led to a particular adaptation.

For more than four decades now we have enjoyed the opportunity to compare many members of the reed warbler group. During that period our work has been a mixture of fieldwork, travel to watch birds in the wild, experimental studies often carried out by students, and systematic and morphological studies using museum collections. In these 45 years, interpretations and methods have changed continuously. At the outset it would have been impossible to imagine just what effective techniques we would eventually have at our disposal in order to answer once imponderable questions. We have been able to employ the full range of modern biological technology only thanks to collaboration with specialists in such areas as genetics and statistics. In all that time we have had the good fortune to be in permanent contact with a host of colleagues around the world, many of whom have produced the most exciting findings. In the course of our work we have continually exchanged ideas and data with them, in close and stimulating friendship, during shared projects, conferences, personal visits, hour-long telephone conversations, and in innumerable e-mails. Without their selfless cooperation and intellectual stimulation this book would not have been possible and we are very grateful to them all.

Photo 1.13 Henderson Island Warblers live in the brush forest of the island after which it is named, one of the remotest in the South Pacific. This species too has helpers at the nest, but in contrast to Seychelles Warblers they are unrelated to the breeding pair (photo Phil Chapman, Nature PL).

2

Systematics – relationships and diversification in the family of acrocephalid warblers

Matching head patterns of two rather distantly related species,
Black-browed (left) and Manchurian Reed Warblers (right).

Every birder knows that identifying reed warblers can be a real headache. Eurasian Reed Warblers and Marsh Warblers are easily differentiated in the field on the basis of their diagnostic songs, but their identification in the hand can be extremely problematic; even experienced ringers will have difficulty distinguishing between them when they are dealing with first-year birds. This is an incredibly tricky problem, and a number of specialists have been working on it for several years (e.g. Dorsch 1979, Leisler & Winkler 1979).

In the case of an ambiguous young bird taken from a mist net by one of our co-workers at the ringing station at Mettnau on the Bodensee (Lake Constance), plumage colouration and wing length are of no help, which is what we expected. We must move to the next step in the in-hand identification process and check where the notch of the second primary falls in relation to the inner primaries. Here our bird falls within the area of overlap between the two species, and we are still no further forward when we calculate the notch/wing ratio – the length of the notch divided by the wing length. Both identification characters are based on the fact that, in general, Eurasian Reed Warblers have shorter wings, but deeper or longer notches, than Marsh Warblers, which undertake more strenuous migration journeys (fig. 2.1).

For those cases where wing structure does not help in the diagnosis, the *Identification Guide to European Passerines* by Lars Svensson (1992) ultimately recommends a complicated score: 'If still stuck, try this method (worked out by G. Walinder). From length of bill to skull (A) is substracted the product of width of tarsus (B) multiplied by width of bill (C)'. Marsh Warblers have a shorter but broader bill, and a slightly wider tarsus diameter (fig 2.2). Thankfully, the value resulting from Svensson's complex calculation does indeed solve the problem: our bird is a Eurasian Reed Warbler. This identification can also finally be confirmed by its slightly larger foot (fig 2.3; Leisler 1972b).

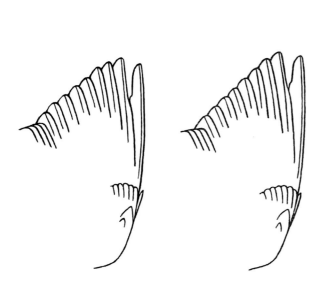

Fig. 2.1 Differences of wing length and shape in Eurasian Reed Warblers (left) and Marsh Warblers (right). Note the deeper notch in the second primary of Eurasian Reed Warblers (from Svensson 1992).

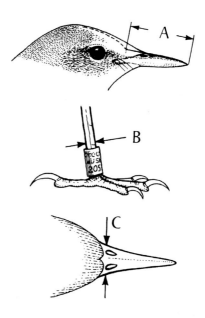

Fig. 2.2 The 'Walinder method': A – (B x C) = ? The score resulting from measurements of bill length, tarsus width, and bill width can be used to separate Eurasian Reed Warblers from Marsh Warblers in difficult cases, usually first calendar year birds (after Svensson 1992).

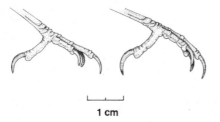

1 cm

Fig. 2.3 Differences in the foot span between Eurasian Reed Warblers (right) and Marsh Warblers (left). The gripping potential between inner toe and hind toe is greater in the former species (after Leisler 1972b).

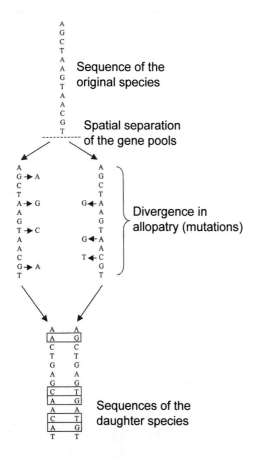

Fig. 2.4 When an ancestral species splits into two daughter species, its gene pool is divided in two, and this initial geographical isolation prevents any exchange of genes between the two groups. Genes in these now separate pools will diverge by the accumulation of mutations. If homologous sequences in the daughter gene pools are compared at a later time, the degree of divergence can be determined using the number of nucleotide differences found (percent sequence divergence; after Helbig 2000).

Mistaken identities and molecular methods

That Eurasian Reed Warblers and Marsh Warblers can present such difficulties in their identification should not really surprise us, because they are *closely related*, and can be expected to *agree* in even more external characters, since they are both descendants of a common ancestor. In a further example, Black-browed Warblers and Manchurian Reed Warblers – whose acquaintance we made on our journey round the globe in chapter 1 – *resemble* each other in their head pattern to a quite astonishing degree (introductory drawing), yet are *not very closely related* and belong to different evolutionary lineages. This led earlier taxonomists incorrectly to consider them as close relatives, or even as subspecies of the same striped reed warbler, despite their considerable differences in physical proportions and voice (Williamson 1968, Wolters 1982, Sibley & Monroe 1990; but see Alström *et al.* 1991). Their similar appearance is not the consequence of a close relationship but the result of *convergent* evolution, which results in species living under similar conditions looking like each other (chapter 14). This phenomenon has always been a major pitfall in the work of taxonomists wishing to gather taxa into groups or clades that contain all descendants of a last common ancestor. Such groups are called monophyletic.

Another example from our world trip in chapter 1 illustrates how systematists can be lured into false conclusions in the reverse direction, namely, when genuinely *related* species evolve under different environmental conditions so as *not to resemble* each other. Until very recently no one would have suspected that the Mountain Yellow Warblers, which we met in the montane forests of Mt. Elgon in East Africa, are closely related to the Booted Warblers of the transition region to the steppe near Novosibirsk. Yet both do indeed belong in the same group, the genus *Iduna*, and the latest DNA-sequence-based phylogenetic analyses by Silke Fregin and co-workers (2009) have proved this beyond any doubt. Even as recently as 2006, in the *Handbook of the Birds of the World* (del Hoyo *et al.* 2006), Mountain Yellow Warblers, together with two other yellow warblers, could still be found in the now redundant genus *Chloropeta*.

Nowadays such methods in molecular genetics help us to avoid the many traps set by divergent and convergent evolution (chapter 5). They date back to the 1970s, when

Charles Sibley and others (Sibley & Ahlquist 1990) recognized the potential of using the similarity of proteins between bird species to assess the degree to which they are related to each other. The essential idea was that molecules also evolve, but some of them – contrary to anatomy and other phenotypic characters – evolve independently of the environmentally-determined forces of selection. With the introduction of pairwise tests to measure DNA similarities between two taxa (DNA-DNA hybridization) Sibley and his colleagues 'finally struck gold' (Mayr 1989). Today's improved methods allow differences in the sequences of DNA nucleotides to be analyzed directly (fig. 2.4; Helbig 2000). The two significant advantages of DNA analysis are that the differences in DNA between two separate taxa develop at a more or less regular rate, driven by random mutation and drift, and are independent of the evolution of species morphology. This 'molecular clock' allows a rough estimate of the time that has elapsed since the taxa split from each other. In this particular mine 'gold' continues to be extracted and deposited in the GenBank, an open access database that now places the nucleotide sequences of more than 100 000 distinct organisms at our disposal.

These new molecular methods have resulted in a revolution in systematics, since it is only thanks to them that it is now possible to recognize the graded relationships of different avian groups – among them the reed warblers – as well as their precise extent and limits. While we can still only construct hypotheses concerning actual relationships, their quality now rests on a much more robust foundation than that of previous taxonomic systems. They provide a solid basis for many of the comparative questions posed in all the subsequent chapters of this book.

Photos 2.1 and 2.2 Although they are counted as two of the classic 'little brown jobs', the closely related Eurasian Reed Warblers (top) and Marsh Warblers (below) can be separated on the basis of their respective brown and olive-coloured upperparts and their differing leg colour. Additionally, their characteristic songs in the breeding season are very different, while their habitats can also be a great help in their identification. However first-calendar-year birds on autumn migration can be extremely difficult to tell apart, even in the hand (photos Dave Bartlett, Oldrich Mikulica).

Photo 2.3 Another species pair that are easily confused are Black-browed (shown here) and Manchurian Reed Warblers, whose head patterns are remarkably similar (see introductory drawing). Despite this they are not closely related (photo Irina Marova).

In this chapter we shall first establish what an acrocephalid warbler actually is – what distinguishes it from other species – before turning to look at the multifarious consequences of speciation, and finally examine the origin and diversification of this family of warblers.

2

What are acrocephalid warblers?

Characterization and boundaries of a uniform family

For very many years the reed warblers and *Hippolais* tree warblers occupied a firm position within the Old World warblers Sylviidae, but in recent years this family has been completely reconstructed in the light of the new molecular techniques. The earlier systematics were based broadly on functional similarities, such as thin insectivorous bills and plain nondescript plumages, but these characters tell us very little about true relationships. Given the paucity of differentiating traits, it has always been difficult to unravel both the relationships within the reed warbler group itself, and also higher-level relationships. Success came one step at a time in a series of studies using increasingly improved techniques of molecular sequencing (at first only with the mitochondrial cytochrome *b* gene, then later with several nuclear genes). Gradually more sophisticated statistical methods of phylogenetic reconstruction, as well as the inclusion of ever more species, enabled an increasingly more precise and complete picture to be formed. Initially lower level relationships were addressed (Leisler *et al.* 1997, Helbig & Seibold 1999, Ottosson *et al.* 2005); later attention turned to those at a higher level (Beresford *et al.* 2005, Alström *et al.* 2006, Johansson *et al.* 2008).

Today the reed warblers and their allies are card-carrying members of the Acrocephalidae, a monophyletic family within the superfamily Sylvioidea as part of the Passerida, the largest avian radiation. Our current state of knowledge suggests that the family encompasses *ca* 53 living species in 6 genera, namely *Nesillas*, *Calamonastides*, *Phragamaticola*, *Iduna*, *Hippolais*, and *Acrocephalus* (Fregin *et al.* 2009). In fig. 2.5 we illustrate with colour coding the phylogenetic tree of the acrocephalids, with the various clades (monophyletic subgroups, subgenera, genera), and we will retain these clade colours throughout the book. Table 2.1 lists the 53 extant species as well as 7 oceanic island species that have become extinct in historical times.

As far as plumage colouration is concerned, the acrocephalids are among the most monomorphic and nondescript groups of passerine birds. All species have rather dull brownish or greyish, sometimes greenish-tinged, upperparts, and whitish buffish-tinged or pale to bright yellow underparts. This external monomorphism is occasionally enlivened by a pale supercilium, lateral crown stripes, or sometimes streaks on the back or breast. Only a few species have developed a juvenile plumage (e.g. Rimitara Warblers; Thibault & Cibois 2006) and in some species the number of tail feathers is reduced (e.g. Dark-capped Yellow Warblers have only 10 and Madagascar Brush Warblers

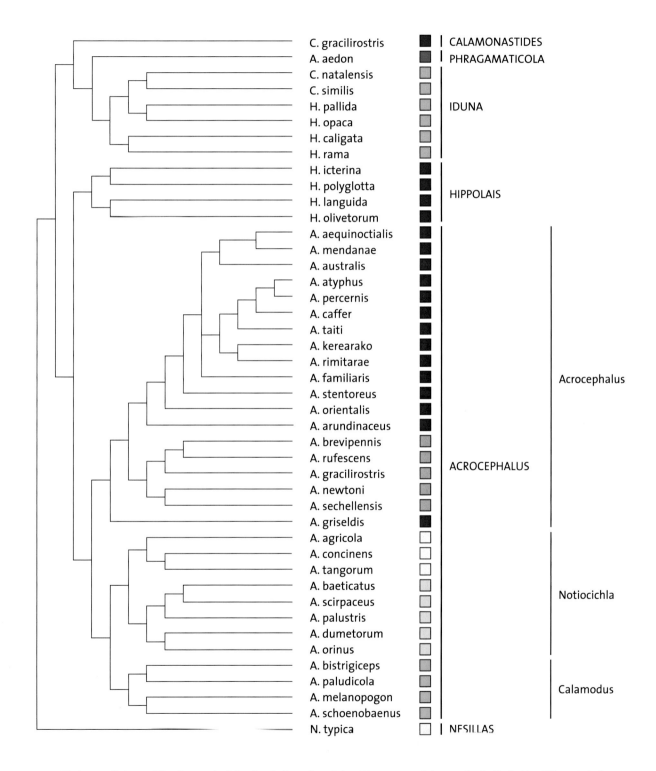

Fig. 2.5 Phylogenetic tree of the Acrocephalidae family based on data of four genes. Colour codes indicate the different clades or genera (after Fregin *et al.* 2009). **Note:** *Acrocephalus aedon* is now *Phragamaticola aedon*; the species from *C. natalensis* to *H. rama* are now placed in the genus *Iduna* (the genus *Chloropeta* being redundant). The colour codes will be retained unchanged throughout the book.

Table 2.1 Taxonomic list of the Acrocephalidae family (vernacular names after Kennerley & Pearson 2010). Extinct species in parentheses. **Note** the abbreviations and colour codes, which will be used throughout the book.

Vernacular name	Scientific name	Abbreviation	Colour code
Madagascar Brush Warbler	*Nesillas typica*	Nt	
Lantz's Brush Warbler	*N. lantzii*		
Grande Comore Brush Warbler	*N. brevicaudata*		
Moheli Brush Warbler	*N. mariae*		
(Aldabra Brush Warbler	*N. aldabranus*)		
Papyrus Yellow Warbler	*Calamonastides gracilirostris*	Cg	
Thick-billed Warbler	*Phragamaticola aedon*	ae	
Booted Warbler	*Iduna caligata*	ca	
Sykes's Warbler	*I. rama*	ra	
Western Olivaceous Warbler	*I. opaca*	op	
Eastern Olivaceous Warbler	*I. pallida*	pll	
Dark-capped Yellow Warbler	*I. natalensis*	na	
Mountain Yellow Warbler	*I. similis*	si	
Upcher's Warbler	*Hippolais languida*	la	
Olive-tree Warbler	*H. olivetorum*	ol	
Melodious Warbler	*H. polyglotta*	po	
Icterine Warbler	*H. icterina*	ic	
Basra Reed Warbler	*Acrocephalus griseldis*	gi	
Great Reed Warbler	*A. arundinaceus*	ar *GRW*	
Oriental Reed Warbler	*A. orientalis*	or	
Clamorous Reed Warbler	*A. stentoreus*	st	
Australian Reed Warbler	*A. australis*	au	
(Guam (Nightingale) Warbler	*A. luscinius*)		
Saipan Warbler	*A. hiwae*	hi	
(Pagan Warbler	*A. yamashinae*)		
Caroline Islands Warbler	*A. syrinx*		
Nauru Warbler	*A. rehsei*		
(Millerbird	*A. familiaris*)	fa	
Nihoa Millerbird	*A. f. kingi*		
Kiritimati Warbler	*A. aequinoctialis*	aq	
Tahiti Warbler	*A. caffer*		
(Leeward Islands Warbler	*A. musae*)		
(Moorea Warbler	*A. longirostris*)		
Tuamotu Warbler	*A. atyphus*		
(Mangareva Warbler	*A. astrolabii*)		
Rimatara Warbler	*A. rimitarae*	ri	
Pitcairn Island Warbler	*A. vaughani*		
Henderson Island Warbler	*A. taiti*	ti	
Southern Marquesan Warbler	*A. mendanae*		
Northern Marquesan Warbler	*A. percernis*		
Cook Islands Warbler	*A. kerearako*		
Greater Swamp Warbler	*A. rufescens*	ru	
Cape Verde Warbler	*A. brevipennis*	br	
Madagascar Swamp Warbler	*A. newtoni*	ne	
Rodrigues Warbler	*A. rodericanus*		
Seychelles Warbler	*A. sechellensis*	se	
Lesser Swamp Warbler	*A. gracilirostris*	gr	

Vernacular name	Scientific name	Abbreviation	Colour code
Manchurian Reed Warbler	A. tangorum	ta	
Paddyfield Warbler	A. agricola	ag	
Blunt-winged Warbler	A. concinens	co	
Eurasian Reed Warbler	A. scirpaceus	sc *avicenniae* sa *ERW*	
African Reed Warbler	A. baeticatus	ba	
Marsh Warbler	A. palustris	pal *MAW*	
Blyth's Reed Warbler	A. dumetorum	du	
Large-billed Reed Warbler	A. orinus		
Moustached Warbler	A. melanopogon	me *MOW*	
Aquatic Warbler	A. paludicola	pld *AW*	
Sedge Warbler	A. schoenobaenus	sch *SW*	
Streaked Reed Warbler	A. sorgophilus		
Black-browed Reed Warbler	A. bistrigiceps	bi	

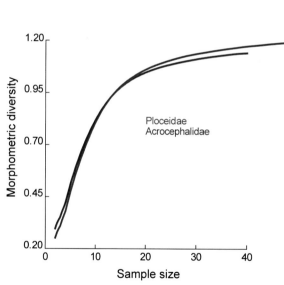

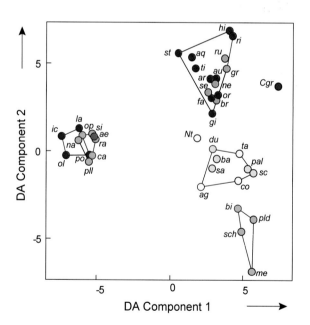

Fig. 2.6 Rarefaction curves showing the dependence of morphometric diversity (mean proper variance) on sample size in weaverbirds Ploceidae and reed warblers Acrocephalidae. The reed warblers are morphologically even more uniform than the markedly uniform weaverbirds.

Fig. 2.7 Discriminant analysis of clades based on morphology, using the combinations of attributes that separate and hence characterize the genera or subgenera. Small feet and short tarsi together with broad bills separate the *Hippolais* and *Iduna* species and *Phragamaticola* from all other groups in the family (*Nesillas, Acrocephalus, Calamonastides*). For differences along component axis 2 see text.

Photos 2.4 to 2.8 The Western Olivaceous Warbler illustrated here is a good example of a species that can be difficult to identify in the hand. Every character has to be carefully scrutinized before making a judgement. In distinguishing the species from Eastern Olivaceous Warbler the stronger bill of the former is diagnostic (photos 2.4 and 2.5 Carlos Zumalacarregui Martinez, 2.6 Jens Hering, 2.7 and 2.8 Wolfgang Mädlow).

have either 12 or 10 rectrices; Kennerley & Pearson 2010). Further common traits include long powerful legs, and possibly their skilfully constructed bowl-shaped nests, the only synapomorphic (i.e. derived) character shared by all species. Their nestlings are naked in contrast to the downy *Locustella* nestlings, and have gapes with two tongue spots as opposed to the three of *Locustella* (Hockey *et al.* 2005, Louette *et al.* 1988). Only in Thick-billed Warblers and Booted Warblers is there occasionally a small dark spot at the tip in addition to the two tongue spots (Cramp 1992, Castell & Kirwan 2005) while in Western Olivaceous Warblers there can be three (Crespo *et al.* 1988).

Are acrocephalids truly uniform?

Such a level of nondescript appearance should make us wonder whether the uniformity of the group might in fact only be suggested by these monomorphic plumages, which, in a sense, might actually mask an otherwise unsuspected phenotypic diversity. In order to answer the question of whether the various reed warblers really are more uniform than other groups, we must first record their overall form or shape statistically, a task often undertaken in the course of answering similar enquiries concerning their anatomy. To simplify a complex issue, the problem is initially resolved by biometrics. We must take linear measurements of various external morphological structures, functional modules such as bill, wing, tail, and hind limb, correct them for body size, and then find the ratios between them. Claramunt (2010) developed a method of calculating a value for the overall shape of a species and, hence, a way of testing the degree of morphological diversification of different lineages. Using other multivariate techniques, combinations of characters are created in the form of axes that span a so-called 'morphospace', in which the positions of each individual taxon can be compared (fig. 2.7; more details in chapter 5).

In fig. 2.6 we compare the morphological diversity of 40 acrocephalid species with that of 70 species of the species-rich weaverbirds (Ploceidae), whose modes of life differ greatly, but whose body shape varies little (Leisler *et al.* 1997). Shape diversity was calculated on the basis of 14 traits measured on extensive series of skins, representing a total of 110 species of both groups. The course of the curves indicate that the acrocephalid warblers are even less differentiated in their shape than the uniform weaverbirds. In other words, they really do constitute a remarkably uniform family.

3 Characterization of the family's genera and clades

In our concept of the monophyletic species groups (clades), and their classification as genera of the reed warblers and associated species, we are following the plausible suggestions of Fregin *et al.* (2009), which have already been adopted by Kennerley & Pearson (2010). This latest systematic arrangement covers a maximum of phylogenetic information and thus most closely approaches the actual relationships between the taxa. In addition, we shall briefly discuss which of the 'classical' characters, such as anatomy, voice, egg characteristics, or distribution range, best support the proposed arrangement.

The new classification based on molecular-genetic data puts us in the happy position of being able to reverse the traditional approach of grouping species in genera according to morphological similarities. Now we can study which morphological traits are characteristic of the clades or genera that have been genetically established. In this we shall quantify their structural properties, using 17 morphological characters together with body mass, and calculate how the genera are maximally separated in a morphospace according to these 18 characters, by employing multivariate methods. Those characters contributing to the differentiation are listed in table 2.1 (see also chapter 5) and identified on the discriminant axes in fig. 2.7. The clearest separation of clades is established by a suite of characters (discriminant axis 1) that encompass traits of the hind limb and bill. Small feet, shorter tarsi, and broad bills clearly differentiate the arboreal clades (*Iduna*, *Hippolais*) from the wetland-dwellers. So the striking congruence in external anatomy in two genera, that in reality are not closely related (*Iduna* and *Hippolais*), is the result of parallel development that obscured their separate origins in such a way that some of their member taxa were wrongly classified in the past. Such adaptive convergences of entire subgroups (and of individual species from different clades; chapter 5), together with an overall weak phenotypic divergence, all contribute to the distinctive morphological uniformity of the group we discussed earlier. The second axis in fig. 2.7 chiefly characterizes an increase in body size, but also in the strength of the legs, in bill length and depth, tail length, and wing width. This suite of traits mutually differentiates all subgroups of the large genus *Acrocephalus* in fig. 2.7.

Clade representatives
1. Madagascar Brush Warbler
2. Papyrus Yellow Warbler
3. Thick-billed Warbler
4. Dark-Capped Yellow Warbler
5. Booted Warbler
6. Olive-tree Warbler
7. Sedge Warbler
8. Marsh Warbler
9. Paddyfield Warbler
10. Clamorous Reed Warbler

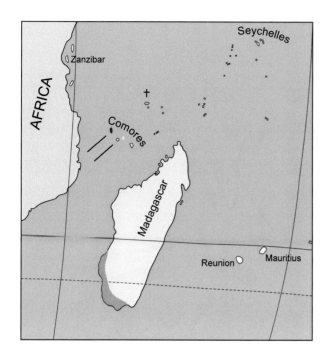

Map 2.1 Distribution of *Nesillas* brush warbler species in the western Indian Ocean (after Louette *et al.* 1988). Madagascar Brush Warbler = yellow; Lantz's Brush Warbler = orange; Grande Comore Brush Warbler = red; Moheli Brush Warbler = violet; Aldabra Brush Warbler (extinct) = green. If two species occur together (e.g. Madagascar Brush Warblers and Lantz's Brush Warblers on Madagascar) then they are separated either by their parapatric (contiguous) distributional ranges or by their ecological niches (e.g. on Moheli, where a subspecies of Madagascar Brush Warbler and Moheli Brush Warblers live together, the former keeping to low vegetation, the latter living higher in trees; from Louette *et al.*1988).

Nesillas (brush warblers)

Much to their surprise, Johansson *et al.* (2008) only recently discovered that the genus *Nesillas* is the sister group to all other groups in the family. This finding has important implications for theories about the origin of the entire family (section 2.5). The four species in this genus (a fifth became extinct 30 years ago on Aldabra) are endemics of the Malagasy Region (Madagascar and the Comoros Islands; map 2.1). Early on they were placed near *Hippolais* on the basis of their egg characteristics (Meise 1976). The brush warblers are medium-sized inhabitants of bushes, undergrowth, and the lower tree canopy, with long graduated tails, although in their overall morphology they resemble *Acrocephalus* species rather than other 'tree warblers' such as the members of *Iduna* or *Hippolais* (fig. 2.7; see also chapter 5). All species are spatially or ecologically separate from each other, as they mostly occur on different islands.

Calamonastides (Papyrus Yellow Warblers)

The position of Papyrus Yellow Warblers on the phylogenetic tree by Fregin *et al.* (2009) at the base of the following clade (genera 3 & 4), is still uncertain. While the species does resemble other yellow warblers in colouration (Dark-capped and Mountain Yellow Warblers, *Iduna* species), and was formerly placed with them in the now redundant genus *Chloropeta*, it is very unlike them in its

fundamentally different morphology (figs 2.7 and 5.6) and in its voice (chapter 7). As a sedentary resident species, confined to a few swamps in central and east Africa, the species is conspicuous within the family for its unusual anatomical adaptations, characterized by long toes and claws, long and powerful legs, and a relatively narrow bill (chapters 5 and 14). The use of a monotypic genus for a species is arbitrary and should only be employed, as in this case, for good reasons such as a pronounced phylogenetic individuality that emphasizes the distinctiveness of the species within the family.

The next clade (genera 3, 4) encompasses seven species in two genera, the first of which once again contains only one species, and is thus monotypic.

Phragamaticola aedon (Thick-billed Warblers)

Thick-billed Warblers are breeding visitors to the dense deciduous vegetation of temperate central and southeast Siberia and northeast China, and winter in India and southeast Asia. They have often been included within the group of the large *Acrocephalus* warblers, but significantly differ from the latter, both genetically (Helbig & Seibold 1999) and also in structure (with a shorter, thick-based bill, domed head, more rounded wings, and longer graduated tail), as well as in nest construction, voice, and egg colour (Marova *et al.* 2005; chapters 7 and 8).

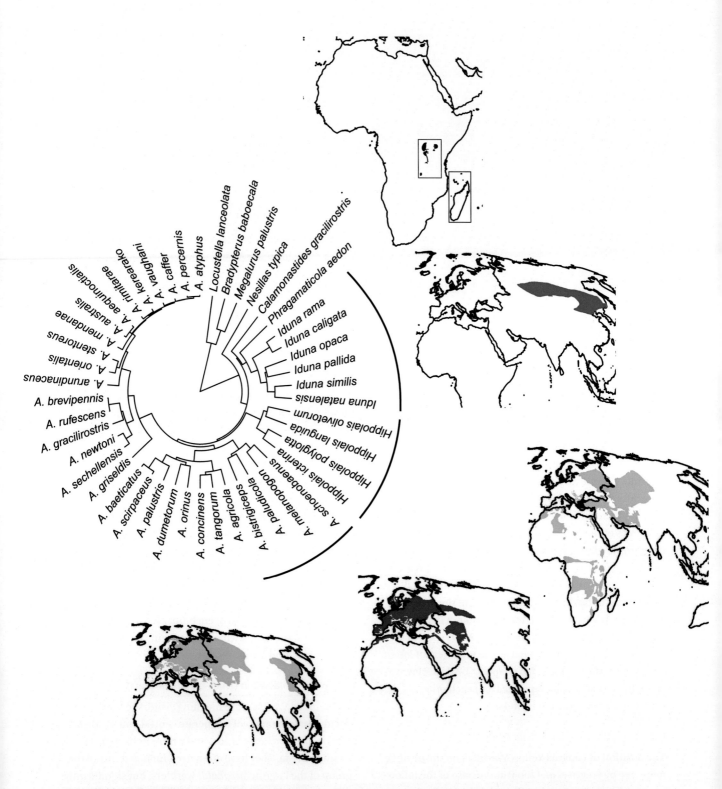

Fig. 2.9 Ladderized molecular tree of the Acrocephalidae and its sister family, the Locustellidae. Branch lengths correspond to time, so it can be established exactly when a divergence (split) occurred. The distribution of early-diverging taxa (deep splits) gives an indication of the origin of the group. In addition, distribution maps of selected genera (clades) are shown: *Nesillas* (white, in box), *Calamonastides* (black, in box), *Phragamaticola* (dark blue), *Iduna* (light blue), *Hippolais* (violet), and striped *Acrocephalus* (green).

Iduna species (bush and scrub warblers)

For a long time, the close relationship between the three species-pairs brought together here remained unrecognized, thus affording us a prime example of parallel and divergent evolution. The two small to medium-sized Olivaceous Warblers (Western and Eastern from north Africa and the Middle East), along with both of the northeast European to central Asian taxa (Booted and Sykes's Warblers) are two brownish-coloured species-pairs which convergently resemble the *Hippolais* warblers in appearance, plumage, and behaviour so much that their independence in a separate genus (*Iduna*) was only confirmed in the first molecular phylogenetic study by the 'Reed Warbler Group' (Leisler *et al.* 1997; note the strongly overlapping or contiguous positions of the two genera in the morphospace in figs 2.7 and 5.6). The parallel similarity between *Hippolais* and *Iduna* species derives from their common adaptations as scrub- and woodland-dwellers. Even more unexpected was the revelation that, as the sister group to the Olivaceous Warblers, two Afrotropical 'flycatcher-warblers' are also firmly placed within the genus *Iduna* (Fregin *et al.* 2009). Dark-capped Yellow Warblers and Mountain Yellow Warblers, together with Papyrus Yellow Warblers, formerly constituted the non-monophyletic genus *Chloropeta*. The two yellow African species diverged markedly in their plumage colour, but remained similar to their present congeners in habitat requirements and hence overall morphology (fig. 5.7)

Hippolais species (tree warblers)

It was long suspected that the tree warblers (*Hippolais;* drawing) and the reed warblers *sensu stricto* (*Acrocephalus*) were very closely related to each other (Voous 1977) and this has now been confirmed by the genetic findings (Fregin *et al.* 2009). *Hippolais* species represent a robustly delimited clade (supported by additional characters, such as egg colour) which contains two species-pairs, one with yellow-green and one with grey plumage, inhabiting bushy growth and woodlands in the south of the Western Palaearctic, and which overwinter in Africa (fig. 2.9). In their anatomy they therefore all unite features of arboreal birds (small perching feet) and migratory species (relatively pointed wings and squarer tails than in *Acrocephalus* warblers).

Acrocephalus warblers (reed warblers sensu stricto)

This is an extensive genus of 37 extant species consisting of two groups, the large and the small reed warblers, which in turn contain respectively three and two genetically clearly differentiated lineages. Their total distributional area encompasses the Palaearctic Region, Africa with Madagascar, the Oriental (Indomalayan) Region, the Australasian Region, and Oceania. As fig. 2.7 makes clear, the reed warblers *sensu stricto* are clearly distinguished from the three previous genera (*Phragamaticola*, *Iduna*, and *Hippolais*) by their longer toes and claws, longer tarsi, and narrower bills, all of which indicate a life style of

Photo 2.9 One of the distinguishing features of the acrocephalids is that their young are naked on hatching and have two black tongue spots (nestling Great Reed Warbler; photo Oldrich Mikulica).

climbing on vertical stems and more frequent gleaning than flycatching (DA1 in fig. 2.7; chapter 4). Within the *Acrocephalus* genus, all subgroups are mutually well differentiated for the most part by traits that represent DA2 in fig. 2.7, and in the following order: the highest values on the second discriminant axis are shown by the large reed warblers, a midway position is occupied by the small plain-coloured birds, and the lowest by the striped species. The genus is also characterized by an attenuated head shape and vocal singularities, such as chattering and chuckling songs.

Large Acrocephalus warblers

Because of their uniform appearance (drab plumage and a common morphology – long powerful legs and bills, long tails, and broad wings; fig. 2.7), the systematics of this group was for a very long period confusing and the object of many lengthy discussions (Stresemann & Arnold 1949, Eck 1994, Shirihai *et al.* 1995, Helbig 2000, Opaev *et al.* 2009). Differences between species were either overlooked or undervalued, such as the particular characteristics of the shape of Basra Reed Warblers compared with other Palaearctic species, or they were exaggerated, such as the rather deviant plumage of Seychelles and Rodrigues Warblers compared with their Afrotropical congeners, which at one time led to them being placed in a genus of their own, *Brebrornis*.

Basra Reed Warblers are specialists of the shallow edges of the saline inundated *Typha* marshlands of Iraq (Kennerley & Pearson 2010). They represent a more ancient sister taxon to all the other large species. Even before any molecular studies appeared, Pearson & Backhurst (1988) emphasized that the species is highly distinct in structure, plumage, and voice.

Amongst the other large reed warblers, the well-supported Afrotropical subgenus *Calamocichla* is made up of the two African Swamp Warblers, Lesser and Greater, together with five species living on islands off the coast of that continent. All are sedentary species, with the corresponding wing structure, and all conform in their song and oological characteristics as well as their rounded heads.

The sister group to these consists of three widespread Palaearctic species – Great, Oriental, and Clamorous Reed Warblers – the last of which have an extensive, if fragmented, breeding range that extends into the Afrotropical, Indomalayan, and Australasian ecozones (drawing). This far-ranging and highly variable species is possibly not monophyletic (Fregin *et al.* 2009). Also belonging to the group are all the species that have colonized Australia and the Pacific islands, which, together with Oriental Reed Warblers, all descend from common ancestors (Cibois *et al.* 2011a). In both of these sub-

Photos 2.11 to 2.15 In the systematic ordering of subgroups, egg colour is a significant differentiating character. The eggs of *Phragamaticola* have a ground colour of pale brownish pink overlaid with hairstreaks; those of *Iduna* also have a pale ground; those of *Hippolais* are tinged pink; those of the large and small plain-coloured *Acrocephalus* have many dark spots and blotches on a pale ground, while the ground colour of the striped species in the same genus is darker (photo 2.11 Thick-billed Warbler; Irina Marova, 2.12 Dark-capped Yellow Warbler; Warwick Tarboton, 2.13 Icterine Warbler; Joseph Hlasek, 2.14 Oriental Reed Warbler; Irina Marova, 2.15 Sedge Warbler; Oldrich Mikulica).

groups of large reed warblers (the subgenera *Acrocephalus* and *Calamocichla*), those species that have settled oceanic islands have become more strongly differentiated in plumage colouration and morphology than their ancestral forms and their closest continental relatives (chapter 12).

Small Acrocephalus warblers

This group is separated from the large reed warblers not only by size but also by long chattering songs (chapter 7) and includes two distinct subgroups: the striped and the plain-backed species.

The *striped* species (*Calamodus*), typified by Sedge Warblers (drawing; fig. 2.9) are a distinct group of five species strongly differentiated from the following group by their streaked plumage pattern, mode of life, and song. They are confined to the Palaearctic. Three of them, Aquatic Warblers, Streaked and Manchurian Reed Warblers, occupy small or disjunct ranges and all of them are wholly migratory, with the exception of Moustached Warblers, which have some sedentary populations. The striped *Acrocephalus* taxa have the lowest values on DA2 (fig. 2.7), and are characterized as inhabitants of undergrowth and very dense vegetation by their smaller body size, short tails, fine bills, and narrow wings. They also have in common their buff-olive coloured and finely mottled eggs.

Photo 2.17 Dark-capped Yellow Warblers are distributed in several subspecies throughout the southern half of Africa. Latest research places them in the genus *Iduna*, together with Mountain Yellow Warblers and four Palaearctic species (photo Warwick Tarboton).

The small *plain-backed Acrocephalus* warblers (*Notiocichla*) are composed in turn of two groups: the shorter-winged eastern 'Paddyfield Warbler' superspecies (i.e. a group of at least two more or less distinctive species with approximately parapatric distributions; drawing), and a group with a distribution further to the west.

The three species of the 'Paddyfield Warbler' superspecies complex – Manchurian Reed Warblers, Blunt-winged and Paddyfield Warblers – possess disjunct breeding ranges from southeast to central Asia, and just touch southeast Europe. Since their ranges are geographically separate, these taxa are allospecies. All three deviate from other small plain species by their longer, more graduated tails. Two species, Paddyfield Warblers and Manchurian Reed Warblers, have head markings similar to their striped congeners (introductory drawing) and thus represent an excellent example of the recurrence of similar plumage pattern elements in taxa which are not closely related. In later chapters of the book, we shall meet even more impressive examples of parallel head-pattern evolution in the even more distantly related inhabitants of reedswamps in the New World (chapter 14).

The second subgroup of plain-backed *Acrocephalus* species is typified by Marsh Warblers and consists of Marsh Warblers themselves, the Eurasian Reed Warbler super-

Photo 2.16 Warblers of the genus *Nesillas* are endemic to Madagascar and the nearby Comoros; Lantz's Brush Warblers (shown here) are confined to low arid regions of sw Madagascar (photo Greg & Yvonne Dean).

Photo 2.18 Basra Reed Warblers are a more ancient sister taxon to all other large *Acrocephalus* species. In their distribution they are largely confined to the Mesopotamian marshes of southern Iraq (photo Mike Pope).

species, and the species-pair of Blyth's and Large-billed Reed Warblers, the former of which inhabit a very large distribution range in the central Palaearctic, while the latter occupy only a tiny one.

4 ▮▮▮▮▮▮▮▮
Species and speciation

Reed warbler systematics formerly created difficulties at both the higher taxonomic levels and at species level. Nowadays, the new DNA-sequencing techniques allow us to measure the genetic divergence between closely related taxa and to recognize speciation processes, thus enabling the position of some species to be newly defined. This is based on the fact that the genetic difference between two taxa correlates with the degree of their reproductive incompatibility. In other words, the development of a barrier to reproduction – decisive for the very definition of a 'species' – is more likely, the further the genetic divergence between them has progressed (see below).

Thus the measuring of genetic differences between different taxa has led to the uncovering of some *cryptic species,* whose morphology is so similar that they could not formerly be perceived as discrete entities. Among them are various Pacific island taxa which, in the course of parallel evolution, have become morphologically almost identical

to each other. An example would be the 'Marquesan Reed Warblers', which in actual fact include two cryptic species (Northern and Southern Marquesan Reed Warblers) belonging to two quite separate lineages (Cibois *et al.* 2007; chapter 12). In other taxa, scientists only started to look for additional differences following the discovery of a genetic distance beween them. It turned out that there were differences in their calls and songs – important signals in species recognition – as well as in their ecology. Examples of this procedure are Eastern and Western Olivaceous Warblers, with a genetic distance of 9.6% (proportion of nucleotide substitutions; fig. 2.4), and Booted and Sykes's Warblers, with a distance of 6.7%. In both cases the species in these species-pairs are differentiated, though only slightly, by the height of the vegetation stratum occupied and by voice. Sykes's and Booted Warblers are not only separable by their songs, but even more so by their calls; that of Sykes's Warblers is a very short, hard 'tsak', while the corresponding call of Booted Warblers can be transcribed as an almost disyllabic 'dsrak' (Svensson 2001, Lindholm & Aalto 2005, Constantine & The Sound Approach 2006).

These fresh insights resulted in an increase in the number of species within the Acrocephalidae family. While Mayr & Cottrell (1986) listed only 40 species, today we put the figure at 53 living species (table 2.1), with the majority of the 'new species' being splits. However, we have not yet reached the end of this particular road. For instance,

Photo 2.19 Lesser Swamp Warblers are a member of the Afrotropical subgenus *Calamocichla* of the large reed warblers which are sedentary species (photo Warwick Tarboton).

Photo 2.20 Madagascar Swamp Warblers also belong to the subgenus *Calamocichla*, which is characterized by resemblances in body structure and song (photo John Mittermeier).

separation of populations: first, when the distributional range of a species is divided or fragmented by climatic or geological events; secondly, when peripheral populations are separated, or thirdly, when parts of a population emigrate. These processes are known as allopatric speciation. The resulting isolated populations then undergo genotypic and phenotypic divergence, because they are subject to different selection pressures in their now differing environments, and because their gene pools change independently through drift and a variety of mutations. Over time the gradual divergence process occurs at different levels – at the genetic level and at various phenotypic levels (e.g. morphological, ethological, physiological, or ecological). Furthermore, this process can take place at different speeds, so that genetic differentiation can run ahead of morphological differentiation or vice versa. Geographic variation with marked differences between populations of the same species (subspecies) will not on its own lead to speciation, so long as the 'elastic band' of gene flow maintains the unity of a population, although it can 'set the points' for speciation (Price 2008).

As time passes, the incidental accumulation of multiple differences between geographically divided populations may generate reproductive incompatibilities. These are primarily a by-product of general divergence, since there is no genuine selection pressure for such incompatibilities between isolated populations. This gives rise, first, to morphological and behavioural differences that influence reproduction and reproductive behaviour, preventing the individuals of one species from mating with those of another (premating isolation) and, secondly, to increasing genetic differentiation which sooner or later creates an intrinsic barrier to reproduction, because it leads to reduced fertility or viability in any resulting hybrid offspring of mixed pairs of the two taxa concerned (postmating isolation; Price 2008).

Alterations to traits that aid species recognition can develop quite rapidly in songbirds, because imprinting and other forms of learning about conspecific individu-

splits can be expected in the disjunctly distributed Clamorous Reed Warbler complex, once genetic techniques are applied to the material. Every decision in systematics remains provisional, with future technology adding ever greater precision to the positioning of taxa.

Speciation – the genesis of reproductive barriers

Before we turn to the question of when it is that two taxa are isolated and must be regarded as separate species, we should briefly look at how speciation in birds takes place. It occurs almost exclusively because of the spatial

Photo 2.21 Australian Reed Warblers represent a previously undiscovered case of reverse colonization, whereby birds settled continents (in this case Australia) starting from (Pacific) islands (photo Karl Seddon).

Photo 2.21 Australian Reed Warblers represent a previously undiscovered case of reverse colonization, whereby birds settled continents (in this case Australia) starting from (Pacific) islands (photo Karl Seddon).

als affects mating decisions in birds. As a result of this, their species recognition system is the complex result of a combination of inborn preferences and various learned experiences. This means that, for example, songs can remain unchanged over long periods of time and large areas on the one hand, yet on the other they can also quickly be changed (Martens 1996; chapters 7 and 12). Thereafter a new song form in one population need only be preferred and handed on for a pairing between individuals from the two populations to be made difficult (premating isolation).

The actual song features, such as syntactical characters, timing, or frequency, which are used in species recognition in individual cases, have not been experimentally analysed in the acrocephalids. However, the precise vocal characteristics which serve to differentiate closely related species have been well studied. We know, for example, that Great Reed Warblers are distinguished from their eastern counterparts, Oriental Reed Warblers, by shorter songs with longer pauses and more repetitions of identical syllables; Oriental Reed Warblers deliver longer songs with shorter pauses, in which the syllables are more variably organized (Opaev et al. 2005). Presumably these vocal differences would suffice to limit hybridization between these two geographically separate species if they came into secondary contact. Furthermore, at 5.9%, their genetic distance – an approximate measure of an intrinsic reproductive barrier – is greater than the rough guideline figure of 5% for stable species (Johns & Avise 1998). In hypothetical 'secondary' contact this would mean reduced fitness of the hybrids produced by mixed pairs, which would help to reinforce the tendency to mate only with conspecifics (postmating isolation). Since the two large reed warblers in question can even be diagnostically separated on the basis of their own derived characters (e.g. Oriental Reed Warblers are smaller and have breast streaks), we are justified in treating them as species (allospecies), even if a gene-flow barrier cannot be demonstrated, because they do not occur sympatrically. Of course birds in particular can undergo dynamic changes in their distribution, so geographically divided populations can meet up again sooner or later and merge or collapse into one if the divergence between them is not too great.

Stages of speciation and 'secondary contact'

In reed warblers we find a variety of different stages of phenotypic and genotypic divergence of taxa, and different gradations of speciation. They are impressive illustrations of two points: first, that a divergence can have progressed to different degrees at different levels, and, secondly, that both reproductive barriers and also ecological differences between two taxa must exist when two formerly separated forms establish themselves in close proximity as species in 'secondary' contact.

Using gene-flow analysis and the genetic structure of present-day populations, Hansson *et al.* (2008) reconstructed the intraspecific diversification and distribution history of Great Reed Warblers. Within the species they unexpectedly discovered that two different clades existed which had been separated during the last glaciation of 26-20 000 years ago; following the retreat of the ice these clades had spread out from different glacial refugia, at different times and along different routes, and then merged again. In recent populations the genetic signature of both clades is present at varying frequencies. However, similar plumage differentiation is not congruent with genetic differentiation, having developed independently in each expanding clade. Moreover, the time period of their separation was too short to produce reproductive isolation in these two clades. Members of the Eurasian-African Reed Warbler complex are

one step further on in their differentiation, though their genetic distances have only been relatively recently investigated (Hering *et al.* in prep.).

Many young species with non-overlapping ranges have undergone relatively little ecological differentiation. When such geographically separate species come into contact there can at times be considerable competition between them. For example, the two *Nesillas* species on Madagascar, Madagascar and Lantz's Brush Warblers, are spatially separated because of an abrupt change between two habitat types, and are now apparently unable to expand their ranges into each other's (map 2.1). The best explanation for the failure of parapatric species or allospecies to penetrate into each other's range seems to be competitive exclusion, coupled with adaptations to environmental factors that give the competitive edge to each species in its own range (Price 2008).

In a situation of complete reproductive and ecological separation, the ranges of two sister species in secondary

Photo 2.22 Grande Comore Brush Warblers are an arboreal endemic of the island. They were previously considered to be a subspecies of Madagascar Brush Warblers (photo Jens Hering).

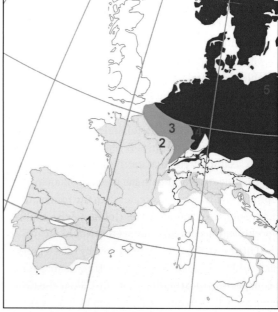

Map 2.2 Current distribution of Melodious and Icterine Warblers in Europe and their shifting hybrid zone. Melodious Warbler = yellow; Icterine Warbler = blue; hybrid zone = green. 1 and 5 = distant allopatry of Melodious and Icterine Warblers respective; 3 = sympatry; 2 and 4 = close allopatry of each species (after Faivre *et al.* 1999 and Secondi *et al.* 2006).

contact may overlap locally without competition. Upcher's and Olive-tree Warblers, which clearly differ from each other both genetically and in their behaviour, occupy differing habitats (thorny bushes and small trees *vs* scrub and higher trees) and altitude zones (500-1800 m *vs* 0-500 m), both allopatrically and in several areas of overlap in Asia Minor, so that in fact they hardly come into direct contact with each other (Roselaar 1995).

In other cases, contact between two species can be more significant: although the gene-flow barriers (premating and postmating isolation) might be strong, they can be incomplete, in which case forms that meet are able to mix to varying degrees. Initial hybridization can then disappear again if correct species recognition is greatly improved due to the reduced fitness of hybrids (reinforcement by increased discrimination; Price 2008).

Hybridizations and hybrid zones

When two taxa hybridize, the evolutionary transition stage is sometimes maintained within what is known as a 'hybrid zone' for long periods of time. Hybrid zones are areas where genetically distinct populations meet, mate, and produce hybrids (Price 2008). They represent a snapshot in the progress towards speciation; individuals from one population mate with individuals from another, but cross-mating does not lead to a collapse of the two populations back into one. The presence of viable and often fertile hybrids in the hybrid zone means that some genes can flow from one population to the other. On the other hand, the maintenance on each side of the hybrid zone of qualitatively different phenotypes, which may often be classified as separate species, implies that many genes fail to cross the zone (Price 2008). One such case is the shifting hybrid zone between

Photo 2.23 Melodious Warblers, breeding in bushy thickets in NW Africa and SW Europe, have extended their breeding range northwards during the 20th century, bringing them into contact with the Icterine Warblers of NE Europe (photo Daniele Occhiato).

Icterine and Melodious Warblers in western Europe. In order to reach a greater understanding of the complicated interactions between both species in the current contact zone, we should first look at their common, but initially separate, past.

The case of Melodious and Icterine Warblers

In time gone by, populations of the common ancestor of both these *Hippolais* warblers were geographically divided. The DNA of two separated taxa drifts apart at a rate of *ca* 2% of its sequences per million years (Weir & Schluter 2008). If we take this '2% rule' as a basis, and the genetic divergence between Melodious and Icterine Warblers as around 3.7%, then this separation must have occurred around 1.9 million years ago (mya). In their separate areas, both a western and an eastern population evolved in different directions, leading to genetic and phenotypic differentiation between the two forms. Differences arose in their morphology (wing length and shape), behaviour (songs and calls), ecology, and probably physiology (adaptations to different habitats and climates). They lived in separate refugial areas during the last glacial maximum around 21 000 years ago (Engler *et al.* 2010) and additional diverging adaptations presumably occurred during the post-glaciation expansion of their ranges northwards. Both species breed in deciduous vegetation: Melodious Warblers in sunny bushy woodland or shrubby areas with scattered trees in dry zones, Icterine Warblers in open or dense deciduous woodland with an understorey. Both pass the northern winter south of the Sahara, but in separate winter quarters reached on different flyways.

Today both species replace each other geographically, occurring sympatrically only in a narrow (*ca* 130-km wide) area of overlap, where they breed locally in the same habitat, forming mixed pairs and hybridizing to a limited extent (Secondi *et al.* 2006; map 2.2). Where they are allopatric, both species are clearly differentiated acoustically, their songs differing substantially in temporal and syntactical features. Icterine Warblers sing more slowly and employ more repetitions than their sister species. However, males of each taxon react to some extent to the song of the other, even when living allopatrically, so that a complete barrier between them does not appear to exist (Ferry & Deschaintre 1974). In the narrow zone of sympatry, interspecific differences fade in that their songs converge, favouring interspecific territoriality as well as the formation of mixed pairs (Secondi *et al.* 2003; chapter 6).

During the past 50 years, Icterine Warblers have retreated from the southwestern edge of their range in western Europe, while Melodious Warblers have simultaneously expanded to the northeast, resulting in a gradual northeastern shift of the zone of overlap by 110 km (Faivre *et al.* 1999). The dynamic of this moving contact zone has been well documented in long-term studies in various locations in northern France, and it appears to have several causes. One is that in some parts of the zone hybridization between the two species has been directly responsible for the local extinction of Icterine Warblers (replacement via reproductive interactions; further details in chapter 6). In other areas a replacement without interaction has taken place and Melodious Warblers have settled an area only after Icterine Warblers have become extinct there.

In a genetic study, Secondi *et al.* (2006) examined many birds in four subsets: first, individuals from the current contact zone of both species (sympatry); secondly, individuals from areas at the front of the contact zone (with Icterine Warblers and dispersing Melodious Warblers: close allopatry I); thirdly, individuals from the former contact zone (presently occupied only by Melodious Warblers: close allopatry II), and fourthly, individuals from distant areas of allopatric occurrence of both species. Throughout the entire contact zone both species mate assortively, that is, preferentially with conspecific partners. Accordingly, parental Icterine Warbler and Melodious Warbler genotypes dominate here. On the other hand, molecular analysis confirmed the rare occurrence of hybrids; backcrosses of viable hybrids to parental species are presumably selected against. They are mainly produced in the stages of co-occurrence in which the availability of conspecific sexual partners is low. Hence mixed pairs arise first in the early stage of sympatry, when only a few individuals of the advancing species are present, and secondly in the late stage, when the receding species is in decline.

Fig. 2.8 illustrates that backcrosses of viable hybrids to both parental species lead to an introgression of nuclear genes that is asymmetric, in that it is stronger in the expanding Melodious than in the receding Icterine Warblers. Even the rarer introgression of Melodious Warbler genetic material in Icterine Warblers has drastic consequences, because it gives rise to morphological changes in the species towards greater similarities with Melodious

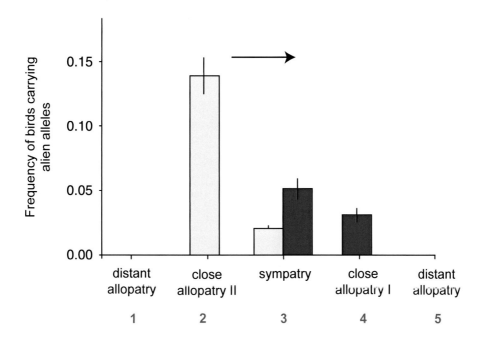

Fig. 2.8 Mean frequency of individuals carrying alien alleles in the distributional context of two hybridizing *Hippolais* warbler species: Melodious Warbler = yellow; Icterine Warbler = blue. Sympatry (3) = both species currently occur together; close allopatry (2, 4) = populations are less than 150 km from the closest sympatric population. Melodious Warbler populations in this category have recently become allopatric since sympatry with Icterine Warblers has not occurred for about 25 years, due to their disappearance from the area. Distant allopatry (1, 5) = populations are more than 150 km from the closest sympatric population and represent reference populations of both species. Note the higher proportion of Melodious Warblers with Icterine genes in close allopatry than in sympatric populations. The arrow indicates the direction of the moving hybrid zone (after Secondi *et al.* 2006; see also map 2.2).

Warblers. Introgression is highest in Melodious Warbler populations that became allopatric only recently, which suggests that alien (Icterine Warbler) genetic material spread mainly at the time just before the receding species became extinct (Faivre *et al.* 1999). Backcrosses are therefore more likely in the increasingly abundant species, Melodious Warblers, and, surprisingly, in the *wake* of the hybrid zone. These phenomena all indicate that the speciation process remains incomplete – stuck halfway, as it were. The formerly isolated taxa have made contact 'too soon', before reproductive incompatibility was reached, which is why such taxa are also called 'semispecies' (Helbig 2000). In them, hybridization results in gene exchange, slowing down divergence in some traits, but also creating novel genotypes that could facilitate evolution in new directions.

Other hybrids

Occasional heterospecific matings and hybridization have been recorded, both between sister species or closely related acrocephalid species within a clade, and also between species belonging to clades distant from each other, such as the large and small reed warblers. A targeted search for hybrids between the closely related Great and Clamorous Reed Warblers in an area of contact in southern Kazakhstan revealed that the two species do in fact hybridize, 1.6% of the population being viable hybrids (Hansson *et al.* 2003c). In this case hybridization is probably the result of general mistakes in mate recognition. In all other known cases, it is a strong asymmetry in abundance between two co-occurring species (related to each other to varying degrees) that has led to mixed pairs and hybridization, which therefore result from a shortage of mating opportunities for the rarer species. This explanation applies to those hybrids that were noted during the expansion of one species into the range of another, and to situations where one of two species has always been rare or has declined sharply in numbers in an area of mixed breeding (table 2.2).

In an area of sympatry it is typically females of the rarer species that engage in heterospecific mating (Wirtz 1999), while the development of an 'interest' in heterospecific females among males is hindered by the strong prefer-

Table 2.2 Reed warbler hybrids. Hybridization generally occurs between a rare and a common species in a region.

Expanding or rare species	Common species	Region	Reference
Blyth's Reed Warbler	Marsh Warbler	Finland Netherlands	Koskimies 1991b Poot *et al.* 1999
Marsh Warbler	Sedge Warbler	Norway	Lifjeld *et al.* 2010
Eurasian Reed Warbler	Marsh Warbler	Belgium	Lemaire 1977
Great Reed Warbler	Eurasian Reed Warbler	Germany Belgium	Beier *et al.* 1997 Hansson *et al.* 2004d

ence for conspecific males among these females, and/or their rejection of heterospecific males. It is not only the shortage of conspecific partners that promotes formation of mixed pairs, but also the occasional appearance of mixed singers in the rarer species. These are males that have been misimprinted as juveniles on the commoner 'wrong' species due to a lack of conspecific models (see above; chapter 7; Helb *et al.* 1985). Furthermore, intermediate songs have also been noted in several hybrids (Koskimies 1991b, Beier *et al.* 1997, Hansson *et al.* 2003c, Lindholm *et al.* 2007).

The viability of a number of known reed warbler F1 hybrids (mostly male) does not appear to have been reduced in a number of studies which have monitored some of them for several years (Koskimies 1991a,b, van Eerde 1999, Lifjeld *et al.* 2010, unpubl. own data). In at least one case, a back-cross of a hybrid female – Blyth's Reed Warbler x Marsh Warbler – with one of the parent species (Blyth's Reed Warbler) was fertile (Koskimies 1991b). Despite such occasionally successful backcrosses, the generally intrinsic loss of hybrid fitness is nevertheless the key factor that contributes to reproductive isolation in speciation (Price 2008).

Map 2.3 Map of the Old World in the Miocene. Note the collision between the Arabian and Eurasian tectonic plates and the position of the inland sea (Second Tethys; sourced from Wikimedia Commons).

5 Origins and distribution of the reed warblers

It is fascinating to ask where a group of birds has originated and where it has diversified. By making use of the new molecular phylogenies, such questions can actually be answered today – something of which earlier evolutionary biologists could only dream. The necessary prerequisite for a reconstruction of ancestral distributions is to understand a group's phylogenetic or evolutionary tree. A version known as an ultrametric tree, calculated so that branch lengths correpond to time, allows us to date when lineages (branches) split from each other (fig. 2.9). Starting with the recent geographical distributions of taxa, and adding the new information supplied by the ultrametric tree, we can reconstruct the ancestral ranges where the splits would have occurred. We can therefore formulate much more precise biogeographical hypotheses than was the case in the past. Computer programs already exist that can be used to reconstruct quantitatively the ancestral home areas of groups with multiple branchings and well-resolved phylogenies (Springer *et al.* in press).

Photos 2.24 and 2.25 Although Eurasian Reed Warblers in Europe are very difficult to imagine without reedbeds, a form of the eastern subspecies *fuscus* commonly lives in the olive and date-palm groves of NE African oases. Others, by contrast, inhabit mangroves along the Red Sea, though they are also able to build their nest in a palm tree (photo Jens Hering).

For the small group of the acrocephalids, an attempt can now be made at a plausible reconstruction of their historical expansion, using the sequence in which splits took place together with the present geographical ranges of the member species (fig. 2.9). The first question to be dealt with is which groups are the closest relatives of the acrocephalids and where do they occur today? The genetic data suggest that the Locustellidae together with the Malagasy warblers (Bernieridae) represent the sister group of the Acrocephalidae family (Beresford *et al.* 2005, Johansson *et al.* 2008). While the Locustellidae (species-rich Old World ground- and low-herbage-dwellers) are widely distributed, the Bernieridae are an endemic songbird radiation of only 10 species, all of which live on Madagascar. They were only recently recognized and

named as an independent group (Cibois *et al.* 2001, 2010 a), whose members were previously distributed between the three families of the babblers (Timaliidae), bulbuls (Pycnonotidae), and Old World warblers (Sylviidae).

The earliest ancestors of the reed warblers would have lived in shrub and forest-edge vegetation in Africa. This assumption can be justified on two grounds. First, according to the latest research, Africa played a central role in the radiation of warbler-like birds (Sylvioidea, Passerida) and the history of this group on that continent has turned out to be older than previously thought (Cibois *et al.* 1999, Fuchs *et al.* 2006, Jonsson & Fjeldsa 2006). Secondly, the oldest branching within the acrocephalids in Africa probably led to their colonization of Madagascar, and the extant genus *Nesillas* then evolved from these Malagasy ancestors. The island's size and relative proximity to continental Africa make it very likely that Madagascar was colonized independently several times by various 'warbler' clades. The next oldest splits, such as those in the ancestors of the *Calamonastides-Phragamaticola-Iduna* and *Hippolais-Acrocephalus* groups, most probably also took place in Africa. The African origin of the former group is also confirmed by the early branching that gave rise to Papyrus Yellow Warblers and accompanied a shift from shrub to swamp habitat.

As early as *ca* 20-15 mya, in the Miocene, it became increasingly possible for landbirds to move from Africa to Eurasia as a result of the landbridges created by the collision of the Arabian and Eurasian tectonic plates in the Middle East (Voelker 1999, Beresford *et al.* 2005; map 2.3). Ancestors of Thick-billed Warblers and the *Iduna* species which presently occur in central Asia must have subsequently moved deep into Asia, with the Asiatic lineage then separating from the forms of the *Iduna* complex that had remained and further differentiated in Africa.

On the other hand, the separation of the *Hippolais-Acrocephalus* group, again associated with a change of habitat from scrub to wetlands, can probably be localized in the Near East. Around 12 mya the large and small reed warblers split from each other (Price 2008). The deep split, hence basal position, shown by Basra Reed Warblers within the large reed warblers underscores the importance of the Near East for the early diversification of the *Acrocephalus* genus. The speciation of the *Hippolais* warblers can also be traced to the Near East, including the western Mediterranean Basin, which has made them the only endemic avian group in the Mediterranean region (Blondel & Aronson 1999).

We have already discussed the further diversification of the large reed warblers into an African subgroup (*Calamocichla*) and a Eurasian one, as well as the colonization of oceanic islands by members of both subgroups (section 2.3; see also chapter 12). The splitting of the small reed warblers into plain-coloured and striped probably happened later than this, most likely in Eurasia, and was once again acompanied by a habitat shift amongst the striped forms to lower, denser vegetation.

A comparison of the maps of species richness and of the Miocene (maps 1.1 and 2.3) illustrates that the area with the highest density of acrocephalid species is western and central Asia. This region is congruent with the remnants of the former inland sea Tethys Seaway (Second Tethys). Vladimir Ivanitskii and colleagues (2005) seem to have been quite correct in their speculation that the centre of *Acrocephalus* diversification lay in the area of the Black, Caspian, and Aral Seas, which are relicts of the prehistoric Tethys Sea. We can easily envisage that the frequent fragmentation and island formation caused by the shrinking or disappearance of inland water bodies in Asia could have resulted in speciation events. A high potential for the isolation of such populations during periods of climate change would also have existed among the other groups that occupied the shrubby habitats of Asiatic fold mountains, such as the ancestors of Blyth's and Large-billed Reed Warblers or Blunt-winged Warblers and Manchurian Reed Warblers.

Nevertheless, the centre of origin of the entire acrocephalid group must be placed in Africa and the Middle East. The acrocephalids therefore represent another impressive example of a situation where the centre of distribution of the ancestral lineages of an avian group is not congruent with the current centre of highest diversity of its evolutionarily younger species (e.g. Mayr 1990, Voelker 1999). A second striking pattern revealed by this reconstruction of the processes of both ancient and more recent splits is that the most closely related species frequently tend to differ in habitat choice. However, in the course of the diversification process there was an early 'commitment', or consistent adherence, to the gleaning form of feeding technique (chapter 4), since it has been successful in all habitats settled by the group. A similar development can be seen in the *Phylloscopus* leaf warblers (Richman & Price 1992).

The change from scrub to reed and vice versa occurred again and again in the course of reed warbler history, and was apparently easily mastered each time. The ability to cope with either habitat is shown today by the young species in particular, such as the reed warblers on oceanic islands or isolated populations of the *A. scirpaceus* superspecies complex in the oases of north Africa (Hering *et al.* in prep.). In both cases we are dealing with environments in which there are no competitors, or at least very few. It has been absolutely no problem for the *scirpaceus* warblers – normally characterized by their close relationship with reeds – to switch to building their nests even in oasis palm trees.

Summary

Throughout this book, the biology and ecology of the various members of the Acrocephalidae family will be repeatedly compared, and this must obviously be based upon precise knowledge of their evolutionary relationships. Consequently this chapter concerns itself with the phylogeny and diversification of the reed warbler group, which – in its plumages and body shapes – appears so remarkably uniform. Its component species have many features in common – powerful legs, skilfully constructed bowl-shaped nests, and naked nestlings whose gapes reveal two tongue spots, to name but a few. For a very long period of time, the boundary between the reed warblers and other groups of Old World warblers was blurred by functional similarities and the lack of unambiguously diagnostic characters. It was only with the advent of cutting-edge molecular techniques, in particular DNA sequencing, that an objective basis was found for the measurement of genetic distances and thence the degree of relatedness of taxa. This resulted in a whole series of revisions to the existing systematics.

Several reed warbler species which were in fact only distantly related had been wrongly placed close together, solely because their similar lifestyles had resulted in acquired physical resemblances (convergence). On the other hand, the genuine close relationships between other species that bore little morphological similarity to each other had gone unrecognized. Furthermore, it was possible to reveal some cryptic species, and also to demonstrate various stages of the speciation process (i.e. the genesis of reproductive barriers between taxa).

The Acrocephalidae are now regarded as a single family within the superfamily Sylvioidea, encompassing *ca* 53 living species in 6 genera. We can assume that the origin of the group lies in Africa, or more precisely in the Malagasy Region, where the earliest splits between ancestral forms of the reed warblers probably occurred. Today, however, the area with the highest acrocephalid species density is central Asia.

Change of habitat from bushes to reeds and vice versa occurred several times during the diversification of the group. Reed warblers are widely distributed and have developed an astonishing variety of lifestyles. More than one-third of the species have settled on oceanic islands, while almost three-quarters of those distributed on continents are migratory.

On the basis of these new insights, the reed warblers can be considered as a clearly defined group, which, given their varying life histories, makes them especially suitable for comparative studies.

3

Habitat characteristics

Cape Verde Warbler in a coffee bush.

When people think of reed warblers, it is quite understandable that they immediately associate them with extensive stands of tall and impenetrable reeds, from which emerges the repetitive song of a Eurasian Reed Warbler or the grating 'karrakeet' of a Great Reed Warbler. Within their range reed warblers are both the most characteristic and the most commonly occurring inhabitants of reedbeds and other damp thickets. However, this does not mean that acrocephalids are to be found exclusively in plant communities associated with silted-up water edges, bottomland, or similar low ground. Members of the family occur in all climate and vegetation zones, from the Arctic to the tropics, including in deserts and high up in the mountains. Irrespective of whether we are talking about reedswamps, bushes, or forest edge, the common denominator in all these habitats is that the vegetation forms have a simple structure (with poor thermal buffering), hold a food supply consisting of large numbers of mobile insects, and are able to satisfy the acrocephalids' essential need for cover. Vegetation types which meet these requirements can be as varied as willow scrub in the Arctic (Sedge Warblers), moist bamboo groves on Tahiti (Tahiti Warblers), or mixed macchia in Morocco (Melodious Warblers).

Two habitat characteristics make it necessary for reed warblers in higher latitudes to leave their breeding grounds in winter: dependence on mobile insects and low-complexity vegetation (Hockey 2005). If we ignore the island forms, a substantial majority of three-quarters of all the acrocephalids are migrants that make a long journey twice a year between their winter quarters and their breeding ranges (chapter 11). In the temperate latitudes of Eurasia, a broad spectrum of vegetation types is utilized for breeding in the northern summer. These migrants have more food resources at their dis-posal for raising their young in the seasonally productive temperate zones than do their resident relatives in the tropics, where the food supply alters less between breeding and non-breeding seasons (Ricklefs & Wikelski 2002). Since their stay in their breeding areas is often only 2-4 months, many long-distance migrant reed warblers, together with *Hippolais* and *Iduna* warblers, spend most of the year on the move or on their wintering grounds, where they frequent thickets of subtropical or tropical vegetation that can be very different from those of the breeding habitat.

Photo 3.1 An aerial view of the Biebrza marshes in Poland, in which all the transitional plant communities of the area can be seen alongside each other. In the few remaining natural riverine plains of central Europe which have meandering rivers, the diversity of these transitional communities between aquatic and terrestrial habitats is still remarkably high. The island-like distribution of reedbeds is well illustrated (photo Norbert Schäffer).

Photo 3.2 Although *Phragmites* stands are sometimes called natural monocultures, the interiors of reed-beds are heterogeneously structured, with a mosaic-like distribution of shallows and deep water. They are frequently and chaotically affected by windbreak and snow, leading to the patchy distribution of crucial resources such as food and nest sites. The many edges between reed and water, increased by inlets and reed islands, have a positive effect on arthropod richness and bird density (Bosnia; photo Elio della Ferrera, Nature PL).

The habitats occupied by acrocephalids can be broadly differentiated on the basis of two ecological factors: decreasing wetness and increasing vegetation height, though there is no necessary correlation between the two. The habitat spectrum settled by the family extends from wetlands with reed or mangrove communities, through marshes with meadow and tall perennial communities, to bushy scrub, edge communities, and woodland vegetation in drier locations. The various plant communities have characteristic vegetation profiles, derived by measuring the horizontal density of cross-sections made at different heights (Cody 1974). On the basis of their average vegetation height and their profiles, these communities can be assigned to five vegetation types (fig. 3.1): swamps and marshes, dry marshes, scrub, shrubby vegetation, and woodlands.

1   Wetlands

About a third of the species in the family Acrocephalidae are closely associated in their distribution with wetland vegetation. Since such plant communities are dependent on the presence of water, or soils with a high water content, and are not tied to the Earth's climatically determined biomes, they are counted among the vegetation types known as azonal and their distribution is worldwide (Ostendorp 1993). Although wetlands are large and continuously distributed in alluvial plains, estuaries, and deltas in some regions of the world, they mainly take the form of habitat islands in the surrounding landscape. This is in strong contrast to the extensive and unbroken forests, agricultural landscapes, and steppes of the temperate zones or tropical savannas. There have been very few studies of the consequences for the birds that occupy

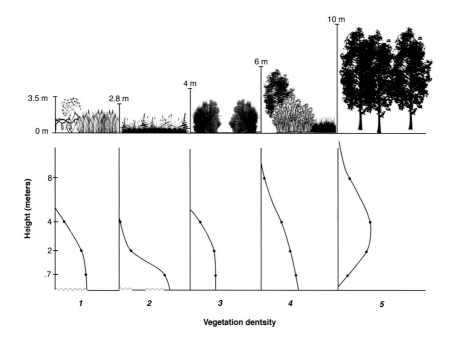

Fig. 3.1 Schematic representation of five vegetation types (top panel) and their rough profiles, i.e. vegetation density at various vertical sections (below panel).

such insular and often transient habitats – for their dispersal behaviour or genetic exchange between populations, for example – even in large non-passerine species, and virtually not at all in the small songbirds (chapter 11).

Depending on the extent of water stagnation, two wetland vegetation types can be distinguished, which commonly overlap: swamps and marshes. A swamp differs from a marsh in that a greater proportion of its surface is open water and it may be deeper than a marsh, with its saturated surface permanently below water level.

Swamps, wet marshes, and mangroves

Swamps and deep water marshes (vegetation type 1 in fig. 3.1) are chiefly composed of silted-up succession commu-

nities, or hydroseres as they are called, with emergents, also known as reedswamps or sometimes swamp-forests, as found in the tropically distributed mangroves. These emergents can be dense-growing tall plants, many of which reach an average height of 3.5 m above the water surface and can characterize entire landscapes.

Reedswamps and tall aquatic emergents

Reedswamps occur at the edges of standing and slow-running water, in marshes and wet depressions, on the tidal flats of the sea coast, at river mouths, in estuaries and brackish marshes, or in man-made places like fish- and settling-ponds. They are generally well-developed hydrosere communities of emergents such as reedmace, club-rushes such as bulrush, great fen sedge, reed sweet-

Photo 3.3 The rising *Phragmites* stems create the vertical, almost two-dimensional character of reedbeds. With a daily growth in *Phragmites* plants of 4.5 cm and a biomass production of 25 g dry matter per m², reedswamps are among the world's most productive ecosystems. Their density of 50-200 stems per m² and height of 2-6 m make reed stands almost impenetrable for the human researcher (Hungary; photo Milán Radisics, Nature PL).

grass, and, above all, reed itself, which is easily the most important in terms of area. The chief reedbed formers belong to the class of monocotyledonous plants and are mainly grasses (Poaceae) or sedges (Cyperaceae). Their rising stems create the typically vertical, almost two-dimensional character of these communities (mono-layer), which frequently occur in extensive pure stands so that they are sometimes called natural monocultures. They represent stable and long-lived communities that are well adapted to their particular location and its climate (Ostendorp 1993).

No ecosystem has sharp boundaries, least of all reedbeds, which form a fluid transition zone between water and land. Their extent and distribution are limited by a number of factors, among which are dryness, lack of nutrients, permanently high salt levels, too great a water depth in the transition to the floating vegetation belt, deep shadow in wet carrs or alluvial woodland, and, finally, alterations made by humans.

Reedbeds reach their greatest extent in the silted-up zones of shallow lakes. Fig. 3.2 shows an example of such a hydrosere, the process of ecological succession by which open water is converted into fen and 'dry' land with

shrubs, following the accumulation of plant remains and an increase in inorganic siltation. At the edge of a lake or pond the pioneer community that develops into the open water consists of submerged plants, some of which may form floating mats. Despite the fact that reedbeds do slowly shift spatially over time, they can be available as a habitat in one location for many hundreds of years. Human influences, such as changes in the water regime or economic exploitation of the bank vegetation, can encourage or hinder succession, for example, the invasion by bushes. This is an issue which has now left hardly one of today's wetlands unaffected (chapter 13).

The most significant shaper of reedswamp, and therefore the most important plant in the habitat of many reed warblers, is the common reed. This cosmopolitan grass is mainly distributed in the northern hemisphere, with related species replacing *Phragmites australis* in several subtropical and tropical areas (Rodewald-Rudescu 1974). Many ecotypes of *Ph. australis* exist, varying in morphology, growth cycle, and response to temperature (Haslam 1973). Common reed is a perennial grass species, ideally adapted to life in the transition zone from water to land. It is made up of stem, leaf-sheath, leaf blade, and panicle or inflorescence, as well as an underground rootstock, the

Photo 3.5 Succession in reedswamps is accelerated in eu-trophic and uncut stands: the proportion occupied by willow and alder scrub increases within a few years (photo Oldrich Mikulica).

Photo 3.4 Great Reed Warblers and other reedswamp acro-cephalids are perfectly adapted to the vertical structure and high density of the vegetation. Here they have a substantial supply of invertebrate prey as well as nest sites (photo Oldrich Mikulica).

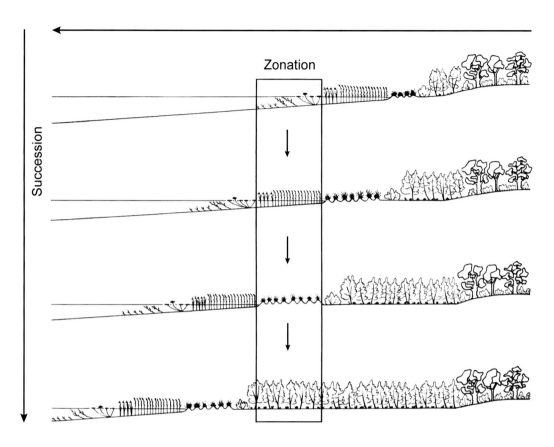

Fig. 3.2 The zonation of a hydrosere, i.e. the transitional area from water to dry land, and its spatial shift over of time (succession).

rhizome body (fig. 3.3). Given a good nutrient supply, the undivided stems grow up to 5-6 m vertically, while the rhizomes can be 1 m deep under the surface of the substrate. In spring, 20-60 buds per m² sprout from the upper part of the rhizome; inside 3 months they grow into long stems, along which the leaf blade unfolds. With their 'growth front' these fresh stems then create a dense vertical zone between the old stalks. The density of stems in a reedbed can be so high that visibility is restricted to less than a metre. This ensures good cover for reed warblers, though it can also make it difficult for those who study them to make any headway!

In temperate regions, panicle-forming and flowering occur in late summer; leaves fall from October onwards and then the stems also die off. Among the adaptations of reed to its habitat are elasticity of the stems to enable them to resist mechanical stresses, deep-lying rhizomes to safeguard against water level fluctuations and desicca-

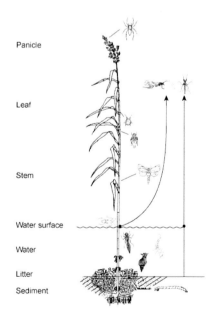

Panicle

Leaf

Stem

Water surface

Water

Litter

Sediment

Fig. 3.3 Vertical sections of the reed plant and the distribution of important invertebrates in the diet of reed warblers (Ostendorp 1993). Note the arrows indicating emergent insects.

Photo 3.6 Heterogeneously structured, extensive reedbeds in larger river deltas and shallow lakes, often with layers of old broken stems, are the typical habitat of Moustached Warblers (Mallorca; photo Peter Castell).

Photo 3.7 In Australia, the habitat of Australian Reed Warblers is also characterized by stands of chiefly *Typha* and *Phragmites*, and can show a close resemblance to that of the Great Reed Warbler in Eurasia (photo Karl Seddon).

tion, rapid stem growth to ensure protection in times of high water, and vegetative reproduction, which bypasses the critical seedling phase if water levels vary too much.

The dense reedbed canopy, with up to seven m² of leaf area per one m² of swamp, serves not only for photosynthesis, but also to shade and suppress competing plants. The enormous growth performance of *ca* 4.5 cm increase in height per day, together with the daily biomass production of 25 g dry matter per m², can be attained by *Phragmites* only with the help of the subterranean storage rhizome. With 2 or 3 kg dry weight per m² and per year of created plant mass, reedswamp is among the world's most productive ecosystems. This figure is higher than that for mature deciduous forest by a factor of almost three, which nearly reaches the values of tropical rainforest or of intensively managed agricultural crops (Ostendorp 1993). The quality of the *Phragmites* plant can be ascertained from stem thickness, growth height, and stand density. Environmental disturbances, such as mowing, fire, water shortage, mechanical and feeding damage, or matting, can lead to a reduction in stem diameter and growth height, though they often result in an increase in the number of stems per unit area. Waterfacing reedbed is mostly higher, thicker, and more sparsely spaced than land-facing reedbed, and these differences are so considerable that the latter is not settled by larger species like Great Reed Warblers or Little Bitterns.

In contrast to most ecosystems, only a very small proportion of the plant mass produced by reedbeds is directly consumed by grazing animals or herbivorous arthropods. Caterpillars, easy prey for songbirds, are therefore uncommon. One reason for this is that silicate deposition results in the reed plant producing extremely hard leaves, in order to protect itself from grazers. Among the primary consumers is the leaf-sap-sucking mealy plum aphid, whose population can increase to infestation levels in some years. Most of the phytophagous insects, such as larvae or caterpillars of the gall-midge, the twin-spotted wainscot, or the reed leopard, are therefore found in the stem interior or in the rhizome, where silicate levels are lower, or they are gall-formers like the shoot-flies of the genus *Lipara* (Chloropidae). Reed warblers find it difficult to reach them, though other songbird neighbours, such as Reed Buntings, can get at them by biting or hammering the stems open. The bulk of the plant mass dies back in the autumn, becoming reed litter that is decomposed to detritus, principally in the water, by fungi, bacteria, etc. Detritus in turn is food for the countless species of aquatic invertebrate grazers and omnivores. These include water slater, amphipods, the larvae of caddis-flies, and snails. Such invertebrates are themselves eaten by the secondary consumers, predatory insects and their larvae. A further important group of aquatic detritus consumers and omnivores lives in the sediment layer of the soil, among them snails, bristleworms, and insect larvae, such as non-biting midges. Many organisms living in the water only become available to most songbirds when they rise to the surface to breathe, hatch, or pupate. Very often in spring and summer, vast numbers of caddis-flies, dragon-flies and damselflies, mayflies, or true flies (Diptera) hatch synchronously, leading to huge emergences of fly-ing insects (see below). Finally, the water surface is inhab-ited by predatory arthropods, such as pond skaters or wolf spiders.

Because reedbeds exercise a considerable braking effect on the wind, a warm microclimate exists in their interi-ors, harbouring rich insect life well into the autumn. Fur-ther arthropod groups, among them spiders and small beetles, migrate into the reed stubble to overwinter (Ostendorp 1993, Chernetsov & Manukyan 2000). These represent an important food source for early-arriving pas-serines in spring (chapter 4).

Photos 3.8 and 3.9 Extensive open sedge-fen mires and wet grassland are home to Aquatic Warblers, whose habitat requirements are very narrowly defined (Belarus and Lithuania; photos Alexej Kozulin, Renatas Jakaitis).

The enormous biomass production of reedbeds in the temperate latitudes is exceeded by that of papyrus swamps in the subtropical and tropical zones. Here the decomposition and transformation of vegetable matter is carried out by specialized worms, even under oxygen-poor conditions, and accelerated by high temperatures (Muthuri *et al.* 1989). Typical marsh dwellers are the large reed

Photo 3.10 Tall herbaceous vegetation bordering on reedbeds in a variety of Asian wetlands is the habitat of Black-browed Reed Warblers (Japan; photo Shoji Hamao).

Photo 3.11 In India, high grass and reedbeds gradually merging into *Acacia* woodland provide both breeding and wintering habitats for Clamorous Reed Warblers (photo Hans Winkler).

warblers, the Eurasian-Africa Reed Warbler superspecies, species of the *agricola* complex, Moustached Warblers, and Papyrus Yellow Warblers.

Mangroves

Mangrove stands are also highly productive communities, which share some features with reedbeds. Both vegetation forms are regularly flooded; in both only a fraction of the primary production is consumed by herbivores, and old plant material, such as leaves, twigs, and seeds, is decomposed and converted by detrital feeders in a generally muddy substrate. However, mangroves differ markedly from reedbeds in their structure, as they consist of woody, leafy vegetation. Mangroves are favoured by the acrocephalids, not only for overwintering (section 3.5), but also as breeding habitat for resident species or subtropical short-distance migrants, such as the local subspecies *avicenniae* of Eurasian Reed Warblers or Clamorous Reed Warblers in the extensive mangroves on the Red Sea, The Gulf, and coasts of India. In addition, various forms of the otherwise scrub-dwelling *Iduna* warblers have specialized in this habitat, such as the subspecies *alulensis* of Eastern Olivaceous Warblers in the mangroves of the Red Sea and northern Somalia (Ash *et al.* 2005), or the resident population of Sykes's Warblers in the coastal mangroves of The Gulf (Pearson *et al.* 2004, Kennerley & Pearson 2010). While these warblers chiefly occupy mangroves of the genus *Avicennia*, Saipan Warblers breed in *Bruguiera* on the Pacific island of Saipan (Mosher & Fancy 2002). Both mangrove forms have emergent aerial roots as important habitat structures.

2 Dry marsh, tall herbaceous, and wet meadow communities

These rather unstable vegetation forms (vegetation type 2 in fig. 3.1) are mostly succession communities landward of deep-water reedbeds. They represent a broad spectrum of low-growing, leafy, rank vegetation, between half a metre and three metres in height, whose most important common features are a very dense lowermost layer and high soil moisture, being truly wet almost exclusively in spring. They would include areas with various kinds of tangled, tussocky plants, such as wet sedge (*Carex*) meadows or other wet grasslands with low-growing scattered bushes, fen mires, luxuriant herbage with or without woody erect stems, and the weedy borders of fields. As opposed to what happens in swamps, a majority of the prey invertebrates develop in the damp soil (e.g. Diptera) and many also occur on the new spring growth of rank vegetation, or are hidden at the base of grassy tussocks. Matted and dry places, or those which are insufficiently flooded, are less productive (Vergeichik & Kozulin 2006a, Tanneberger *et al.* 2008). A characteristic of certain tall perennial communities, such as nettles, is that they wither rather quickly in late summer. A similarly pronounced seasonality is also shown by tropical marshes dependent on the timing and intensity of the rains. Typical inhabitants of such vegetation forms are the striped acrocephalids, such as Black-browed Reed Warblers, Sedge and

Aquatic Warblers, but also Marsh Warblers, while among the *Iduna* species Booted Warblers are attracted to rich ground cover.

Although it is the case that most wetlands are in lowland areas, damp places at higher altitudes also provide a habitat for some reed warbler species. Moustached Warblers inhabit mountain valleys up to 2000 m in the Caucasus, Marsh Warblers occur at 1500 m in Switzerland and up to 3000 m in Georgia, and Lesser Swamp Warblers have been recorded in Africa up to 2700 m (Urban *et al.* 1997, Kennerley & Pearson 2010). Some species inhabit both low-lying wetlands and high-altitude sites, where they shift to scrub; examples would be Blunt-winged Warblers in the Himalayas up to 3000 m or endemic Madagascar Swamp Warblers (Kennerley & Pearson 2010). Generally Madagascar Swamp Warblers occur in lowland or mid-elevation aquatic habitats, but a population was found quite recently by Goodman *et al.* (2000) above the tree-line in a dry ericoid vegetation zone.

3 Wetlands in sharper focus for passerines

As far as locomotion and nest building are concerned, reed-swamp actually represents an extreme habitat for birds; plant diversity is low, reed and associated plants have a uniform vertical growth form, and this kind of vegetation is often less than 4 m in height. In addition, the structural simplicity and homogeneity mean that the number of niches is limited. This explains why only a few bird species occur there, for instance a maximum of 20 out of a total of 150 breeding species in central Europe, though those that do inhabit reedbeds are often found at high individual densities (chapters 6 and 13). Population figures for insectivorous songbirds in marshes are higher than those associated with the same vegetative mass in forests (Fjeldså 1980). The adaptations necessary for a life in reeds, whether for climbing or for nest construction, are so great that they are dealt with at length in chapters 5, 8, and 14.

The varied nature of wetlands

Close relationships beween vegetation structure, food supply, and species distribution inside reedbeds have been discovered in the Nearctic icterids as well as the Palaearctic acrocephalids. Peculiar to swamp and marsh ecosystems are the insect emergences, masses of swarming mosquitoes, midges, mayflies, or dragonflies and damsel-flies (fig. 3.3). They are important as a continuously renewable food source for reedbed passerines, and explain both the prolonged breeding seasons and the frequently high population densities of some reed warbler species (chapters 6 and 13). The scale of emergences can vary considerably and is relatively unpredictable (Imhof 1972). Seasonal

Photos 3.12 and 3.13 In the Russian Ussuri region, tall herbaceous vegetation, *Phragmites* stands, and scrub are the habitat of Thick-billed Warblers (below illustrated here), Black-browed, and Oriental Reed Warblers (photos Irina Marowa, Jonathan Martinez).

Photos 3.14 and 3.15 Dry marsh and wet meadow herbaceous plant communities, which may be already wilting by August, are preferred by Marsh Warblers. In many regions this vegetation can be found on low-intensity farmland (Bulgaria; photos Bård Stokke, Joseph Hlasek).

2000). Whether they are water-facing or land-facing, the edges of a reedbed, with their higher number of plant species, are consequently richer in food than the centre (section 4.2). Furthermore, *Typha* and *Phragmites* reedbeds offer more food resources than *Scirpus* reedbeds, since they provide structural features, such as sheathing leaf bases or clumps of old plant material, that afford more foraging opportunities (Orians 1980). With increasing reed density the food supply declines (Taylor 1993, Hoi *et al.* 1995, Poulin *et al.* 2000). Dense, old *Phragmites* stands are therefore much poorer in food than places where thick-stemmed reed grows more sparsely, because the availability of light and oxygen-rich water in such dense stands is very restricted. In the reed belt around shallow lakes, such areas can even be anoxic at times in summer. Similarly, the food supply is reduced in stands of reed whose substrate has been dry for a long period (Poulin *et al.* 2002). These food-poor sections of reedbeds are usually occupied by the reed warbler species that can best cope with them, Moustached Warblers (alongside Bearded Reedlings), whereas the most productive parts attract the larger Great Reed Warblers (chapter 6). This negative relationship between density and productivity will be considered in more detail in chapter 4.

If, in this 'two-dimensional' habitat, the reed-water boundary is increased by inlets and reed islands, the ever-changing pattern of cover and open water has a positive effect on both arthropod species richness and breeding bird density (the 'edge effect', as it is known; Leisler & Dyrcz 1988). Reedbed edges are therefore not only attractive for foraging, but also for nesting birds, in spite of the fact that predation at the edge is greater than in the centre of the reeds (Poulin *et al.* 2000, Baldi & Batáry 2005, 2007; section 8.4).

In addition, the interiors of reedbeds are not necessarily homogeneously structured, but rather show a mosaic-like distribution of shallows and depths and are frequently chaotically affected by windbreak and snow, particularly in old stands. This has important consequences for the biology of many birds, since crucial resources like food and nest sites are unevenly distributed in space and time as a result of these effects (chapters 8 and 12). Hence reedbed quality can vary considerably in a small area, and 'good' areas can be monopolized for nest sites, while 'good' feeding places are not so easy to monopolize because of their low predictability. This means that many acrocephalids

and locational fluctuations occur, not only between regions, but also within them, depending on reedbed type or the presence of fish stocks, which may reduce aquatic larvae numbers. In general, songbird density in reed-swamp rises with the increasing availability of food in the form of insect density (Poulin *et al.* 2002).

This food resource is dependent on plant diversity; the number of potential prey animals rises with the increasing number of plant species and vegetation structures, resulting in more feeding and hiding places for arthropods (Orians 1980, Leisler & Dyrcz 1988, Poulin *et al.*

Photo 3.16 Low shrubs scattered in the wide steppe are the favoured habitat of Booted Warblers in western central Asia (Kazakhstan; photo Klaus Nigge).

Photo 3.17 A series of acrocephalids live in shrubby and bushy vegetation, sometimes in arid regions or at high altitudes. In the dry landscapes of western central Asia, tamarisk and saxaul bushes are the characteristic habitat of Sykes's Warblers (Kazakhstan; photo Peter Castell).

must frequently fly from their nest to alternative foraging sites elsewhere (Poulin *et al.* 2000; chapter 8).

The water-depth gradient

Although wetlands may superficially appear to be homogeneous habitats, closer observation will reveal a multitude of structural differences to which their inhabitants are well adapted. The water-depth gradient is the one eco-

logical factor that determines all else. The steadily decreasing water column, from the edge of open water to the landward side of a reedswamp, has a decisive influence on the composition of the avian community in both temperate and tropical habitats (Britton 1978, Henry 1979, Leisler 1981). For example, in the hydrosere of the Neusiedler See, where Aquatic Warblers are now extinct, the territories of the six *Acrocephalus* species which

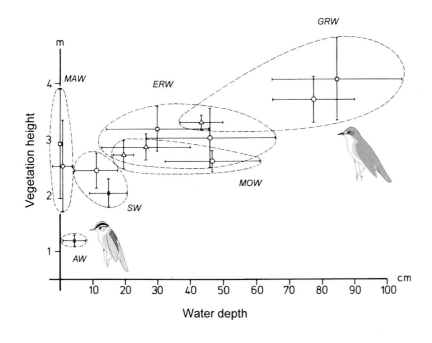

Fig. 3.4 Distribution of the habitats of six European reed warblers in relation to water depth and vegetation height. The means, with variation from different study areas, are indicated (after Leisler 1981). Niche separation and overlapping are thus visible.

AW   Aquatic Warblers
ERW   Eurasian Reed Warblers
GRW   Great Reed Warblers
MAW   Marsh Warblers
MOW   Moustached Warblers
SW   Sedge Warblers

Photo 3.18 Dry rocky hillsides with scrub vegetation are the habitat of Upcher's Warblers (Turkey; photo Peter Castell).

occured there are characterized by the close relationship between water depth and vegetation height (fig. 3.4). The largest species, Great Reed Warblers, occupy the tallest reedbed vegetation in the deepest water. At the landward end of the gradient we find either Marsh Warblers, in tall herbs and shrubby vegetation, or Aquatic and Sedge Warblers, in very low growth with sedges or grasses. In between these two areas, Moustached Warblers, Eurasian Reed Warblers, and Sedge Warblers all find their place, although in the littoral zones of eastern Europe and central Asia, their position is taken by Paddyfield Warblers. We can therefore observe that several coexisting species prefer the differing sections of the hydrosere succession which result from the silting-up process and plant invasion, and are consequently horizontally separated from each other (fig. 3.2).

Water level is of varying direct or indirect significance to different species. For Great Reed Warblers it can even be a proximate indicator of habitat suitability. If the level in the lakes of the Kalmuk steppes of southern Russia falls so much that the bottom becomes visible, they will leave the area, even if they have already built nests and laid eggs (Ivanitzkii *et al.* 2002). By contrast, Paddyfield Warblers, which breed more on the landward side, remain unaffected by that situation.

It is probably because reedbeds can be colonized by only a few species with specialized adaptations, and which consequently have no competitors, that the reed warblers were able to diverge in such a marked fashion in this hab-

itat. In chapter 14 we shall meet representatives of other groups that have 'taken the plunge' of colonizing reed-swamps where reed warblers are absent.

4 Scrub, shrubby, and woodland vegetation

A number of acrocephalids live in shrubby, bushy, and edge communities, brushwood and parkland, even in gardens, often far from water and sometimes at high altitudes. In these vegetation forms (vegetation types 3 and 4, which merge into each other: fig. 3.1) woody plants, bushes, and trees become more prominent. 'Scrub' (vegetation type 3) is characterized as being mainly below 5 m in height and on drier soils, whereas vegetations of type 4 are on average taller (*ca* 6 m) and always possess a luxuriant undergrowth of either shrubs, seral stages of marshes on wet substrate, or forest edge vegetation.

Scrub is principally occupied by a number of *Iduna* and *Hippolais* species and various island reed warblers, as well as *Nesillas* species (e.g. Kiritimati and Tuamotu Warblers, Millerbirds, Madagascar Brush Warblers, Lantz's Brush Warblers). It extends from markedly barren, open, sandy or rocky sites with low scrub, thorn thickets, and tussocky grass to copses and groves. In the dry landscapes of central Asia, Sykes's Warblers favour saxaul and tamarisk bushes (Svensson 2001, Castell & Kirwan 2005). In these semidesert regions they are able to exploit the seasonal food supply in damper places, alongside Asian Desert Warblers. Upcher's Warblers breed in Pakistan on dry hillsides with *Artemisia* scrub at 1800-2400 m (Kennerley & Pearson 2010), while Large-billed Reed Warblers have been found breeding in scrubby riparian bushland of sea buckthorn (*Hippophae*) at 2000-3200 m in northeast Afghanistan and adjacent Tajikistan (Timmins *et al.* 2009, Aye *et al.* 2010, Koblik *et al.* 2011).

Vegetation type 4 (fig. 3.1) comprises shrubs surrounded by grassland, reedy places among scattered trees or at forest edges, undergrowth in gallery forest or old plantations, clearings or burnt areas becoming overgrown with pioneering scrub, and bushy glades. Along with several *Iduna* and *Hippolais* species (e.g. Melodious Warblers) and some insular reed warblers (e.g. Saipan Warblers, Cape Verde Warblers), the typical inhabitants of such plant communities are Blyth's Reed Warblers and Thick-billed Warblers.

Of these two rather more arboreal reed warblers, Blyth's Reed Warblers have penetrated as far as the coniferous taiga forests of the Eurasian boreal zone. Here they have settled bush-rich forest edges with a dense understorey, even shrub tundra in the mountains of Siberia (Rjabitzev 2001). Despite the long cold winters in this region, the brief growing period suffices for the development of deciduous woody plants and herbs, and hence many arthropods. The Thick-billed Warblers of east Asia utilize dense thickets of deciduous bushes, such as hazel, oak, bush clover, or false spiraea, and forest edges. However both forms avoid the interior of larger, denser stands; like other bush dwellers, they prefer pioneer vegetation that is clearly more productive (Ivanitzkii *et al.* 2005).

Some species which prefer vegetation type 4 can reach substantial altitudes; for example, Melodious Warblers can reach 2200 m in the High Atlas (Thevenot *et al.* 2003) and Dark-capped Yellow Warblers occur up to 2600 m in Ethiopia (Kennerley & Pearson 2010).

Only a few acrocephalids are true forest birds (vegetation type 5 in fig 3.1). Of these, more are resident species (especially island endemics) than migratory. On the steep volcanic islands of eastern Polynesia, Tahiti Warblers and two species of Marquesan Warblers live in natural forests up to at least 1200 m; they avoid the higher-altitude wet montane rainforest, but also occur in secondary forests and the latter two on scrub-covered hillsides (Bruner 1974). Indian Ocean species (Seychelles and Rodrigues Warblers) and other Pacific taxa occupy low brushy forest, or are found in agricultural forest, an ancient, man-made plant community of lowlands found throughout the tropical Pacific (Pratt *et al.* 1987; chapter 12). The most arboreal of the *Nesillas* species, Grand Comore and Moheli Brush Warblers, frequent tall forests, the former actually occupying all vegetational zones to over 2000 m. Both these species use the upper strata when foraging (Louette *et al.* 1988). Although Mountain Yellow Warblers are characteristic inhabitants of the east African mountain forests up to 3400 m in Kenya and 3700 m in the east of the Democratic Republic of the Congo, they actually prefer rank growth and bushes at forest edges and clearings, or the undergrowth of lichen-draped primary forest trees.

Among the migratory acrocephalids, Icterine and Olive-tree Warblers are distinctly arboreal; these two *Hippolais* species favour the canopies of well-spaced trees.

Photo 3.19 In the desert regions of NW Africa, Western Olivaceous Warblers and a race of Eastern Olivaceous Warblers come into contact. The latter inhabits *Acacia* woodland and high tamarisk trees, while the former is restricted to oases and palm groves (Morocco; photo Volker Salewski).

Occupying the highest vegetation of all the continental acrocephalids, Icterine Warblers prefer trees with small, open foliage (as do Tahiti Warblers), while Olive-tree Warblers are found in open woodland of both evergreen and deciduous trees. Both these species exploit the short-lived food resources of the northern temperate and Mediterranean zones to the full.

Without exception, all acrocephalids avoid the interior of dense and dark forests. Some species seem to have profited from ancient and modern human interference leading to the opening-up of forests, as well as from a proliferation of edge effects or arrested ecological succession, but this will be considered further in chapter 13.

5 Habitats for migration and in winter

Habitat preference in some species can change as early as the immediate post-breeding period while they are still on the breeding site. When birds were netted over several years in a fragmented Hungarian wetland, it became clear that Marsh Warblers shifted to more open habitats after

Photos 3.20 to 3.23 Habitats of the island acrocephalids range from secondary growth, rank vegetation, bushy thickets, or dense undergrowth in woodland to native forest in damp or dry locations. The subspecies *longicaudata* of Madagascar Brush Warblers occurring on Anjouan (or Nzwani; Comoros, top right) has even been recorded in dense forest. Cape Verde Islands (top left), Marquesas Islands (below left) photos Peter Castell, Franz Bairlein, Jens Hering.

breeding, while Eurasian Reed Warblers remained in the habitat they had chosen to breed in (Preiszner & Csörgö 2008). When several resting species were trapped in various habitats over many years during an autumn migration stopover at the Mettnau peninsula on Lake Constance (Bodensee) in the foothills of the Alps, Eurasian Reed Warblers showed a constant species-specific pattern of habitat selection over 15 consecutive years (fig. 3.5; Berthold 2001). Further south, most of the long-distance migrant acrocephalids on passage – including Eurasian Reed Warblers – were found in a wider range of habitats.

In their winter quarters, birds may be less habitat-tolerant than when they are on migration, though they are still more tolerant than when they are on the breeding grounds (e.g. Black-browed Reed Warblers and many other species; Kennerley & Pearson 2010). More or less fundamental correspondences seem to exist between the breeding and winter niches of species for general ecological characteristics, such as a specific climatic regime, (e.g. aridity) and associated overall vegetation structure. Hence differences in the habitat requirements of various species-pairs in the breeding season are retained on the wintering grounds. For example, Melodious Warblers are more inclined to make use of lower vegetation than Icterine Warblers, which prefer taller, mature, green woodland, and Olive-tree Warblers exploit the canopy of tall acacia woodland, in contrast to Upcher's Warblers, which occupy low, dry scrub (Lack 1985, Urban *et al.* 1997, Kennerley & Pearson 2010).

Some Palaearctic acrocephalids are to be found in drier locations in their winter quarters than in their breeding areas, for example Eurasian and Great Reed Warblers in Africa and Blyth's Reed Warblers and Booted Warblers in the Indian subcontinent (Urban *et al.* 1997, Grimmett *et al.* 1998). Such reversals of habitat preference when compared with breeding sites are especially striking in Sedge Warblers and Eurasian Reed Warblers. In Africa the latter are commonly found away from water in a variety of dry thickets and acacia (Kennerley & Pearson 2010). Conversely, wintering Sedge Warblers are more confined to waterside situations than Eurasian Reed Warblers, and even more strongly than during their own breeding period. In the Sahel they occupy open, often ephemeral or seasonally inundated wetlands, or floating vegetation, not occupied by resident species (Zwarts *et al.* 2009, Kennerley & Pearson 2010).

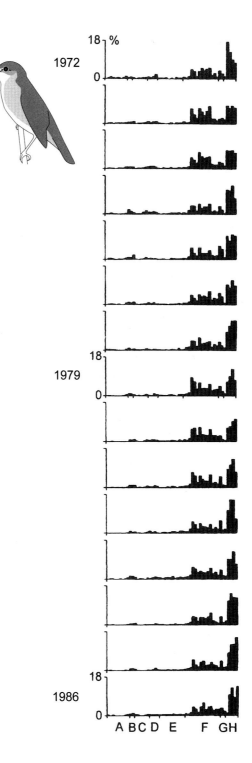

Fig. 3.5 Distribution of trapped Eurasian Reed Warblers over eight different habitats (A-H) on the Mettnau peninsula (Lake Constance/Bodensee) over 15 years (after Berthold 2001). Note the long-term constancy of habitat choice in a migration stopover site.

Other species, such as Thick-billed Warblers in east Asia, retain their close association with damp thickets on their overwintering sites. Specialist species show the strongest correspondence between breeding and winter habitats; Aquatic Warblers are very closely associated with low flooded vegetation at all times, even on migration and in their west African wintering grounds (Flade *et al.* 2011). Basra Reed Warblers are also tied to dense, though emergent, aquatic vegetation in east Africa, similar to that found in their summer quarters in Iraq (Walther *et al.* 2004, Kennerley & Pearson 2010, G. Nikolaus, pers. comm.).

The habitat spectrum utilized by acrocephalids in their wintering areas extends from semidesert scrub, stands of tall grass, savanna, and broadleaved woodland (*Hippolais/Iduna* species) to a variety of habitats associated with water, including inundation zones, seasonally flooded tamarisks and rushes along lakes and rivers, overgrown stream beds and riverbanks, and also mangroves, whose significance as winter habitat has long been underappreciated. There are probably several million Eurasian Reed Warblers concentrated in the mangrove zone of Guinea-Bissau (Altenburg & van Spanje 1989, Kennerley & Pearson 2010). Acrocephalids also exploit man-made habitats, and to a greater extent than in their breeding quarters; arboreal species are found in orchards, palm groves, and gardens, while those which prefer tall herbs and shrubs occupy secondary growth, sugar cane, maize, paddies, tea and coffee plantations and other cultivations.

Habitats, especially in Africa, may not remain suitable for the entire stay of a migrant in its wintering area, and this forces some species to move on in midwinter, a phenomenon that Moreau (1972) described as 'itinerancy' (chapter 11). In Africa the wetland species remain at least partly in the northern tropics. These species are less obliged to go south than those dependent on seasonal vegetation, since

Photos 3.24 to 3.26 While the vegetation types in the breeding and winter quarters of some Palaearctic acrocephalids are very different from each other, the preferences of habitat specialists show a strong correspondence between the two. For instance, Aquatic Warblers are very closely associated with low flooded vegetation at all times (breeding habitat in western Poland, migrational stopover site in Bretagne/France, wintering site in the Djoudj wetlands, Senegal; photos Franziska Tanneberger, Arnaud LeNeve, Volker Salewski).

some wetlands persist in north Africa throughout the winter (Pearson & Lack 1992). In contrast, species like Marsh Warblers that depend on green, leafy thickets will winter further south.

During moult, which carries high energy costs, habitats with rich food resources closely resembling the breeding areas are preferred, and territories are defended. Examples would be Marsh Warblers and Blyth's Reed Warblers (in bushy scrub with rank grasses and herbaceous vegetation in high rainfall regions; Gaston 1976, Kelsey 1989), Great Reed Warblers and Sedge Warblers (in reeds and moist thickets; Bensch *et al.* 1991, summarized in Pearson & Lack 1992), and Oriental and Manchurian Reed Warblers (in reedbeds; Nisbet & Medway 1972, Round & Rumsey 2003). For further discussion of these topics, see chapters 11 and 14.

In conclusion, when we review the spectrum of all the habitats utilized by the acrocephalids, an interesting pattern emerges. As migrants, they use a variety of simply structured, seasonal vegetation types, not only in the northern temperate zone but also in southern latitudes, such as in sub-Saharan Africa, where, for example, Eastern Olivaceous Warblers (subspecies *laeneni*) are the commonest Afrotropical 'gleaner' in the central Sahel (Wilson & Cresswell 2010). Acrocephalids are often confined to plant communities which are pioneers, and/or only briefly productive. As residents they have apparently only been able to establish themselves in tropical swamps that have very few other coinhabitants (with the exception of Dark-capped Yellow Warblers in edge vegetation). Only on islands have they successfully penetrated forests where other species are absent (chapter 12). All these factors indicate that the acrocephalids are a poorly competitive group compared with a lineage of *Sylvia* warblers of similar size, containing Blackcaps, Garden Warblers, African Hill Babblers, and Dohrn's Thrush-babblers.

Summary

All habitats favoured by acrocephalids have much in common: the vegetation forms are simply structured, hold a food supply of large numbers of mobile prey, satisfy the need for cover, and harbour few competing species. Almost three-quarters of the continental acrocephalids are migrants breeding in temperate zones, where they make use of seasonally productive plant communities, mostly successional or pioneer communities. In the tropics, the remainder of the continental species live chiefly in swamps, while the island-dwellers have successfully penetrated forests.

Habitats can be differentiated on the basis of two simple ecological factors: decreasing wetness and increasing vegetation height. The habitat spectrum embraces swamps and marshes, dry marshes, scrub, shrubby vegetation, and woodland. Only a third of the species are closely associated with wetland vegetation. Reedswamps are an extreme habitat for birds, given their low plant diversity and uniform vertical growth of thick stems. It is probable that the reed warblers were able to diverge into this habitat in such a marked fashion precisely because reedbeds could only be colonized by a few specialist species. Wetlands are among the most productive ecosystems, with the almost unique phenomenon of insect emergences offering a continuous renewable food source. This accounts for the high population densities seen in some reedbed passerines and their frequently prolonged breeding seasons. Although wetlands appear to be homogeneous habitats, they actually contain a multitude of structural variations. The water-depth gradient is the ecological factor that determines all else.

Several coexisting species prefer different sections of a hydrosere, and are consequently separated from each other by differences in horizontal habitat selection. A series of acrocephalids live in shrubby and bushy habitats, far from water and sometimes at high altitudes. Scrub is principally occupied by *Iduna*, *Hippolais* and various island reed warblers. Only a few species are true forest birds – some island endemics as well as migratory Icterine and Olive-tree Warblers – yet all of them avoid the interior of dense forests.

Habitat preferences can alter after the breeding period. Some inhabit drier locations than in their breeding habitats, with the exception of extreme habitat specialists such as Aquatic Warblers. Most of the long-distance migrants are found in a wider spectrum of vegetation on passage, while in their winter quarters they may be less habitat-tolerant than on migration

Their habitat requirements, and their often pronounced specialization, indicate that the acrocephalids were able to diversify in environments where competition pressure was relaxed.

Foraging, diet, and habitat use

A pair of Seychelles Warblers foraging. Insects are gleaned from the undersides of the leaves.

When searching for food birds must constantly make decisions about what they should eat, where they should look for it, and which part of their habitat is likely to be the most profitable source. When members of the family Acrocephalidae forage they are constantly on the move; none of the species uses a 'passive' sit-and-wait technique with long periods of immobility, as would be typical for flycatchers or shrikes. Although the spectrum of their search-and-capture feeding method shows few modifications, it is very surprising just how much the techniques used by Moustached Warblers and Eurasian Reed Warblers differ in the same habitat. In an old undisturbed reedbed a Moustached Warbler utilizes the lowest layer, the tangled jam of broken reed stems, to search for tiny, immobile, hidden invertebrates, while higher up a Eurasian Reed Warbler actively hops from stem to stem in the rapid pursuit of larger, mobile prey, which it seizes in skilful 'leap-catches'.

A variety of 'search-and-capture' tactics

The reedswamp-dwelling Acrocephalus species, the more bush-loving Iduna, and the arboreal Hippolais warblers all feed opportunistically on arthropods gleaned from leaves, stems, and twigs or snatched in flight. Some typical photographs (photos 4.1 - 4.17) graphically demonstrate the spectrum of capture techniques in the group. Such differences are often only fine, as are the corresponding variations in morphology (chapter 5), where the bill form can indicate such species-specific differences (Glutz & Bauer 1991, Cramp 1992).

The most widespread foraging tactics are the so-called 'near-perch' manoeuvres, when a target food item can be reached without the bird moving from its perch (Remsen & Robinson 1990). These include various forms of gleaning, including picking prey from a surface while remaining at rest ('stand-picking'). A number of species feed methodically on the ground (e.g. some island reed warblers, Upcher's Warblers), or on mats of floating vegetation where they hop about, reminiscent of chats (Moustached Warblers). Also common are 'reaches', where the legs or neck are completely extended upwards, outwards, or downwards, sometimes accompanied by fluttering of the wings. In this way prey animals are reached from a perching position by a rapid stretching of the body, or by a reaching upwards to glean them from the underside of a leaf, or by hanging downwards as exemplified by Seychelles Warblers (introductory drawing).

In addition, the plain-coloured reed warblers and most Hippolais species will seize passing aerial prey from a standing position. Leap and aerial manoeuvres merge into one another in these techniques. During 'leap-catching', brief flutters of the wings are employed to snatch a flying insect or one just in the act of taking off in escape. In 'flycatching', airborne insects are taken following a brief flight, accompanied in some species by an audible snap of the bill. Flycatching techniques include the 'sally', in which the bird flies from a perch in pursuit of an insect and then returns to a perch – a manoeuvre occasionally executed by tree warblers (Hippolais and Iduna) as well as by small plain reed warblers – and the 'hover', generally in the form of 'hover-gleaning' by an airborne bird seizing a stationary prey item (Glutz & Bauer 1991, Cramp 1992). This manoeuvre is frequently performed in evergreen tamarisk by Eastern Olivaceous Warblers of the north African subspecies reiseri, for example, and is very similar to that employed by Coal Tits in conifers.

When 'pouncing', a bird flies to the ground for either stationary or mobile prey. Acrocephalid substrate manoeuvres consist of 'probing' and 'flaking' of material, where loose substrate is pushed or flicked away with a sideways motion of the bill in order to reach hidden prey, as sometimes seen in Moustached Warblers. Other more substrate-oriented and more complex extractive techniques have been developed by insular species in particular (chapter 12). These include leaf-turning and 'prying or gaping', whereby the bill is opened while it is inserted in substrate to expose prey. Use of the feet in treatment of prey is also known from some island reed warblers and Australian Reed Warblers have also been observed to use their feet as though to disturb prey (Higgins et al. 2006a). Tree warblers (genus Hippolais) use

Photos 4.1 to 4.3 Wetlands – the habitat of many acrocephalids – harbour a high density of invertebrates. Mass reproduction of a variety of taxa in spring and summer (known as emergences) and a high renewal rate guarantee short-term peaks as well as a long-term steady food supply. (Sedge Warbler and swarming chironomids (Diptera) in the Netherlands, mayflies at Tisza/Hungary (below), flying insects in early morning light in Poland; photos Marten van Kempen, Milan Radisics, Nature PL; Arnauld LeNeve).

their habitat in what seems a surprisingly clumsy way to our eyes; their 'awkward' entry into or exit from foliage is especially striking when compared with *Sylvia* or *Iduna* species

The diet of the family is chiefly made up of small invertebrates, mainly arthropods (insects and spiders, but also crustaceans such as freshwater shrimps and crayfish), as well as molluscs and even small vertebrates, such as newly metamorphosed frogs, newt larvae, or fish fry (Cyprinids, Perciformes), which are mainly taken by the robust species Great, Clamorous, Oriental, and Australian Reed Warblers. The Pacific insular forms are even capable of catching small lizards and geckos (chapter 12). When raising young, several species, including

Aquatic Warblers, occasionally gather large numbers of hairy caterpillars.

Plant food is rarely taken by *Acrocephalus* species, probably because the supply of berries in their habitats is fairly limited, although *Salvadora* berries are eaten by several forms. However, the *Hippolais* and *Iduna* tree and scrub warblers do make use of plant material to supplement their predominantly animal diet. A wide variety of fruits (e.g. figs, berries) and nectar is eaten, and these become particularly important in late summer prior to migration. In their winter quarters and on spring migration nectar uptake can play an important role in the diet of these genera (Salewski *et al.* 2006), although still to a lesser extent than in *Sylvia* warblers (Swilch *et al.* 2001).

Habitat and food supply – a comparison of six wetland species

The juxtaposition of even the smallest details can quickly reveal just how different both dietary preferences and foraging strategies can be within the acrocephalids. Studies of the six reed warbler species which live in varying combinations in the reedswamp hydrosere in different parts of central Europe can serve as representative examples. The model species are Great and Eurasian Reed Warblers, Moustached, Sedge, Marsh, and Aquatic Warblers – we have met them already in chapter 3. Such comparative studies also throw light on the interactions between habitat structure and food supply, as well as on the ability of individual species to exploit this supply, thus allowing us some insights into habitat selection in each species.

When we analyse the food spectra of these six European *Acrocephalus* species, what is initially striking are the differences that result from the regional variations in food resources. Hence the Great and Eurasian Reed Warblers, in an area of eutrophic (nutrient rich) lakes in Poland,

have larger prey animals at their disposal than birds in oligotrophic (nutrient poor) lakes in Switzerland, where both taxa have to make do with smaller invertebrates (Dyrcz 1979). Additionally, there is a well-defined seasonal variation in the food available. In reedbeds, spiders form part of the diet early in the year, as do beetles (small Carabidae, Staphylinidae, Cryptophagidae, Curculionidae, and marsh beetles Helodidae) when they leave the reed stubble that has served them as hibernation shelter and gather in warm places (Chernetsov & Manukyan 1999, 2000). Some time later Hymenoptera, Diptera (true flies), caterpillars, and cicadas are present in large numbers. Beetle larvae and aphids are available only at the end of the breeding season (Leisler 1970, Bibby & Thomas 1985, Dyrcz & Zdunek 1996).

In a comparative study of the ecological separation of the six species, measurements were taken at four different European wetland sites, not only of water depth and vegetation height (section 3.3), but also of habitat characteristics, such as density of emergent elements (e.g. stalks and erect leaves) and vegetation profile (Leisler 1981).

Photo 4.4 All acrocephalids feed opportunistically on arthropods gleaned from the vegetation or snatched in flight. Immobile prey, such as aphids, are taken while the bird, here an Oriental Reed Warbler, remains at rest – a technique called 'stand-picking' (photo Owen Chiang).

Photo 4.5 Larvae or caterpillars are also taken. With perfect aim this Olive-tree Warbler picks a caterpillar from a cocoon (photo Mike Pope).

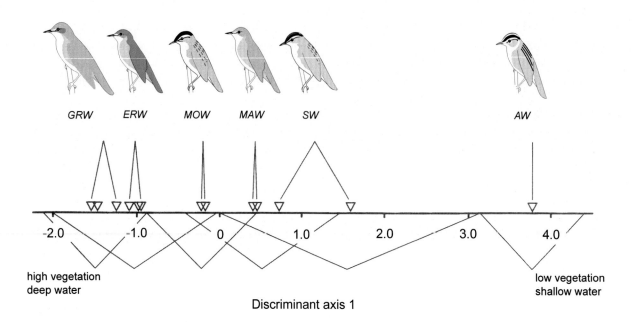

GRW ERW MOW MAW SW AW

-2.0 -1.0 0 1.0 2.0 3.0 4.0

high vegetation
deep water

low vegetation
shallow water

Discriminant axis 1

Fig. 4.1 Separation of six reed warbler species on the basis of a discriminant analysis of various habitat characteristics. The species are clearly separated from each other by a combination of several of these characteristics. The means from different study areas are shown together above the line (triangles), the variations in the discriminance values for each species below the line. For example, the values shown here for Aquatic Warblers do not overlap with those of other species, while those of Sedge Warblers overlap with both Marsh and Moustached Warblers. Ecologically this complex axis of differentiation roughly corresponds to the real gradient of decreasing vegetation height and wetness (after Leisler 1981).

Photo 4.6 This Sedge Warbler has interrupted its song to seize an item of aerial prey, a passing dipteran, by rapidly stretching its body forwards from the perching position (photo Simon White).

Using a multivariate analysis, the properties most effectively differentiating the microhabitats exploited by the warblers could be isolated. It was found that the species were mainly separated by combinations of characteristics of the vegetation that were independent of its floristic composition. Fig. 4.1 shows how the resulting discriminant axis (of the multivariate analysis) separates the species. Apart from their clear separation, what is also apparent is that some species pairs overlap more markedly than others (chapter 6). Ecologically this complex axis of differentiation roughly corresponds to the real gradients of decreasing vegetation height and wetness, as shown in Fig. 3.4 (chapter 3). The axis mainly describes a decrease in vegetation density in the upper layers, but also a decrease in vegetation height, number of emergent elements, and water depth, as well as an increase in undergrowth.

Making a living in the centre or on the edge

When this information is correlated with food supply studies (Schulze-Hagen *et al.* 1989, Schulze-Hagen & Flinks 1989, Hoi *et al.* 1995, Poulin *et al.* 2000), a general

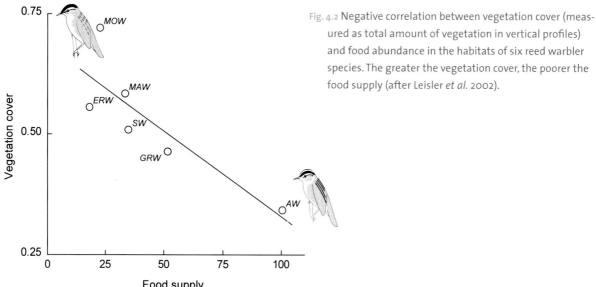

Fig. 4.2 Negative correlation between vegetation cover (measured as total amount of vegetation in vertical profiles) and food abundance in the habitats of six reed warbler species. The greater the vegetation cover, the poorer the food supply (after Leisler *et al.* 2002).

relationship between habitat structure and food abundance can be recognized. Thus, although the habitats of Great Reed Warblers and Aquatic Warblers are furthest apart on the discriminant axis, both are characterized not only by being more food rich but also by harbouring larger arthropods than those of the other four congeners. What unites the two is that they are both ecotones – boundary zones between two different communities. Great Reed Warblers live at the water-reed interface, while Aquatic Warblers favour the landward edge of reedswamp or fen mires. These fringes are particularly rich in food, either because the vegetation is rather sparse, as favoured by Great Reed Warblers, or because it is the most open, as preferred by Aquatic Warblers; in both cases greater invertebrate productivity is the result (Leisler 1985, Leisler & Catchpole 1992; chapter 3). The higher plant diversity of ecotones presumably also leads to a greater food biomass.

By contrast, in the habitats occupied by Eurasian Reed Warblers, Moustached, and Marsh Warblers, much vegetation cover is concentrated between a height of 1 and 1.5 m, which absorbs much of the light. These habitats are also composed of only a very few plant species, resulting in a decrease in the invertebrate food resource. Overall, a significant negative correlation could be shown between vegetation cover (measured as total amount of vegetation in vertical profile) and food abundance in the habitats of the six species (fig 4.2). Since the density of vegetation cover in a reedbed tends to increase from the edges towards the centre, the area of poorest invertebrate productivity is located in the thick and dark interior (Taylor 1993, Hoi *et al.* 1995, Poulin *et al.* 2000). That relationship also holds for more terrestrial habitats, such as stands of herbage (Ille & Hoi 1995). Sites where willow scrub or oak stands border immediately on the reeds are also preferred by reed warblers because of their higher arthropod density (Dyrcz & Zdunek 1996).

There are not only quantitative, but also qualitative differences in food supply between the centre and the edges of a reedbed. For instance, Moustached Warblers, Eurasian, and Great Reed Warblers find reedswamp with far more Diptera in the inundated centre of the Neusiedler See than do Sedge and Marsh Warblers in their habitats on the landward fringe (Hoi *et al.* 1995).

Because productivity declines as vegetation cover increases, a breeding season trade-off between nest site and food-rich areas arises for some species. Although places with dense vegetation offer nest sites safe from predators, they are poor in food resources; conversely, areas of open vegetation are rich in food, but nests there are vulnerable to predation. This conflict affects Moustached and Marsh Warblers in particular, since their nests are placed in dense vegetation (chapter 8). A habitat shift to a more productive area is only easily achieved once the young have departed from the nest or when the breeding season is over. At this time Eurasian Reed Warblers and Sedge Warblers can often be found where plum aphids (always an

Photo 4.7 This Eurasian Reed Warbler takes its target prey, an aphid, in a similarly agile fashion from its perch. The reach attained in such acrobatic lunges from the perch is impressive (photo Rene Winter).

Photo 4.8 If the prey cannot be caught from a perching position then reed warblers will try to reach it by jumping up, like this Eastern Olivaceous Warbler. The distances covered are short, mostly under 20 cm (photo Mike Pope).

unpredictable resource) most commonly occur, even when such reed patches are standing in deep water. It appears that birds in reedbeds can easily estimate the food richness of a site, either directly, or by using habitat cues like wetness, vegetation type, or density (Chernetsov 1998).

One question we must ask is why are food-rich zones like Aquatic Warbler habitat not used as breeding sites by other acrocephalids? As Aquatic Warblers' closest neighbours, we might think that Sedge Warblers could also exploit such areas. One answer is that the vegetation in these sedge meadows is too pliable for Sedge Warblers, with too few rigid elements to be successfully utilized. When they climb on such structures, Sedge Warblers lack the evolutionary adaptations to use thin-stemmed grasses for locomotion. The respective 'required' morphological endowments of the various species are considered in chapter 5.

3 The food spectrum

There have been many studies of the composition of the breeding season diet and of the food given to their nestlings by the six wetland species we have been discussing (e.g. species accounts in Glutz & Bauer 1991, Cramp 1992). It is generally true of songbirds that adults show greater foraging diversity when feeding their young, or, to put it another way, they expand their species-specific adaptive mode of finding food. This applies to the type of prey taken, its size, and to the foraging methods employed (e.g. Root 1967). With young in the nest, adult foraging behaviour and prey choice become less selective because of the increased workload; for example, Moustached Warblers and Eurasian Reed Warblers forage for prey for their young in a significantly broader spectrum of sites than when seeking food just for themselves. By contrast, the generalist Great Reed Warblers do not increase the normally high diversity of foraging sites used at this time (Leisler 1970). However, differences between adult diet and nestling food are not so great as to mask considerable interspecific differences, which are known as diet specificities (Bussmann 1979, Henry 1979, Kazlauskas *et al.* 1986, Taylor 1993, Poulin *et al.* 2000, Glutz & Bauer 1991, Cramp 1992).

Diet specificities and food selection

Although each of these six European reed warblers has a broad diet, some of them specialize in particular groups

Photos 4.9 and 4.10 If the required distances are greater than 20 cm, the wings will increasingly be employed for assistance, so that 'leap-catching' and aerial manoeuvre merge into one another. Photos of Eurasian Reed Warblers illustrate the transition to 'flycatching', in which airborne insects are taken following a brief flight (photos Greg Morgan, Keith Cochrane).

of prey, both in the breeding and the non-breeding seasons. While spiders and flies fulfil an important role in the diet of all six species, Eurasian Reed Warblers are conspicuous in having a special preference for Diptera in general, and for hover flies (Syrphidae) in particular (Bairlein 2006, Grim 2006, Kerbiriou *et al.* 2011). Marsh Warblers concentrate on less mobile Diptera and aphids. The latter are also an important food resource for Sedge Warblers, but, like Great Reed Warblers, this species does not limit the choice of prey to certain systematic groups of arthropods, but rather takes any abundant small items, including beetles (Coleoptera), chironomids (Diptera), and hymenopterans, and will do so in the non-breeding season as well, providing the prey are plentiful in one place (Fry *et al.* 1970, Chernetsov & Manukyan 2000, Zwarts *et al.* 2009, Kennerley & Pearson 2010). A variety of insect larvae are high on the menu for Great Reed Warblers and Moustached Warblers, with the latter generally preferring relatively inactive orders such as Coleoptera. Items taken by Aquatic Warblers can hardly be characterized as belonging to any particular arthropod groups, since the species feeds on a large range of main prey, depending on availability. When supplying their young the females react opportunistically to the changing insect

population surges that are typical both of the breeding habitat and the wet grasslands on migrational staging and wintering sites. From year to year, and varying with the seasons, the species will also exploit infestations of caterpillars or Orthoptera, or mass emergences of Tabanidae or Tipulidae (Diptera) and caddis flies (Schulze-Hagen 1991a, Kerbiriou *et al.* 2011).

When only the broad classes of prey are compared, the overlap in the diets of individual species appears at first unrealistically high (e.g. Henry 1979, where an ecological overlap index for Eurasian Reed Warblers and Sedge Warblers of 0.93 is given). However, identification of prey to genus or species level soon reveals the extent of the differences between the diets of the various *Acrocephalus* species (e.g. Bussmann 1979, Leisler 1991a, b).

Prey size

There is often a considerable prey size-class overlap in the diets of the species. The majority of prey items taken by the small species, Moustached, Sedge, and Marsh Warblers and Eurasian Reed Warblers, are 2-6 mm in size (Henry 1979, Leisler 1985, Glutz & Bauer 1991), though the first of these feeds on extremely small arthropods, including mites and springtails (Collembola) (Bibby 1982, Leisler 1991a). On the other hand, the similarly-sized Aquatic Warblers appear to prefer larger prey (8.5 to over 9 mm), and three particular groups of large prey, namely Odonata, Araneida, and Orthoptera, significantly contribute to the total biomass consumed by this species (Schulze-Hagen *et al.* 1989, Tanneberger *et al.* 2010, Kerbiriou *et al.* 2011). Even when we allow for the larger body size of Great Reed Warblers, it is striking how they select larger items than their smaller congeners. For example, they bring only the larger females of orb-web spiders to their nestlings, while Eurasian Reed Warblers in the same area feed their young only with the smaller individuals of these spiders (Leisler 1991b).

A comparison of arthropod size distribution in the food supply based on sweep-net samples with that of the items actually given to nestlings (Schulze-Hagen & Flinks 1989, Schulze-Hagen *et al.* 1989, Dyrcz & Zdunek 1996) confirms not only the obvious fact that small arthropods are much commoner than larger ones, but also that the adults of the majority of acrocephalid species often feed their young with above-average-sized prey for 'economic' reasons. By collecting larger prey and carrying it in 'bundles', they can save energy by reducing the number of otherwise costly flights to the nest (Leisler *et al.* 2002). An extreme example can be found once again in Aquatic Warblers, where the average size of items brought to the

nestlings was substantially greater than in the sweep-netted samples (average 8.4 *vs* 2.7 mm; fig. 4.3). Amongst the other small acrocephalids this difference was less pronounced (e.g. Sedge Warblers – the closest neighbours of Aquatic Warblers) or there was no difference at all (e.g. Marsh Warblers; Schulze-Hagen *et al.* 1989, Tanneberger *et al.* 2010).

One advantage of the habitats of Aquatic Warblers and Great Reed Warblers is that they are fundamentally so rich in food that the rarer, larger arthropods are common enough to be effectively exploited (section 4.2). By contrast, the prey of those species which live in relatively food-poor habitats is mostly small and has to be carried to the nest by both parents in bundles (multiple prey loading). Species which inhabit more productive habitats, and are able to capture larger prey, can afford to bring them in fewer or single feeding bouts (single prey loading; for the negative relationship between prey size and prey load see fig. 4.4). As a result, they are sometimes in a position to raise a brood unassisted. This has far-reaching consequences, not only for the development of the breeding systems of Great Reed Warblers and Aquatic Warblers, but also for the selection of migration stopover sites by the latter, a globally threatened species (chapters 10 and 13).

Climbing the ladder of success in vertical habitat

The exact identification of prey animals, obtained via collar-sampling, washing-out of stomachs, or faecal analysis, allows us to determine their way of life and habitat, and hence the foraging sites of their predators. Since many arthropods are tied to defined zones in the vegetation, information can be gained by qualitative analysis as to whether the birds caught them on the ground, in water or from its surface, in lower or higher vegetation layers, or in the air. Because the warblers are more or less impossible to observe for long periods in the dense growth of their habitats, such indirect information is of considerable significance. Indeed, on the basis of their food spectra, it can be demonstrated in a direct comparison that each of these six European *Acrocephalus* species visits a specific vegetation height or zone when foraging.

Photo 4.12 Fluttering with the wings prevents sliding on smooth substrates. Tahiti Warbler (photo Jens Hering).

A more direct method of establishing the distribution of birds throughout the differing heights in a habitat (their

Photo 4.11 Here a Basra Reed Warbler has just gleaned an insect from the water surface and is returning to its perch (photo Mike Pope).

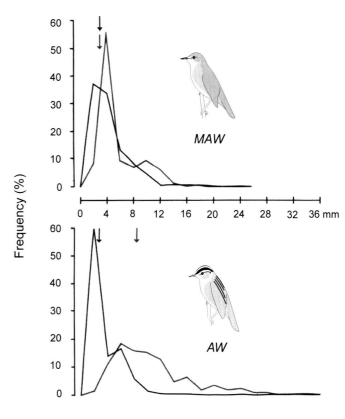

Fig. 4.3 Frequency distribution of size of potential arthropod food supply (black; sweep net samples) and of prey taken (red) in Marsh and Aquatic Warblers. Arrows indicate mean values. In Aquatic Warblers the gap between the means of potential and selected prey items is greater than in Marsh Warblers (after Leisler & Catchpole 1992).

stratification) can be carried out by trapping them in nets under standardized conditions. In this procedure the distribution of the captured birds in the four vertically-arranged pouch panels of a mist net is evaluated (fig. 4.5). This simple method of data collection has been carried out throughout Europe via the numerous standardized trapping schemes in operation in various parts of the continent, both during and after the breeding season and at migration stopover sites (e.g. Bairlein 1981, Baccetti 1985, Spina 1986, Pambour 1990, Mädlow 1992, Taylor 1993). Although such data need to be heavily qualified before evaluation, a good rough guide to species stratification can be obtained in this way. In general, the three striped species (Sedge, Moustached, and Aquatic Warblers), which rely more on a picking strategy to catch relatively inactive invertebrates, spend more time in lower strata than the three plain-coloured species (Great and Eurasian Reed Warblers and Marsh Warblers). Aquatic Warblers are most consistent in their use of vegetation layers close to the ground, where Sedge Warblers are also frequently found, especially when breeding. Both during and outside the breeding season, Moustached Warblers take about half of their food from the lowest stratum – the inundated part of a reedswamp with its layer of tangled, broken stems – or even fish directly from the water (Baccetti 1985). This is significantly different from the behaviour of sympatric Eurasian Reed Warblers: 40% of spiders fed to the young by Moustached Warblers come

from the water or its surface, whereas for Eurasian Reed Warblers this figure is only 7% (Leisler 1991a). However, Moustached Warblers are also capable of foraging in the upper strata, picking spiders from inflorescences or aphids from leaves. In this endeavour they employ their excellent climbing skills on such erect features, which serve as 'tramlines' for movement from the base to the top of the reed stems (section 5.1).

In essence, the three plain-coloured species (Great and Eurasian Reed Warblers and, to a lesser extent, Marsh Warblers), which all employ active hunting techniques to catch mobile prey, keep to higher zones, or use all layers equally, more so than their three striped congeners. As in

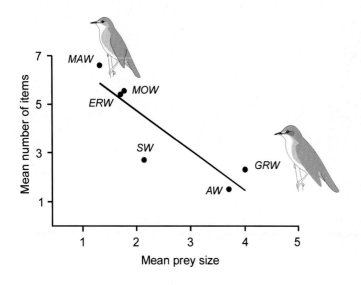

Fig. 4.4 Negative relationship between mean prey size (corrected for size) and number of items per feeding trip in six reed warbler species. Left are multiple prey loaders (Marsh Warblers, Eurasian Reed Warblers, and Moustached Warblers). Right are single prey loaders (Aquatic Warblers, Great Reed Warblers; after Leisler & Catchpole 1992).

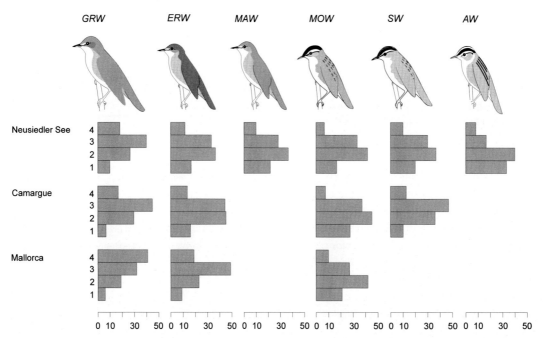

Fig. 4.5 Distribution of trapped reed warblers in mist net panels (1=bottom, 4=top) in three study areas (for references see section 4.4).

their food spectrum, Great Reed Warblers are generalists, do not prefer any particular stratum, and opportunistically take some of their food from the water, including moribund and washed-up insects, which are extremely numerous on large steppe lakes exposed to the wind, for example. Eurasian Reed Warblers collect their prey most frequently in the upper vegetation layers or snatch it out of the air. Marsh Warblers use all strata more or less equally when foraging.

5 Feeding techniques

In the final analysis only data on habitat exploitation and direct observation of prey capture permit a precise classification of the feeding ecology specializations of individual acrocephalids, as well as of their prey spectrum overlap, foraging methods, and feeding sites. The majority of arthropods are taken in flitting, creeping, or hopping locomotion through vegetation, with or without the aid of the wings. All six of our study reed warblers move in these ways when foraging, though to very variable extents. Simple jumps can thus be very short (less than 15 cm), while leap-catches can be from 15 to 60 cm, and flight-stretches from 60 cm to several metres. Eurasian Reed Warblers, and Marsh and Sedge Warblers occur sympatrically in many reedbeds; thus, provided the growth is not too dense for observation, we can see how Marsh and Sedge Warblers mainly glean their prey from the vegetation while perched, whereas Eurasian Reed Warblers more often use their wings in support. As the distances covered in chasing insects demonstrate, Marsh Warblers move the slowest when foraging (table 4.1).

Seizing prey in hovering or pursuit flight is rarely employed by any of the six species. As they are not so specialized, Great Reed Warblers have a large number of foraging tactics at their command – chasing flying insects, gleaning prey from surfaces, extracting it from cavities, or picking it out of the water with equal facility. The species also flies more freely than other *Acrocephalus* warblers. Aquatic Warblers forage as secretively as Grasshopper Warblers, almost exclusively gleaning items while scurrying and threading their way through the vegetation. Moustached Warblers pick their prey from leaves and suchlike (almost always while remaining perched and without using their wings), and frequently examine openings and crevices in a manner reminiscent of wrens. In experimental situations, captive Moustached Warblers succeeded in pulling invertebrates from 9 cm-deep water,

while Eurasian Reed Warblers could only manage it from 1.5 cm of water (van den Elzen in Leisler 1991a).

Such differences become particularly clear under controlled conditions. If Moustached Warblers and Eurasian Reed Warblers are kept in individual cages with a few perches and food *ad libitum*, it soon becomes apparent just how fundamentally different their foraging behaviours are. Since both birds are the same size, there is no difference in their food requirements or rate of intake, both consuming an average of one item every 50 seconds. However, the picking rate per hour is significantly higher in Moustached Warblers (89 *vs* 54) and the number of 'search stares', or instances of 'peering', is as much as twice as high (164 *vs* 77). These stares are directed at the food or water container or at the aviary floor, showing unambiguously that search behaviour has its own underlying drive, independent of the degree of satedness (B. Leisler unpubl.).

Like Moustached Warblers, other birds that specialize on small, less mobile, and hidden prey, such as wrens, treecreepers, kinglets, or treerunners, also show intense searching behaviour. Conversely, this behaviour is absent in Eurasian Reed Warblers, which stalk optically conspicuous and rapidly fleeing prey species. This foraging tactic can lead to individual Eurasian Reed Warblers interfering with each other in areas of high population density by scaring off their respective targets while feeding (interference competition). For this reason Eurasian Reed War-

Table 4.1 Comparison of feeding techniques in Eurasian Reed Warblers, Marsh and Sedge Warblers. Percentage feeding techniques (Green & Davies 1972) and size of movement when feeding (Davies & Green 1976, Schulze-Hagen 1991b,c). Each of the three species has a different feeding technique.

	Eurasian Reed Warblers	Marsh Warblers	Sedge Warblers
Stand-picking	26	81	94
Leap-catching	68	16	6
Flycatching	6	3	0
Movement			
< 15 cm	9	83	23
15-60 cm	73	9	67
> 60 cm	18	7	10

blers drive away both conspecifics and other species and defend temporary feeding areas. These overlapping short-term territories are even maintained at stopover sites during migration and on the African wintering grounds (Bibby & Green 1981, Kennerley & Pearson 2010).

6 Diurnal niches and weather-dependent strategies

Given the broad range of prey choice and the plasticity of foraging techniques among the acrocephalids, it might be expected that they would also differ in their diurnal activity patterns, in order to reduce feeding competition and increase foraging efficiency. In a very extensive study, Brensing (1989) analysed the diurnal foraging patterns of various songbirds in the course of a long-term trapping project, which included acrocephalids in the post-breeding period and on migration, and which compared their behaviour patterns with those of captive conspecifics. The study showed that these patterns were almost identical, from which we can conclude that the foraging behaviours of different species reflect their species-specific, probably endogenous, locomotory schemata. Hence Eurasian Reed Warblers, which pursue highly mobile prey, are still active much later in the day than the other three

congenerics (Marsh, Sedge, and Moustached Warblers), which catch more sluggish arthropods (Brensing 1989, Trnka *et al.* 2006). In addition, the discovery that Eurasian Reed Warblers bring their nestlings smaller portions of more inactive animals early on cold mornings, when flying insects are scarce, than later in the day, when they are plentiful, indicates that their foraging techniques during cold weather are less efficient than in higher temperatures (Kazlauskas *et al.* 1986). The accurate daily timing of being in the right place at the right time is therefore of adaptive value. Further evidence comes from the fact that, on the breeding sites, Eurasian Reed Warblers can be as negatively affected by long-lasting periods of cold and rain as Barn Swallows, which are aerial feeders (Nowakowski & Wojciechowski 2002). Interestingly, (and this has also been demonstrated in Barn Swallows) longer-winged and larger Eurasian Reed Warbler individuals are better adapted to survive such food crises (section 5.3).

The differing diurnal niches of the various species can therefore be viewed as closely related to the activity periods of their prey and to their foraging tactics; species that take their prey when it is largely immobile can be active earlier in the day than those that are better equipped to pursue fast-moving arthropods.

7 Learning the whys and wherefores

For an analysis of the development of prey-catching in juvenile birds, the most suitable study species are the two acrocephalids that employ the more complex, though distinct, foraging techniques, namely Moustached Warblers, which mostly use 'hop-picking' and probing, and Eurasian Reed Warblers, which often catch prey in short leaps.

The way in which innate elements of foraging behaviour and habitat selection can be influenced by varying the experiences of the young bird (natal experiences) has been studied in Moustached Warblers by Raach & Leisler (1989). For this purpose, a group of experimentally deprived juveniles was hand-reared in an extremely impoverished environment, in aviaries containing only climbing perches. In this way the birds were denied the opportunity of investigating structured objects. By contrast, a control group ('experienced birds') was offered

Photo 4.13 Species with smaller and finer bills, such as this Booted Warbler in India, can also capture large prey animals, such as an ant-lion (Neuroptera; photo Clement Martin).

Photos 4.14 and 4.15 With their powerful bills, the large warblers are even able to take vertebrate prey such as small fish or amphibians. The large island warblers can capture small lizards or geckos (Great Reed Warbler, photos Oldrich Mikulica).

objects of differing structures, such as cardboard tubes, wood shavings, wood-wool, or string, imitating the lowest reedswamp layer of broken-down reed stems, whose crannies and passages are of crucial importance to Moustached Warblers as nest sites as well as for foraging. A third group of freshly caught wild birds of about the same age was also part of the study. Once the hand-reared birds were independent, all three groups were tested for their fear of new experiences (neophobia), their agility and skills, and their curiosity. The deprived birds showed the highest level of neophobia, measured as a reluctance to approach a food dish in the presence of a novel object over it. They also displayed less aptitude for procuring food from structures made of wire than experienced or wild birds. Although the birds raised in the stimulation-poor environment were more fearful and less skilful than both control groups, they did just as well in the curiosity tests.

The experiments demonstrated that explorative behaviour is deeply embedded as an important component of Moustached Warbler species-specific foraging technique and is insensitive to any effect deriving from the experi-

ence of the young bird. This ability allows them to track down immobile and hidden prey successfully, so that, in contrast to other more migratory acrocephalids of the temperate zone, they can overwinter in northern frost-free areas and can start their breeding cycle earlier. An additional double-choice experiment tested the birds' preferences for either of two test chambers with the corresponding rearing situation (with or without an artificial undergrowth). Here, the experienced group and the wild birds showed a clear preference for the enriched (though not optimal) chamber, while the deprived birds were indifferent. High neophobia in the deprived birds could be responsible for the lack of preference for the new, more appropriate situation, and may restrict individuals to those microhabitats which they explored when they were still learning. However, such behaviour can be reversed, when a habitat experienced later corresponds better to the innate habitat schema, as illustrated by experiments with Coal Tits (Grünberger 1992).

Young Eurasian Reed Warblers show a presumably innate interest in Diptera at a very early age (Impekoven 1962). Even eleven-day-old nestlings observe flies and follow

Photo 4.16 Fruits play only a marginal role in the diet. This Eurasian Reed Warbler in Portugal is feeding on figs (photo Manuel San Miguel Bento).

their flight with their bills. In an experiment on the development of prey-catching behaviour in these birds, nestlings were raised under two different regimes: one group gained continual experience with flies, the other was presented with them only at the age of 30-40 days. In subsequent tests the naïve birds performed no worse at catching Diptera than the experienced ones. However, they were worse at prey handling, and learned this more slowly than the control group. This illustrates that, in the development of foraging skills, flycatching matures independently of specific experience, while mandibulating prey is learned within a short period (Davies & Green 1976). The authors also found that juveniles that were still being fed by their parents mostly used 'stand-picking'. Although the proportion of this foraging method in their overall performance declined as they grew, it was still employed to the same extent as 'leap-catching' right into September. Experienced adults more often indulge in 'leap-catching', which is more energy demanding than 'stand-picking'.

The trapping patterns found in the bird-ringing schemes mentioned above also show age-dependent differences in habitat utilization by our six species when the nets are placed so that they cut across different habitats. The young of these species show a wider range in their habitat choice than the adults; in fact, juvenile birds in general more frequently use habitats other than those that are species-specific. Young of Great and Eurasian Reed Warblers, and Moustached Warblers, all inhabitants of deep-water reeds, were more often trapped in dry parts of the hydrosere than were adults. The reverse held true for juvenile Marsh and Sedge Warblers, birds typical of the drier vegetation edges, as they increasingly visited the reedswamp zones (Bairlein 1981, Zwicker & Grüll 1984, Mädlow 1992, Preiszner & Csörgö 2008). Furthermore, in their first calendar-year, Sedge, Moustached, and Marsh Warblers, together with Eurasian Reed Warblers, all 'prefer' lower vegetation layers more than do adults (Bairlein 1981, Spina 1986, Mädlow 1992). Three reasons, not necessarily mutually exclusive, have been put forward to explain the apparently widespread phenomenon of juveniles using less species-specific, and possibly suboptimal, habitats. First, the exploratory behaviour that begins at fledging might lead to a differential separation of juveniles and adults; secondly, to avoid competition with adults; thirdly, the still inexperienced young birds move to areas in which they are safer and/or in which they can catch prey more easily. In many species, young birds have rounder wings and lower wing loadings, making them more manoeuverable when making use of their habitat and escaping predators (Alatalo et al. 1984, Norman 1997).

All these examples show that several acrocephalids, occurring in the same habitat, have developed a variety of 'lifestyles' that can only be distinguished when closely examined.

Summary

All reed warblers, whether reedswamp-dwellers, bush-specialists, or arboreal, feed mainly on arthropods gleaned from vegetation or snatched in flight. Their foraging techniques are mainly 'near-perch' manoeuvres, or are derived from such. Passive sit-and-wait strategies, so typical of flycatchers for instance, are absent from their repertoire, and only those species that have colonized oceanic islands have developed more complex feeding tactics. However, much closer inspection reveals that each species actually possesses its own foraging method and consequently has a different prey spectrum.

A comparison of six European wetland species illuminates the relationship between habitat structure and food supply, which, together with foraging strategy, determines how a species utilizes its habitat. There is a negative correlation between vegetation cover and food abundance. In the zonation of a hydrosere, the zones at the edges have a sparser vegetation cover, which lets in more light and is thus richer in food, whereas the denser central growth absorbs much of the light and is correspondingly food poor. The species in the food-rich habitats can exploit larger prey when raising their young, with far-reaching consequences for the development of their breeding systems.

Along the hydrosere, the species occupy different sections, which effectively means that they are ecologically separated from each other. They are less separated vertically, although the plain-coloured taxa, which employ active hunting methods to catch more mobile prey, make greater use of higher strata than their striped congeners. The diurnal niches of the taxa are closely related to the activity periods of their prey. Species that preferentially take immobile prey can be active earlier than those specializing in fast-moving dipterans.

A study of hand-reared young Eurasian Reed Warblers and Moustached Warblers, whose feeding strategies are almost antithetical, reveals exactly which components of their foraging skills are inherently fixed during juvenile development and which are flexible. The young of several species show a wider range in their habitat choice than adults.

By examples such as these, we are able to reveal that several apparently similar acrocephalids, occurring sympatrically in the same broad habitat, have nevertheless developed a variety of lifestyles that can only be accurately distinguished when closely examined.

Photo 4.17 Some *Iduna* and *Hippolais* species consume nectar, such as this Booted Warbler in India (photo Clement Martin).

Integrated ecomorphology –
challenges and solutions

From right to left. Three different modes of clinging on vertical stalks. Bearded
Reedlings straddle, Eurasian Reed Warblers grasp the vertical stem with a supporting
leg and a pulling leg held far apart, and Penduline Tits climb actively and energetically.

Whilst on board his ship, moored off the island of Sulawesi (then called Celebes) during an expedition to the South Seas in 1900-1901, Oskar Heinroth (1871-1945), one of the founders of ethology, was given two newly-fledged birds. Although he was unable to identify them, he was experienced bird breeder enough to be able to feed them up and restore them to health. He soon suspected that they must be some kind of reed-dwelling warblers, although at that time no *Acrocephalus* species was known from the island. He decided to try a little experiment and built a cage with both horizontal and vertical perches. As soon as he placed the birds in the cage they clambered with ease up and down the vertical perches. Based on this observation, and an examination of the anatomy of the fully grown individuals, he concluded that the birds could only be reed warblers, and proceeded to name them as a new species (Heinroth 1903). Today they are classed as a subspecies (*celebensis*) of Clamorous Reed Warblers.

Heinroth had intuitively combined two ideas, as is now common practice in ecomorphology. First, that the demands of a habitat are reflected in the body structure of a bird and, secondly, that this interrelationship manifests itself by some linking behaviour, in this case the specialized climbing method. In ecomorphological studies, data gained from observations in the field or aviary are integrated with morphological parameters – usually a series of measurements from museum skins or skeletons – using complex statistical procedures such as principal component analysis (PCA) or multiple regression. In this way, relationships can be identified, that would otherwise remain hidden, between ecology and both body structure and behaviour (which develop in tandem in the course of their evolution). In studies involving several species the additional question arises as to which traits can be explained by adaptations to environmental demands and which have their origins in a phylogenetic relationship.

Even the apparently uniform acrocephalids, whose members appear to lack any species-specific characteristics, can be separated from each other in their ecological and phylogenetic aspects by employing ecomorphological methods (section 2.1). Not only do they show very little variation in basic shape (see below) and colouration, but with a weight of 8-20 g they mainly belong to the small to medium-sized warblers, with Blunt-winged and Booted Warblers being the smallest. Only a quarter of the species are larger at 25-35 g (the large reed warblers), and just a few island species are larger still, such as Tahiti Warblers, which weigh in at more than 40 g (chapter 1). Several factors can shape the size or body mass of a species; interspecific size differences are determined by, for example, phylogeny, physiology, habitat productivity and structural complexity, habitat use,

and competition (e.g. Polo & Carrascal 1999). Despite the special effect of phylogeny on body size, which varies only slightly between closely related taxa (section 5.4), there are exceptions where sister species actually do differ considerably in body size. We are thinking here of the species pairs Upcher's and Olive-tree Warblers or Cape Verde and Greater Swamp Warblers. The insular forms also display a certain plasticity within the group for either decreasing or increasing body size, depending on ecological conditions (chapter 12). On the other hand, size difference between the sexes remains slight.

Using the results of ecomorphological studies, this chapter will illustrate how a range of ecological challenges, such as climbing on vertical stems or long-distance migration, were solved, and which structural characters were required. These examples are good illustrations of both the working methods of integrated ecomorphology and the fact that birds in the wild and in laboratory conditions behave and act precisely according to their morphologically defined capacities. A further issue with far-reaching implications is how young birds explore their own species-specific and individual abilities. Finally, by analysing the morphology of the different species and clades in the light of their phylogeny, conclusions can be drawn regarding a possible diversification in this superficially homogeneous family.

1

The insectivorous reedbed passerine guild

The morphological adaptations of a species to its environment sometimes become clearer when we look at other members of a guild in the same habitat, rather than

Photo 5.1 Reed warblers show a wealth of adaptations to their habitat that are expressed in their shape, such as in this Paddyfield Warbler in Kazakhstan: pointed head and spindle-shaped body (photo Klaus Nigge).

Let us look first at the members of this guild as they search for food. Eurasian and Great Reed Warblers are mostly seen feeding on vertical vegetation, with a marked ability to climb up and down reeds, hopping and jumping from one stem to the next. Both when they hold their body-line horizontally and when they are perching upright on vertical stems, the upper leg and foot are concealed by the body, while the lower leg is fully stretched. On the other hand, Bearded Reedlings hop or walk on the ground and clamber just as agilely on single reed stems, in a movement best described as vertical 'jump-climbing'. A particular characteristic is their horizontal straddling of two adjacent vertical stems, with the soles of the feet turned outwards, and legs more or less widely spread (see introductory drawing). Savi's Warblers tend to creep and climb alternately through the layer of broken-down reeds. In this thicket of stalks they not only sit frequently in a straddle position, but also move upwards in a shuffling straddle in which, in contrast to Reed Buntings, the grasping feet are slid forward on the vertical stems one after the other. Reed Buntings themselves use a hopping and walking gait on the ground, and move easily between horizontal and vertical perching in reeds. They progress upwards and downwards on vertical stems with great agility, often in alternating jumps and with frequent flicking of the tail and wings. Even more skilful and occasionally positively acrobatic movements are performed by Penduline Tits when 'jump-climbing' on vertical stalks, whereby first one foot and then the other is uppermost and the wings are also actively employed (see introductory drawing). Thus various reedbed inhabitants use at least three completely different techniques to negotiate vertical stems: grasping a vertical stem with a supporting leg and a pulling leg held far apart (the acrocephalid technique described in more detail below), straddling, and forceful climbing.

Acrocephalids do it their way in a vertical world

In this first ecomorphological example we shall examine which morphological adaptations are exhibited by six wetland *Acrocephalus* species (whose ecological requirements we have already noted in chapters 3 and 4) in order to cope with their extreme habitat. Initially, body shape differences between the six species were recorded using a series of biometric measurements (32 external and skeleton characters of bill, flight apparatus, and hind limbs). If we then correlate a range of ecological data from their habitats, such as vegetation height and density – or discriminant

comparing the species in question with its closest relatives. The term 'guild' was introduced into ecology by Root to define a group of species which, independent of their taxonomic position, exploit the same class of resources in a similar way (Root 1967). Originally the guild concept was principally intended to encompass the potential competitors of a species. In a reedswamp, apart from the various *Acrocephalus* species (whose foraging techniques were described in chapter 4), many other insectivorous passerines use this habitat to raise their young. Western Palaearctic examples are chiefly Savi's Warblers, Reed Buntings, Bearded Reedlings, and Penduline Tits. In order to be considered a member of the insectivorous reed passerine guild the major portion of a species' diet has to consist of arthropods gleaned from the vegetation.

values calculated from them – with aspects of their morphology, we find them to be closely connected (Leisler & Winkler 1985). The most revealing correlation indicates that species living in *high* vegetation (Great and Eurasian Reed Warblers) possess, above all, long legs for efficient grasping, longer bills, and broader wings. By contrast, species inhabiting *low* vegetation with few vertical reeds on the landward side of reedswamps (Sedge and Aquatic Warblers) have legs that are less effective for grasping, shorter bills, and narrower wings. The correlation evidently results from three different relationships: most important of all, the differing methods of locomotion on foot, followed by variation in diet, and differing use of the wings.

Since this correlation does not reflect any direct causal relationship, Leisler *et al.* (1989) looked more closely at climbing behaviour as a possible mediator between body structure and habitat. In double-choice experiments, individuals of the six species were tested for the extent to which they favoured vertical or horizontal climbing in their movements. The birds were in a neutral enclosure and could chose between two chambers equipped with either only vertical or only horizontal perches. The relative share of locomotor activity served as a measure of

preference for one chamber over the other. Eurasian Reed Warblers used vertical perches the most, Aquatic Warblers the least. The climbing data obtained from the six species (percentage share of vertical climbing) were then correlated with both habitat features and morphological values. We can see from fig. 5.1 that the deep-water marsh species (Eurasian and Great Reed Warblers, Moustached Warblers) climb vertically more often than the landward vegetation dwellers.

To better understand the relationship, it is important to tease out those individual traits in a bird's morphology and environment that combine to form a functional unit via its climbing behaviour. Using specialized statistical methods, the length of the hind toe was identified as the structural feature that best explains the differences in climbing behaviour (fig. 5.2). The more a species climbs vertically, the longer will be its hind toe. In the bird's environment, the number of upright stems is the best predictor of climbing technique. So individual species do not only have a species-specific morphology, but also specific locomotor traits that are precisely adapted to certain habitat structures and play a decisive part in coping efficiently with their environment.

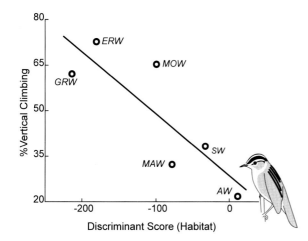

Fig. 5.1 Negative correlation between frequency of vertical climbing (in double choice aviary experiments) and discriminant scores of habitat variables in six reed warbler species. Inhabitants of deep-water marshes (Great and Eurasian Reed Warblers, Moustached Warblers) prefer to climb vertically (after Leisler *et al.* 1989).

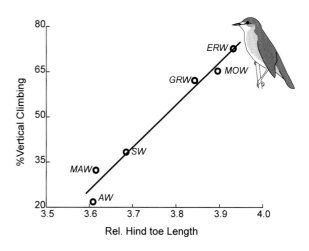

Fig. 5.2 Positive correlation between frequency of vertical climbing (in double choice aviary experiments) and relative hind toe length in six reed warbler species. The more a species climbs vertically the longer its hind toe (after Leisler *et al.* 1989).

As they are birds that cling to vertical stems without the aid of a stiffened supporting tail (unlike treecreepers (Certhiidae) for example), most *Acrocephalus* species employ a special technique. They place their feet widely apart with the lower leg maximally extended, so the lower foot is the main support against gravitational force, while the upper leg is fully bent. At the upper grasping point, a pulling force has to be compensated for, and this task is mainly performed by the hind toe. It clasps fully around the stalk and must therefore be of sufficient length to allow this. The toes of the lower foot mainly have the function of preventing slippage. To achieve this, the inner toe is spread widely, and the central pads of the front and hind toes provide enough friction to overcome gravitation. Subtle differences in the length of the inner toe are even diagnostic in separating the vertical-climbing specialist Eurasian Reed Warblers from Marsh Warblers (chapter 2). There is no indication that reed warblers use a strong friction-generating grip at either point of sup-

port, in the manner of the acrobatically clinging Penduline Tits, for example. A further precondition for this climbing technique is that the legs must be long and capable of being folded like a jack-knife.

Juvenile development

How fixed or how plastic is such specialized locomotion in reed warblers? Is it chiefly genetically programmed, or can it be altered through experience in the course of juvenile development? Willy Ley has performed many experiments with Eurasian Reed Warblers to try to clarify the extent to which a preference for vertical structures is innate (Ley 1988). He chose the species because it is the most pronounced of all the vertical climbers. He raised one batch of nestlings on horizontal perches only (group 1), another on vertical perches only (group 2), while a third (group 3) was reared in an artificial mixed habitat of both types and rewarded with additional food when they used the horizontal perches. How would the rearing envi-

Photos 5.2 to 5.4 An important adaptation in reed warblers is their ability to climb on vertical stalks. These Clamorous Reed Warblers in Egypt and on the Philippines demonstrate the special *Acrocephalus* technique of placing the feet widely apart with the lower leg extended, so the lower foot is the main support against gravitational force, while the upper leg by contrast is fully bent. At the upper grasping point a pulling force has to be compensated for, and this is performed by the well-developed hind toe. The main function of the toes of the lower foot is to prevent slippage (photos Jens Hering, Romy Ocon).

ronment influence their later choice of perch? To test this, the birds were offered the choice between horizontal and vertical perches in the same set-up. The outcome was rather unexpected. While the birds that were raised in a vertical environment were evenly divided in their choices (group 2; fig. 5.3), those in groups 1 and 3 preferred the species-specific vertical perches. That the group 1 birds did not favour their familiar horizontal substrate, but chose vertical perches instead, supports the idea of an innate preference for vertical elements (fig. 5.3.). The division within the vertical batch (group 2) can be explained by the novelty effect of the horizontal structures coming into conflict with an inborn preference for vertical ones. Even the food reward as a positive reinforcement to choose horizontal perches in group 3 could make 'horizontal' only slightly more attractive (not shown in fig. 5.3). All groups increasingly used the vertical perches from the first to the third day during the three-day test period. This increase could only be based on individual experience gained in the comparison of different substrates.

Similar results were obtained for Marsh Warblers, a species that prefers horizontal perches (own unpubl. data). Birds raised on horizontal stalks preferred the vertical habitat during the first days of the test, and vice versa. The results in the two species may be tentatively explained in the following way: during juvenile development, a strong drive to explore seems to guarantee that a bird gets the necessary information regarding its abilities in as many situations as possible. The experiments with Eurasian Reed Warblers specifically show that habitat choice is influenced, first by an innate preference for vertical structures and, secondly, by exploratory behaviour controlled by novelty. Learning leads to the final adjustment of habitat choice.

Straddling the reedbed

Although *Acrocephalus* species, with their relatively long legs, employ few of the powerful climbing capacities shown by Penduline Tits and other birds in reed stands, some do use the straddling technique fairly frequently. We have also undertaken quantitative studies of this method of locomotion in standardized captive conditions (own unpubl. data). In contrast to the climbing experiments, where reed warblers could voluntarily choose between vertical and horizontal components, in this study various reed-dwellers were 'compelled' to adopt a straddle stance. Table 5.1 shows how long the study spe-

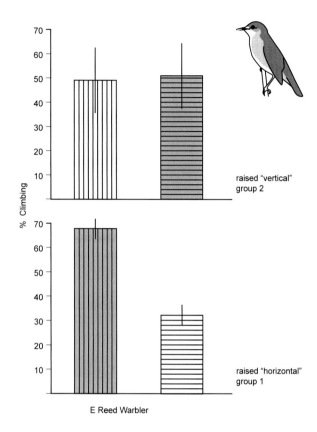

Fig. 5.3 Relative utilization of vertical (vertically hatched columns) *vs* horizontal (horizontally hatched columns) perches in Eurasian Reed Warblers raised under different conditions (groups 1 and 2; double choice experiments; yellow indicates a novel situation for the test group; from Ley 1988). Together with an innate preference for vertical structures, the novelty of a structure has an important influence on the choice (see text).

cies spent in this position. Bearded Reedlings are best at mastering straddle-perching, spending more than half of the observation time in the position. Amongst the acrocephalids, Aquatic Warblers were observed to be specialists in the technique, which enables them frequently to hang between several stems. Moustached Warblers will also often straddle-perch, while other species, such as Eurasian and Great Reed Warblers, are completely unable to adopt the posture. When we search for the morphological attribute corresponding with this locomotory feature, we find that it is the proportions of the pelvis that are decisive here: three aspects of pelvis width explained 93% of the variation in straddle-perching in the 11 study species.

Table 5.1 Percentage of time spent straddling in 11 wetland passerines; average values of several individuals in each case. The best straddlers are Bearded Reedlings, the worst are Sedge Warblers.

Species	% time
Bearded Reedling (*Panurus biarmicus*)	57.4
Aquatic Warbler (*Acrocephalus paludicola*)	41.7
Savi's Warbler (*Locustella luscinioides*)	30.3
Moustached Warbler (*A. melanopogon*)	26.5
Grasshopper Warbler (*L. naevia*)	13.7
Reed Bunting (*Emberiza schoeniclus*)	7.6
Penduline Tit (*Remiz pendulinus*)	0.3
Sedge Warbler (*A. schoenobaenus*)	0.2
Great Reed Warbler (*A. arundinaceus*)	0
Eurasian Reed Warbler (*A. scirpaceus*)	0
Marsh Warbler (*A. palustris*)	0

These two examples are impressive illustrations of how important it is in morphological studies to define the various behaviours, or sets of movements and behavioural decision rules, by which morphology and ecology are connected in individual birds (Winkler 1988, Ricklefs & Miles 1994; see also chapter 14 for ecological equivalents).

2 Other forms of locomotion and habitat use

Those acrocephalids which live either in soft grass-like vegetation (e.g. Aquatic Warblers) or in bushes and even high trees (e.g. Icterine Warblers) face completely different challenges when moving around on foot. Aquatic Warblers are the most terrestrial of all the species. Sometimes they run very quickly over the ground between grassy vegetation and sedge tussocks (which offer few perching opportunities); sometimes they climb with great agility onto emerging structures. When they move on foot they are reminiscent of grasshopper warblers (*Locustella*), whose feet have adaptations for walking: the inner and

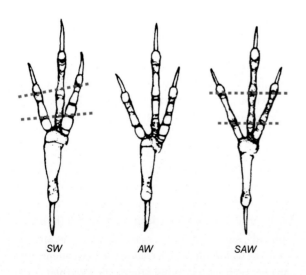

Fig. 5.4 Soles of the feet of Sedge (sw), Aquatic (aw), and Savi's Warblers (saw). Note the angle of the axes between the phalangial joints (after Leisler 1975). In a walking foot, the red lines run at 90° to the foot axis, indicating that the foot rolls off a broad surface. The Aquatic Warbler foot has a form midway between the walking foot of Savi's Warblers and the climbing foot of Sedge Warblers.

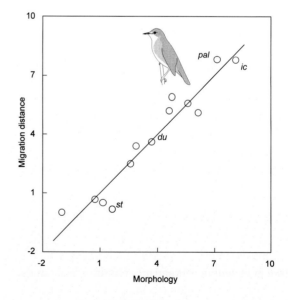

Fig. 5.5 Agreement of migration distances predicted from three morphological traits (length of the notch on the primary feathers, tail graduation, and tarsal diameter) in 13 acrocephalid species (x-axis) with the actual length of their migrations (y-axis, in 1000 km). The three morphological features explain 92% of the variance in migration distances (from Leisler & Winkler 2003).

outer toes are of equal length, so that the phalangial joints lie on a line at a right-angle to the direction of locomotion (Leisler 1975; fig. 5.6). Hence the foot can always roll off a fairly broad support. In reed warbler species such as Sedge Warblers, which often climb more vertically than do Aquatic Warblers, the inner and outer toes differ in length, and the joint axes are oblique (fig. 5.6). When the sole of an Aquatic Warbler foot is compared with that of the other two species it is obvious that it combines the characters of both a walking and a climbing foot.

Those taxa that move through twigs and branches, often holding their body in a horizontal position, are the furthest removed from this very terrestrial life style. We are thinking in particular of the tree warblers (*Hippolais, Iduna*), which have a small, perching foot and relatively short legs that keep their centre of gravity close to the often moving substrate (Leisler 1980). These inhabitants of the canopy and upper storey hunt faster-moving insects than do birds of the denser ground vegetation and undergrowth. Consequently their bills are broad, flat, and have well-developed rictal bristles, which – as in flycatchers – have a tactile function and play a role in prey-handling (Cunningham *et al.* 2011). Conversely, the species that catch their prey more by gleaning or probing techniques

possess narrower bills with short facial bristle feathers. The most flycatcher-like members of the acrocephalids are the two African yellow warbler (*Iduna*) species, birds which truly combine the body characters of both the flycatcher and warbler forms.

The arboreal species show additional morphological adaptations. In the genera *Hippolais* and *Iduna*, which inhabit vegetation of varying height, it is quite straightforward to demonstrate which external features correlate with which vegetation height – a simple ecological measurement. Of 17 body characters studied by us (in flight apparatus, bill, and hind limb), relative wing and tail length and also bill width correlate well with vegetation height. Species with longer wings and tails and broader bills inhabit taller vegetation (Icterine and Olive-tree Warblers), but traits of the legs and feet do not play a very important role. Tree warblers are thus more wing- than leg-orientated forms. This is the reverse case of the six *Acrocephalus* species living in various wetland vegetation layers, whose foot and leg characters correlate with vegetation height (see above). The explanation for this difference is that *Acrocephalus* taxa cover vertical distances mainly by climbing, thus gaining some freedom to adapt their wings to other needs.

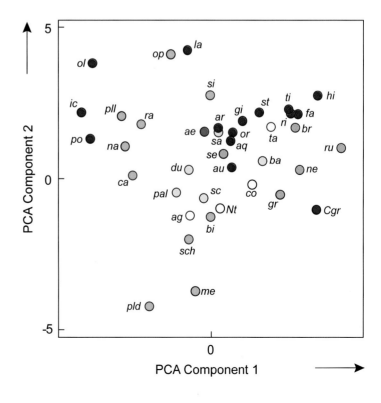

Fig. 5.6 Principal component analysis (PCA) of overall morphology in 40 acrocephalid species. From the many character measurements (which of course are correlated with each other), the PCA extracts those which best account for the overall variance of the species under study. *PCA component 1* mainly combines characters of forms possessing long powerful legs, blunt wings, graduated tails, and long bills. At the high end of the axis these are the resident swamp dwellers such as Greater Swamp Warblers, Saipan Warblers, and Papyrus Yellow Warblers. The contrary character expression is found in the *Hippolais* species.
PCA component 2 mainly combines characters such as large wings and prominent bills. At the low end of the axis we find the striped reed warblers and at the high end the *Iduna* and *Hippolais* taxa. In this way differences in overall body shape can be shown in acrocephalid morphological space.

Photo 5.5 The vegetation density of reedbeds can be very high, with up to 200 stems per square metre. In this habitat 'grid' of vertical and horizontal structures (leaves, bent stems), reed warblers move with great agility, often using a combination of feet and wings like this Eurasian Reed Warbler (photo Greg Morgan).

3

Adaptations to migration

So far we have looked at examples in which foraging and habitat use, so closely interwoven with each other, are reflected in morphology. In migratory birds, the environment experienced in the course of a year encompasses two pronounced migration events (chapter 11). Migration is a powerful factor, which shapes not only the flight apparatus but also overall morphology in birds (Winkler & Leisler 1992). Leisler & Winkler (2003) tested how well the migration distances of 13 acrocephalids could be predicted from their morphological features by regressing external and skeletal traits on migration distances, in a similar manner to the above example with vegetation height. The best three morphological predictors explained 92.3% of the variance in migration distances

(fig. 5.7). The longer the migratory distance in these birds, the shorter was the notch on the inner web of the primary feathers, and the smaller were both tail graduation and tarsal diameter.

First, and unsurprisingly, we see that those parts of the body most strongly correlating with migration distance are those concerning the external flight apparatus (a direct relationship). These are traits of the wing tips, which assist in propelling the bird, and which are thus active in speed and economical forward flight. The wings of migrants in general are longer (larger aspect ratios) with more pointed tips and primary feathers which are less separated (shorter notches) than those of non-migratory subspecies or the species most closely related to them (for general results see Lockwood *et al.* 1998, Mönkkönen 1995; chapter 2).That individual morphological differences in wing shape actually do influence flight efficiency has been directly confirmed in migrating American thrushes (*Catharus*), by measuring their heart rate and wing beat frequency (Bowlin *et al.* 2004).

Secondly, tails are important in guiding the airflow around the bird, but as they get longer they may lessen flight efficiency by creating drag through friction. On the other hand, long and graduated tails assist in slow and manoeuvrable flight. Tails have often been selected for functions other than flight, as signals, or as balancing

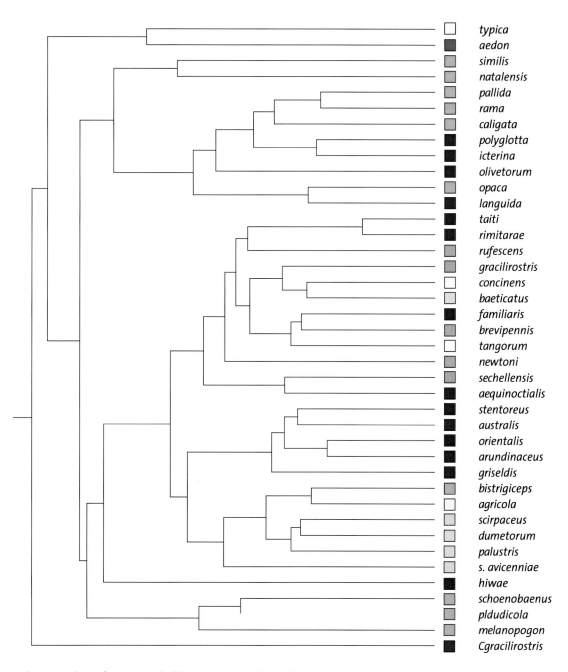

Fig. 5.7 Cluster analysis of 40 acrocephalid species using 18 morphological characters. This illustrates shared similarities on the basis of morphological traits and shows only slight differences from the phylogenetic tree constructed from the genetic distances (see fig. 5.5). Note the importance here of the symbol colours.

aids in other types of locomotion, for example. In both acrocephalid warblers and other passerine groups studied by Leisler & Winkler (2003), tails tended to be shorter and more square-ended in long-distance migrants. Thus migration seems mainly to affect wing tips and tails as a coevolving suite of characters (see also Fiedler 2005).

However, the development of pointed migratory wings is countered by opposing demands for increased manoeuvrability, since shorter rounder wings are ideally required for moving around in a habitat and escaping from predators. This trade-off is thought to have different outcomes, not only in different species but also in different age classes of the same species, or in populations of partial-migrants with varying migratory distances. Such age- and distance-related variations in wing characteristics within a species have been thoroughly studied in Eurasian Reed Warblers (Nowakowski 1994, 2000, Merom *et al.* 1999, Peiro 2003, Hall *et al.* 2004). The substantial amount of data accumulated from the trapping, retrapping, and ring-recoveries of reed warblers of known age allow the following pattern to be constructed: adult birds have longer wings and a more elongated wing shape than juveniles in the same population. This difference can be explained by postulating that juveniles, which are natu-

rally more vulnerable to predation than adults, profit more in early life from rounder wings, at the expense of a somewhat impaired migratory performance (Alatalo *et al.* 1984). Subsequent movements to and from the winter quarters considerably reduce wing length and form variability; stabilizing selection eliminates extremes of wing development. Depending on weather conditions, it could be shown that in some years shorter-winged, and in others longer-winged individuals were favoured in returning birds. These swings of the wing-length pendulum explain the relatively broad peak of favourable wing lengths found in single populations. In addition, there is still a tendency towards rather longer wings in older birds. Thus Eurasian Reed Warblers encounter different selection pressures during migratory and non-migratory periods. In this species, both wing length and wing shape vary according to the differing migratory distances in different populations, as has been shown for other passerines (e.g. Mönkkönen 1995).

Thirdly, let us turn to an indirect relationship. Migration influences not only the flight apparatus, but also has many indirect effects on other functional complexes, such as skull, bill, and hind limbs. There is a considerable trade-off between the muscle mass that can be allocated to either the fore limb or the hind limb. This negative relationship leads to certain modes of locomotion requiring strong hind limb muscles that are incompatible with migration (Leisler & Winkler 2003, Winkler & Leisler 2005). In the acrocephalid warblers the negative correlation between tarsal diameter and migration distance is best explained in terms of this problem – the migrants among them possess low-mass hind limbs.

A further unexpected negative relationship has also emerged, namely between migration distance and skull height or brain size (Winkler *et al.* 2004). This seemingly improbable result has also been confirmed in many other groups: migratory birds appear to have relatively small forebrains compared with sedentary taxa (Sol *et al.* 2005, Pravosudov *et al.* 2007). Explanations suggest that changes in brain size might result from different life styles, because of the differing selection pressures on forebrain size in nomadic and sedentary birds. For instance, unlike resident species, nomads do not require a large forebrain for behavioural flexibility and cognitive tasks, including the maintenance of social networks; on the other hand their brain sizes are subject to strong energetic and devel-

Photo 5.6 In dense vegetation a rounded wing is an advantage, as can be seen in this sedentary African Reed Warbler in Senegal (photo Volker Salewski).

However, details of morphological evolutionary changes have been little studied to date. Leisler & Winkler (2009) analysed the changing fate of the flight apparatus in various Old and New World songbird groups, for which changes in their migratory behaviour had been documented more than once. Using discriminant analysis (DA), they computed a 'migration-score' that best characterized and separated long-distance migrants from residents and short-distance migrants in all 234 species studied (calculated in the same way as the 'papyrus-score' in section 14.1).

Wing length and primary projection contribute positively to this score, while tail-length and tail-graduation enter negatively. The resulting scores of a taxon can then be used to study its evolutionary change within a lineage. To do this, Leisler & Winkler (2009) used a method devised by Schluter *et al.* (1997), which allowed them to take data from their phylogenetic trees and calculate probable ancestral states from this score. Rates of change of the score were computed by taking branch length into account. Both 'tree warbler' clades (*Hippolais* and *Iduna*) show a steady trend towards a migratory morphology. Of the striped reed warbler species, Sedge and Aquatic Warblers tend to have high migratory scores. This trend is reversed in the branches of Moustached and Black-browed Warblers, species with less pronounced migratory behaviour. Much stronger reversals occur in other clades. The transitions in the Eurasian-Africa Reed Warbler superspecies are the most extreme cases, where the flight apparatus underwent extensive alterations in the inordinately short evolutionary time period indicated by the very short branch lengths.

Photo 5.7 The longer the migration route of temperate acrocephalids, the longer and more pointed will be their wings. However, a long wing can occasionally be an annoyance in dense vegetation, as in the case of this Icterine Warbler (photo Jiri Bohdal).

opmental constraints. Thus brain size reduction could have evolved to a certain degree as a precursor for the development of long distance migration (Winkler 2010).

New molecular phylogenetic studies have shown that migrants evolved from (tropical) sedentary species and also that non-migratory populations may originate from migratory ones (chapter 11). This implies that evolutionary changes may occur in either direction – from traits more suited to a sedentary existence to those required for migration, and vice versa.

4 Measuring diversification in the acrocephalids

In the above examples, we have seen how differences in avian body form reflect differences in mode of life (ecology). Let us now turn to the variation in morphology and shape within the Acrocephalidae family as a whole, which

Photo 5.8 The wings of the sedentary island reed warblers are rounder, more slotted, and broader than those of their continental relatives, as shown by the flight outline of a Tahiti Warbler (photo Jens Hering).

Let us look first at the distribution of the species in the morphospace (fig. 5.6) in relation to their position on the first component (horizontal x-axis) only. The highest values actually appear in tropical marsh inhabitants and island reed warblers (Greater Swamp Warblers, Papyrus Yellow Warblers, Saipan Warblers). The low end of the axis consists of wing-emphasized forms with small perching feet, in other words the arboreal migratory *Hippolais* taxa (Icterine, Melodious, and Olive-tree Warblers). The bulk of the intrafamilial morphological variation lies between these two extreme life styles, where we find inhabitants of shrubs and bushes, tall herbs, and undergrowth.

The vertical axis in fig. 5.6 is dominated by a contrast between prominent bills and small delicate bills. Parallel to this development in bill structure runs a similar one in the flight apparatus, tending to broad, long wings and rather long tails. Species that show the highest PC 2 values (Upcher's and Western Olivaceous Warblers) forage very actively by hopping or flitting through dense or open foliage and twigs of bushes and trees. These habits coincide with their bill form, which is perfectly suited to the capture of larger active prey. It is noteworthy that these two warblers come from two different groups (*Hippolais* and *Iduna*; section 2.3). At the lower end of the vertical axis in

we have already partly outlined in chapter 1, and consider how it should be interpreted. To this end we shall again employ biometrics and measure various external traits in the three functional modules of bill (length, breadth, depth), fore limbs (wing tip, wing breadth), and hind limbs (tarsus length and diameter, toe and claw length). These must include as many features as possible with a known function, such as the thrust-providing wing tip. All linear measurements must then be corrected for size, as in the previous examples. An important next step is to use a multivariate statistical technique to reduce this character variation to only a few principal components (PC) or axes, from which we can then construct a morphospace (fig. 5.6). We are able to interpret these axes functionally, using the strength of their correlations with the values of the original characters (table 5.2). The first axis in fig. 5.6 combines mainly the hind limb characters, and represents large feet (long toes and claws) and a long robust tarsus. Other strongly correlated traits are rather blunt wings, a graduated tail, and long bills, which represent a suite of characters we have already found in various reed-dwelling passerines that do not use make much use of their wings (chapter 2).

Photo 5.9 Within their normal habitat reed warblers only usually make short flights, in stark contrast to their migrational movements. Eurasian Reed Warblers rarely cross gaps between two reedbeds more than 50 m apart. Great Reed Warblers like to stretch their wings a little more, like this individual flying low over the vegetation (photo Oldrich Mikulica).

Table 5.2 Important correlations of the original morphological characters of 40 spp. with the first 2 principal components (PC). Also indicated is whether a trait shows a significant phylogenetic signal (+) or not (–). For details of the test procedure see Blomberg *et al.* 2003.

Character	PC 1	PC 2	Phylogenetic signal
Wing		0.54	+
Tail		0.48	+
Tarsus	0.74		+
Body mass			+
Bill length	0.51	0.58	+
Bill width		0.74	–
Bill depth		0.64	+
Hind toe	0.91		+
Middle toe	0.71		+
Hind claw	0.86		+
Middle claw	0.82		+
Tarsus diameter 1	0.75		–
Tarsus diameter 2	0.78		+
Wing width		0.76	–
Primary projection	-0.58		+
Rictal bristles			–
Notch length			+
Tail graduation	0.55		+
% variation explained	33.67	17.27	

Photo 5.10 The strength of the bill corresponds with habitat structure and the prey supply found there. Reed warbler species which take larger arthropods in food-rich habitats have more powerful bills, as in this Great Reed Warbler. Other large species, such as Olive-tree Warblers, also possess a 'tooth', which helps to hold larger items (photo Oldrich Mikulica).

fig. 5.6 are inhabitants of very dense, lower-growing vegetation. These consist of most of the striped reed warblers, and combine complementary characters such as a short delicate bill and narrow short wings, morphological attributes that enable them to pick and glean smaller, more inactive invertebrates from dense vegetation.

Another way of representing shape similarities between species is by cluster analysis, which we shall also consider in chapters 7 and 8. This shows us resemblances that occur throughout the entire character space but cannot be interpreted ecomorphologically, such as those in fig.5.7. In this figure, the clusters are more or less congruent with systematic units – genera, subgenera – which

are also differentiated ecologically. For instance, *Nesillas* and the monotypic genera *Phragamaticola* and *Calamonastides* are also morphologically distinct. All bush- and tree-dwelling species (*Hippolais*, *Iduna*) are brought together. The other main cluster in fig. 5.7 consists of three subgroups: the first comprises the striped species (with the exception of Black-browed Reed Warblers), the second (from *stentoreus* to *avicenniae*) is overwhelmingly made up of migratory wetland dwellers (the small, plain-coloured and also the large *Acrocephalus* species), while the third subgroup (from *taiti* to *aequinoctialis*) contains more sedentary species found in a variety of different habitats, including all the island species (except Saipan Warblers, which hold an isolated position), as well as representatives of the subgenus *Calamocichla* and the *agricola* complex.

Although it is primarily the congruences between morphological and phylogenetic similarities that are made visible by the cluster analysis, there are also some disparities. An especially striking example of these would be the forms of the Eurasian-Africa Reed Warbler superspecies, which, although genetically only weakly differentiated, are morphologically readily distinguishable and are placed in different groups. These and other 'false' allocations (e.g. *bistriceps* close to *agricola*) challenge us to further clarify exactly which ecological conditions have led to such morphological similarities.

A final crucial question that must be answered if we are to understand the morphological diversification of the Acrocephalidae family is: which of the external body traits vary with the degree of relatedness of the species to each other? In other words, which traits contain a phylogenetic signal and which are independent of phylogeny and hence attributable to adaptation? It is only very recently that appropriate statistical procedures have become available to address this particular problem, since they require a phylogeny based on genetic data (e.g. Blomberg *et al.* 2003). Most of the 18 structural characters measured on *ca* 40 species used in these analyses show a phylogenetic signal; they vary with phylogeny (*cf.* table 5.2.). In the acrocephalids, the traits that depend most strongly on relatedness are tail graduation, body mass, and tarsus length, all of which we have already shown to be important in the characterization of the genera and subgroups (e.g. of the *Hippolais* and *Iduna* tree warblers or of the subgenera within *Acrocephalus*) in the multivariate procedures (PCA and DA; figs 5.6 and 5.7). By contrast, bill breadth, rictal bristles, large tarsal diameter, and wing breadth are all independent of phylogeny. Therefore, within a lineage, bills can become slightly broader or narrower, rictal bristles correspondingly longer or shorter, wings larger or smaller, and tarsi thicker or thinner, all as adaptational responses to a changing environment. However, this does not mean that the characters that possess a significant phylogenetic signal (see table 5.2) might not also be adaptive, as suggested by Walter Bock (1980) when he drew attention to this double meaning of the very word 'adaptive'.

From the many different results presented in this chapter, we can perhaps draw two general conclusions. First, on closer examination, the members of the Acrocephalidae family turn out to be far more morphologically diverse than first impressions suggested. Secondly, it would appear that seemingly small morphological differences can in fact be of considerable ecological significance and consequence (see also chapter 14).

Summary

Ecomorphological studies investigate how the ecological demands made on an organism via its environment are reflected in its body structures. Climbing on vertical stems, foraging techniques, and long-distance migration are among the particular ecological challenges faced by the apparently uniform group of acrocephalids. Various reedbed inhabitants use completely different climbing techniques; in these, acrocephalids are characterized by grasping a vertical stem with a supporting leg while the pulling leg is held far apart. A comparison of six central European reed warblers that inhabit the zone of succession between water and land demonstrates that species living in deep-water marsh are more likely to climb on vertical stalks than those living in the landward vegetation. In this regard, Eurasian Reed Warblers are the champions!

The more a species climbs vertically the longer will be its hind toe. Individual species do not only possess a species-specific morphology but also specific methods of locomotion that are precisely adapted to negotiating specific habitat structures. Experiments with Eurasian Reed Warblers raised among differing physical structures have shown that both an innate preference for vertical structures and a strong drive to explore new situations are important for their later choice of habitat. In lower vegetation straddling is an effective form of locomotion, and this is most impressively adopted by Aquatic Warblers.

Those taxa, such as the 'tree warblers' (*Hippolais*, *Iduna*) that live in scrub and in trees, possess shorter legs and smaller perching feet; their bills are broad and flat with well-developed surrounding rictal bristles to facilitate catching the flying insects that constitute their prey.

Photo 5.11 When hunting for faster-moving insects long rictal bristles are advantageous, as shown by this Manchurian Reed Warbler. They have a tactile function and play a role in prey handling (Thailand; photo Phil Round).

Photo 5.12 Other inhabitants or users of reedswamp have developed their own techniques for moving around in vertical structures; these are different from those employed by acrocephalids, as well illustrated here by the acrobatically clinging Penduline Tit and Blue Tit, which use more of a friction-generating grip at either point of support (photos Oldrich Mikulica).

Migration is a further powerful factor in shaping not only the flight apparatus, but also the overall morphology of an acrocephalid, even including its brain size. The longer the migratory distance flown by a warbler in this family, the less separated are the primary feathers of its open wings and the smaller its tail graduation and tarsal diameter. In order to fulfil the requirements of long-distance migration the necessary changes in the wing tips and tail were developed quickly during the evolution of the taxa, and also reversed just as quickly as a result of any decline in migratory behaviour. Today's opposing selection pressures towards longer or shorter wings are currently being studied in Eurasian Reed Warbler populations.

A close look at the ecomorphological diversification of the family demonstrates that the genera and subgenera defined on the basis of genetic characters can also be distinguished by their overall shape – a congruence that is also confirmed by cluster analyses of the species based on morphological characters. At the extremes, the key morphological variation divides leg-emphasized and wing-emphasized forms. Thus a variety of body shapes can be

found within the family, ranging from the resident inhabitants of tropical marshes and oceanic islands, with their large feet, powerful long legs, rounded wings, graduated tails, and long bills, to the tree- and bush-dwellers (the migratory *Hippolais* and *Iduna* species), with their opposing versions of these characters. Between these extremes are the inhabitants of scrub and tall herbs. A further morphological contrast exists between taxa with short wings and fine bills (birds of undergrowth and dense low thickets, such as the striped reed warblers), and forms with broad, long wings and prominent bills (such as Upcher's and Western Olivaceous Warblers).

Using the most up-to-date phylogenetic and morphometric techniques, we are now able to estimate which traits contain a phylogenetic signal and which are independent of phylogeny, and hence attributable to adaptation. In a nutshell, ecomorphological studies show 1) that the members of the family turn out to be far more morphologically diverse than first impressions suggested, and 2) that small morphological differences can have great ecological significance and far-reaching consequences.

6

Competition and coexistence

Aggressive posture of a Moustached Warbler.

Joseph Beier had been studying the breeding biology of Great and Eurasian Reed Warblers in Franconian carp ponds for decades. He soon came to suspect that the two species somehow did not seem to 'like' each other. As he monitored their nests in the reed fringes, he noticed that he never found both species breeding at the same time in close proximity. So from 1981 onwards he began to record both the breeding biology data from the nests and also the distance between them. Sure enough, the nests of those Great and Eurasian Reed Warblers which were actively breeding at the same time as each other were at least 20 m apart. Only when the Great Reed Warbler young had fledged and moved away from the immediate nest surroundings were the Eurasian Reed Warblers able to move into the space vacated in order to start breeding (fig. 6.1; Beier 1993). Beier had thus noted a strong indication of pronounced interspecific competition between the two species (section 6.2).

This chapter deals with particular aspects of territoriality, and specifically with the question: 'Under which conditions is it advantageous to be aggressive towards both conspecific individuals and those of other species?' Exactly which resources are being defended by territory owners: a particular nest site, fertile females, a predator-free space in which to breed, or an area in which to forage? Important foundations in this debate were laid in the 1950s, at a time when the great societal and political themes for humans also revolved around competition and coexistence. After the Second World War, the market economies of the West were booming, and economic competition became one of the export commodities of the USA, while the Cold War conflicts with Eastern block countries became ever more intense. It is surely no coincidence that ecological research at that time was preoccupied with problems of competition and coexistence. Among the most hotly contested issues was the way in which species and individuals divided up their living space, rather like cutting a cake into portions, and how it is that all organisms are at the same time in a network with each other in ecological communities.. The modern pioneers of ecological competition research in the 1950s were two ornithologists, David Lack and Robert MacArthur.

Photo 6.1 The world of the reed warblers is a small one. What this Eurasian Reed Warbler is defending against the supposed rival from the tape recorder is only the few square metres of reedbed that make up his core territory and contain the nest. The adjoining herbaceous vegetation and bushes serve as foraging areas, since reed warblers frequently search for food outside the reedbed proper. The mean size of Eurasian Reed Warbler territories is only around 250 m² (photo Bob Johnson).

1

Intraspecific territoriality and densities

Territorial behaviour allows competing animals to achieve spacing, the apportioning of the local habitat among various individuals. All acrocephalids show territorial behaviour, especially at the start of the breeding season. In varying degrees of quality and quantity, the areas defended by them harbour resources such as food or access to

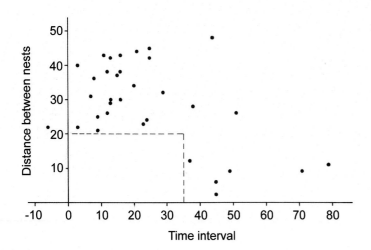

Fig. 6.1 Spatio-temporal separation of nests of Great and Eurasian Reed Warblers. The x-axis shows the time interval in days between laying of first Great Reed Warbler egg and first Eurasian Reed Warbler egg; the y-axis shows the distance in metres of Eurasian Reed Warbler nests from Great Reed Warbler nests (after Beier 1993). Eurasian Reed Warblers are prevented from breeding at the same time as Great Reed Warblers if their nests are too close to those of the larger species.

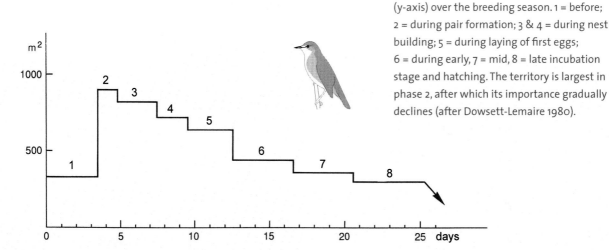

Fig. 6.2 Changes in Marsh Warbler territory size (y-axis) over the breeding season. 1 = before; 2 = during pair formation; 3 & 4 = during nest building; 5 = during laying of first eggs; 6 = during early, 7 = mid, 8 = late incubation stage and hatching. The territory is largest in phase 2, after which its importance gradually declines (after Dowsett-Lemaire 1980).

females. However, as a rule these resources are patchily distributed, and thus differences in the attractiveness of habitat segments often exist at a small-scale level. Where to choose a territory is therefore of great significance for the reproductive fitness of an individual warbler (Pulliam & Danielson 1991, Rodenhouse *et al*. 1997).

What is territoriality for?

In general, territoriality can serve several different functions, including for instance, the attraction of females, ensuring paternity, securing a suitable or safe nest site, or securing food resources. These functions may be combined (multipurpose territory), or each of them in itself can become stronger or weaker over the course of the reproductive cycle. In the early days of reed warbler research, some of these functions were studied by Henry

Eliot Howard (1920), Philip Brown and Gwen Davies (1949), and in a comparative study by Clive Catchpole (1972).

Our comparison of four reed warbler species (Marsh, Moustached, Eurasian Reed and Great Reed Warblers) living side by side at the Neusiedler See can be used here to illustrate the differing functions of territoriality. In the course of our research, playback experiments using species' own songs were made throughout the entire breeding season, with territorial aggression being 'measured' by how close a bird approached the loudspeaker (Laußmann & Leisler 2001). While Marsh Warblers and Eurasian Reed Warblers reacted most aggressively to playback during the nest-building phase, Moustached Warblers and Great Reed Warblers showed their strongest reactions earlier, during establishment of the territory and while searching for a mate. This indicates differing emphasis regarding nest site and mate attraction (greater in Moustached Warblers and Great Reed Warblers) or assurance of paternity (greater in Marsh Warblers and Eurasian Reed Warblers). Behavioural studies further sharpened the picture and indicated that each species has its own priorities in the function of territoriality (Leisler 1975, Dowsett-Lemaire 1980, 1981a,b, Catchpole *et al.* 1985, Ezaki 1990, Hasselquist & Bensch 1991, Bensch & Hasselquist 1992, Borowiec 1992, Laußmann & Leisler 2001).

First, attraction of females is the primary function of territoriality in all acrocephalid warblers. Male Marsh Warblers devoid of a territory are unable to acquire a female (Dowsett-Lemaire 1981a). Female Great Reed Warblers do not choose a partner until they have visited the territories of several owners (Bensch & Hasselquist 1992). This raises the question of whether it is the quality of the territory or of the male, as characterized by his song, for example, that crucially determines female choice. Older males (with larger song repertoires) usually gain the most attractive territories (chapter 10). An indication that territorial quality plays a direct and significant role in mate attraction can be seen in Marsh and Sedge Warblers, whose females prefer males with larger territories (Dowsett-Lemaire 1981a,b, Buchanan & Catchpole 1997).

Secondly, sperm competition theory suggests that, in general, high incidences of extra-pair copulations (EPC) strongly affect territorial behaviour. A defined territory makes it easier to detect and expel invading male conspecifics during the female's fertile period, the time when intrusions most often occur (Møller 1990). Territory size in Marsh Warblers is greatest during the females' fertile period, and territoriality in Eurasian Reed Warblers persists throughout the entire fertile phase, only to collapse soon afterwards (fig. 6.2; Dowsett-Lemaire 1980, Borowiec 1992). An additional strategy for avoiding EPCs is mate-guarding, which can be observed during the fertile phase in all four warblers, and which is especially pronounced in Eurasian Reed Warblers (chapter 10).

Thirdly, areas containing suitable or safe nest sites in reedbeds are, on the whole, heterogeneously distributed (chapters 10 and 14). Nest predation is especially high in reedbeds (section 8.4), and Catchpole (1972) found some

evidence that grouped Eurasian Reed Warbler nests suffered greater predation than nests which were more spread out. This species is very aggressive towards both conspecifics and heterospecifics such as Moustached Warblers and Great Reed Warblers (section 6.2) until the end of the egg-laying period, a tendency that works against high nest densities (though the advantageous effects of social aggregations are considered below). Another possible counteradaptation of Eurasian Reed Warblers in avoiding high predation pressure is the extension of the breeding period into such times as the pressure is lower (Laußmann & Leisler 2001).

Fourthly, food supply in reedswamps is also very unevenly distributed (chapters 3 and 4). The lack of any sort of territorial defence at the very time of incubation and raising the young indicates that securing food resources is of minor significance in all our four species, as is also the case in Sedge Warblers (Catchpole 1972). All of them, but especially Eurasian Reed Warblers, have adapted to the situation of patchily distributed food, often foraging far beyond their former territory boundaries. However, better quality territories may still be related to food availability if certain territories are closer to good feeding areas, as shown in Great Reed Warblers (Dyrcz 1986, Bensch *et al.* 2001; section 4.2).

In communal feeding areas aggressive intra- as well as interspecific contacts are common, though these may be more to ward off disturbing influences while hunting for mobile prey than a direct defence of food resources (Davies & Green 1976; chapter 4). Such agonistic behaviour is also shown by leaf-gleaning acrocephalids (Eurasian Reed Warblers, Melodious and Olivaceous Warblers) against other similar species outside the breeding season, particularly at migration stopover sites, where many individuals might congregate. At a stopover in Mauritania during spring migration, Salewski *et al.* (2007) recorded body-size-dependent hierarchies, in which larger birds were dominant and 'inferior' individuals were temporarily deprived of food. Ultimately it is the distribution of their food that determines the spatial behaviour of migrants at stopover sites. Classical territoriality, such as that seen in robins and wrens defending areas with homogeneously distributed arthropods at stopover sites, has not been observed in reed warblers (Eurasian Reed Warblers and Sedge Warblers). Reed warblers are more dependent on spatiotemporally unpredictable resources,

Photos 6.2 and 6.3 Male reed warblers draw attention to themselves by their song, which serves as courtship-display to the females and to mark the boundaries of their territory against competing males. Prior to pair creation they sing – like these Eurasian (top) and Great Reed Warblers (below) – almost continuously from dawn to dusk. It remains to be investigated whether the strikingly red mouth and throat lining, probably created by carotenoids in the diet, has any kind of signal function (photos Johan Stenlund, Oldrich Mikulica).

often emerging arthropods, and will share a site with conspecifics, moving almost randomly through a stopover site, as demonstrated in the recapture and radio-tracking studies of Chernetsov & Titov (2001) and Chernetsov (2005). However, a whole series of species – Eurasian and Basra Reed Warblers, Marsh, Melodious, Olive-tree and Eastern Olivaceous Warblers – will defend overlapping or exclusive territories (frequently used as moulting sites) for several weeks in their winter quarters, sometimes using song (Kennerley & Pearson 2010; chapter 11).

The case of Great Reed Warblers and Sedge Warblers

Each year, on Lake Kvismaren in central Sweden, those Great Reed Warbler males that return first to the breeding area are also the first to settle in the most attractive territories. They then achieve higher fitness, with higher mating success, more secondary females, and more fledglings than the occupiers of less desirable territories (section 10.3). The attractiveness of territories is remarkably constant between years, despite their male owners changing almost annually. For instance, one site, known as territory A, was continuously the first or second to be established over a period of 17 years. It is interesting to note that attractive and less attractive territories are often located side by side (Bensch & Hasselquist 1991, Bensch *et al.* 2001).

How can it be that the same territories are always more appealing than others over such a long time period? Many studies have failed to identify the factors behind territorial attractiveness, often due to the complexity of habitats such as forests. However, habitat quality in reedbeds can be measured more easily, as they are effectively two-dimensional. On their Swedish long-term study sites, Bensch *et al.* (2001) measured a number of food supply and nest-site-quality parameters in territories of differing occupation order (fig. 6.3). The slight differences they found in food availability between territories did not correlate with territory attractiveness. On the other hand, measures of habitat structure, such as reed edge length, as well as nest predation rate, were correlated with territory attractiveness, albeit not very strongly. As an additional explanation the authors suggest that territory attractiveness in Great Reed Warblers might also be a result of tradition, when inexperienced birds copy the territory preferences of experienced conspecifics (section 10.5).

Preemptive site selection is the tendency of individuals to move to sites of higher quality as they become available and has been demonstrated in Sedge Warblers in the Nida wetlands of southern Poland (Zajac *et al.* 2006). Located between two side arms of a river, this study plot consists of temporarily inundated areas of sedges, reed-

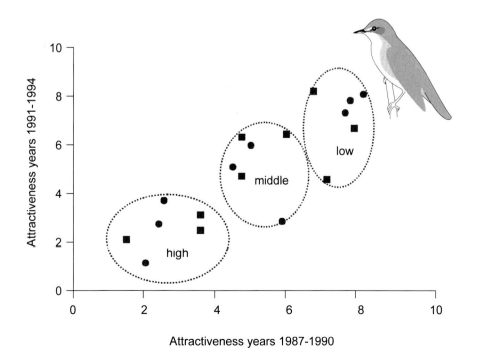

Fig. 6.3 The relationship between mean order of male occupation of territories (figures on the axes) during two different time periods for two marshes (circles and squares) showing the constancy in territorial attractiveness over long periods (Bensch *et al.* 2001). The degree of territory attractiveness can be very consistent over long periods of time.

Fig 6.4 Attractive and less attractive Great Reed Warbler territories side by side (Bensch *et al.* 2001). Territory numbers indicate the order of occupation (photo S. Bensch).

mace, and reed. Here, the most attractive sites are those with the largest amount of tall wetland vegetation (reed-mace and reeds). These sites are the first to be occupied year after year by the arriving males. Just as in Great Reed Warblers, the order in which the territories are settled remains strikingly constant over periods of time which exceed the life span of individuals. A preference for larger patches of emergent wetland vegetation probably evolved for two reasons: first, they allow females to choose between several males ('hidden lek'; see below) and secondly, they directly increase reproductive success by providing fledglings with more cover than smaller patches as protection against predators. Thus males occupying larger patches of tall vegetation had greater chances of finding a mate and of producing more recruits to the breeding population in succeeding years. The importance of territory quality for birds' fitness has been recently demonstrated in Seychelles Warblers by Janske van de Crommenacker and co-workers (2011). Birds in territories of lower quality were more oxidatively stressed because they had to invest more energy in finding their arthropod prey. Variation of fitness with site quality should therefore constitute a strong selection pressure, as it would favour the males' ability to arrive at and occupy sites early, thus preempting them. As males became

older and more experienced, they did indeed choose sites of increasingly higher quality, which indicates that the ability of males to assess the suitability of a site seems to improve with increasing experience throughout their lives (Zajac *et al.* 2006).

Both the varying suitability of sites for reproduction and preemptive site occupancy play an important role in the density-dependent regulation of populations (Rodenhouse *et al.* 1997). These two features in concert generate a negative feedback when progressively less suitable sites are used, whereby population size increases and breeding success decreases.

Getting together socially

Territorial songbirds generally use song to defend sites and attract mates, but both hearing song or just the presence of conspecifics itself may also serve as a cue in judging if a particular breeding site is attractive. This can lead to the settlement of additional males in the immediate neighbourhood of already occupied territories, resulting in aggregations and higher breeding densities on certain patches per unit area, as opposed to lower densities at other sites that are not necessarily of lesser habitat suitability. Social aggregations of this kind are common in most acrocephalid species. For instance, some examples of particularly impressive aggregations include 26 simultaneously occupied nests of Clamorous Reed Warblers in 120 m², which is the area covered by a house (Bairlein 2006), 6 nests of Eastern Olivaceous Warblers in 0.1 ha (Hagemeijer & Blair 1997), or 45 nests of Eurasian Reed Warblers active at the same time in 1.17 ha of *Phragmites* (Bibby & Thomas 1985).

In the breeding season, the use of conspecific attraction (and hence the preference for breeding aggregations) is selectively advantageous because individuals are more likely to find a vacancy in an aggregation as compared to a solitary site (Pärt *et al.* 2011). The benefits include easier discovery of better habitats, more effective mate choice, higher potential for extra-pair copulations (the 'hidden lek' hypothesis; Wagner 1998), and more efficient predator deterrence. Reedswamp-dwelling acrocephalids utter their warning calls communally when a predator appears, and the intensive mobbing of Common Cuckoos by Eurasian Reed Warblers is a very successful anti-parasitization technique. Furthermore, even though there is a positive correlation between nest density and predation rate, synchronous nesting by many breeding pairs can contribute to a reduction of the individual risk of predation or parasitization (the 'dilution effect'), as can be seen in Great Reed Warblers (Hoi &

Winkler 1988, Moskát *et al.* 2006, Welbergen & Davies 2009; see also below and chapter 9). Since conspecific attraction has a clear impact on both settlement patterns and population dynamics, it is also of special significance in conservation biology (chapter 13). Conspecifics can also act as attractors to sites outside the breeding season; nocturnally migrating Eurasian Reed Warblers and Sedge Warblers can be encouraged to land by playing their song, an established procedure in trapping projects (Kennerley & Pearson 2010, Mukhin *et al.* 2008; chapter 11).

Habitat structure and population density

The vegetation density and structure of reedbeds, together with their productivity, make for high population densities amongst the birds which frequent them. Hence the 'radius of tolerance' between Eurasian Reed Warbler males in dense *Phragmites* stands is less, and their territories correspondingly smaller, than those of the Sedge Warblers that inhabit more open vegetation (Catchpole 1972). No matter whether in temperate or subtropical regions, the densities of acrocephalids living in wet *Phragmites* beds are among the very highest for non-colonial songbirds (tab. 6.1). Comparatively high values are also reached in the deep-water marshes of the New World (chapter 14). Conversely, and in spite of the fact that it may still be of high vegetation density, perennial herb

Reed beds *(Phragmites)*	Average territory size (m²)	Maximum density/ha
Eurasian Reed Warbler	250	30
Oriental Reed Warbler	860	18
Moustached Reed Warbler	550	10
Great Reed Warbler	1 400	3
Herbaceous vegetation and grassland		
Marsh Warbler	1 000	8
Booted Warbler	?	5
Blyth's Reed Warbler	900	3
Sedge Warbler	1 000	3
Aquatic Warbler	6 000	1
Shrubs and saplings		
Melodious Warbler	1 100	1
Icterine Warbler	1 400	1

Table 6.1 Territory size and maximum densities in some acrocephalids according to habitat (data from Glutz & Bauer 1991). The term 'territory size' is also applied to Aquatic Warblers. The highest densities occur in reedbeds.

vegetation in drier locations sees average bird population density falling significantly, reaching its minimum in grassland and stands of shrubs and bushes or young trees. One interesting phenomenon arises when we compare the large species, Great and Oriental Reed Warblers. The maximal population density of the latter in China and Japan can be 10 times that of the former in Europe

(Dyrcz & Flinks 2000). While the Oriental Reed Warblers of the Eastern Palaearctic are often the only acrocephalid species found in reedbeds, in the Western Palaearctic Great Reed Warblers almost always occur together with Eurasian Reed Warblers. Adding the abundance values of both these species together results in a biomass density corresponding to that of the Oriental Reed Warblers.

Photos 6.4 and 6.5 Intraspecific aggression is a frequent occurrence, given the high population densities and mostly small terrritory sizes. As these photos of Eurasian Reed Warblers illustrate, the participants threaten each other and the situation can escalate into physical fighting (photos Albert Steen-Hansen, Mike Grimes).

Interspecific aggression and territorial overlaps

It soon became clear to the early students of acrocephalid territorial behaviour (Howard 1920, Brown & Davies 1949, Springer 1960, Catchpole 1972, 1973) that individuals of different coexisting species occasionally engaged in interactions or even maintained non-overlapping territories. Most birds defend their territories primarily against other members of their own species, but some also defend against individuals of other species. Since territoriality is a means of guarding resources or mates, interspecific aggressive actions are usually directed against heterospecific individuals that closely resemble the territory holder, either because they are closely related species, or they have similar resource requirements (Cody 1974, Rice 1978), or they would be most likely to attempt copulation with the mate of the occupier. Thus males of Melodious and Icterine Warblers, for example – two very similar, but mutually exclusive sister species – defend their territories in the very narrow zone where they occur sympatrically and where hybridization is possible, not only against male conspecifics but also against males of the sister species (section 2.4).

Further potential for interspecific territoriality exists where the habitat is so simply structured that it does not permit the divergence of environmental exploitation between two competing species, or where the habitat allows only incomplete niche segregation (Orians & Willson 1964). This is the case on some Scottish islands, for example, where the woodlands are structurally simpler and smaller than on the adjacent mainland. Here Great Tits and Chaffinches defend territories against one another, even though they are representatives of totally different families which do not even resemble each other and have completely different songs. Yet in the richer woods of the nearby mainland the two species live in overlapping territories alongside each other without any problem (Reed 1982). A parallel situation pertains in simply structured reedswamps, where the common members of the local bird guild, the reed warbler species, are even closely related, so that they resemble each other in appearance and voice (section 5.1 and chapter 7).

Marshland warblers keep their distance

In chapters 3 and 4 we saw that syntopically occurring reed warbler species have differentiated themselves along three different niche axes or resource dimensions in order to stay out of each other's way. The mechanisms of such a separation are the selection of different horizontal or vertical habitat segments and preferences for different food types. Temporal separation – in breeding phenology for instance – also contributes to ecological separation. Many studies have arrived at the same conclusion: the strongest niche separation among syntopic reedbed warblers results mostly from the utilization of different *horizontal* microhabitats and not so much from differences in feeding technique, or diet, or a partitioning along the *vertical* component of reedbed structure (Catchpole 1973, van der Hut 1986, Kagawa 1989, Saino 1989, Schulze-Hagen & Sennert 1990, Ivanitzkii & Keseva 1996, Vladimir Ivanitzkii *in litt.* 2009).

However, the ecological segregation of these species through the selection of varying horizontal habitats is far from total. This is evidenced by various degrees of overlap and at times by intense direct interactions between members of individual species-pairs: Oriental Reed Warblers were observed to be strongly aggressive towards Black-browed Reed Warblers, but not towards Thick-billed Warblers, by Ivanitzkii and co-workers (2002, 2005). In northern Israel, Great Reed Warblers formerly bred as a summer visitor alongside the resident Clamorous Reed Warblers, but separated by habitat, the former exclusively in reedbeds, the latter in a mixture of papyrus and reed. After drainage, the papyrus habitat was largely destroyed and widely succeeded by reeds, which may have increased habitat competition, and subsequently the Clamorous Reed Warblers supplanted the Great Reed Warblers (Shirihai 1996). In Sweden, later arriving Eurasian Reed Warblers were observed to oust Sedge Warblers from sites that were optimal for the former and suboptimal for the latter (Svensson 1978). In the Netherlands, the territorial overlap of Sedge Warblers with Eurasian Reed Warblers was generally only 17% (van der Hut 1986). Hoi *et al.* (1991) and Laußmann & Leisler (2001) were not able to determine consistent dominance relationships between Eurasian Reed Warblers and Moustached Warblers at their Neusiedler See study site.

The outcomes of these interactions are very variable, depending on the prevailing ecological and seasonal situation. Dominance-mediated relationships range from expulsion of a subordinate species (forcing it to breed in a suboptimal habitat) to partial exclusion or tolerance of a lower-ranking species. For example, the

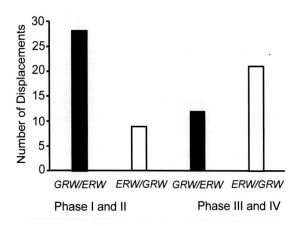

Fig. 6.5 Frequency of territorial displacements in Great *vs* Eurasian Reed Warblers (black columns) and Eurasian *vs* Great Reed Warblers (white columns) during the early (I and II) and late (III and IV) Great Reed Warbler breeding phases. In the early phases, Great dominate Eurasian Reed Warblers, while the situation is reversed in the later phases (Hoi *et al.* 1991).

lower-ranking Aquatic Warblers and Black-browed Reed Warblers, with their skulking habits, are tolerated by their respective dominant neighbours Sedge Warblers and Oriental Reed Warblers, and can therefore coexist with them in partly overlapping territories (Leisler 1988b, Kagawa 1989).

The case of Great and Eurasian Reed Warblers

Several authors have shown that the larger, dominant Great Reed Warblers exclude their smaller relatives with similar requirements from their habitat (Saino 1989, Hoi *et al.* 1991, Rolando & Palestrini 1989, Honza *et al.* 1999). The issue that has been most intensively studied, including by experimentation, is ecological overlapping and interference competition in the species-pair Great Reed Warblers and Eurasian Reed Warblers (fig. 6.5). Both species partially share limited resources such as nest sites and food, both suffer the same nest predators and brood parasite (Common Cuckoos), and they may even hybridize under certain circumstances (section 2.4). The active interference of Great Reed Warblers with the nesting activities of Eurasian Reed Warblers has been confirmed, both by direct observations of individual birds destroying nests containing eggs (Hoi *et al.* 1991) and also, indirectly, through experiments in which artificial eggs were placed in Eurasian Reed Warbler nests sited at various distances from active Great Reed Warbler nests. The reactions of the Great Reed Warblers ranged across the entire spectrum from acceptance (no eggs disappeared, all manipulated nests remained intact) to extreme rejection (all eggs disappeared or all manipulated nests destroyed; Honza *et al.* 1999). There was a strong correlation between the 'aggression level' of individual Great Reed Warblers and the distance to the nearest Eurasian Reed

Warbler simultaneous breeding attempt, so that the distance between nests increases as the aggression of an individual Great Reed Warbler increases (chapter 10).

Where nest site requirements are similar, the timing of breeding of the competitively weaker Eurasian Reed Warblers is delayed, in spite of their arrival times being the same, because the impact of the dominant Great Reed Warblers early in the breeding cycle is at its strongest. Eurasian Reed Warblers can only nest successfully when Great Reed Warblers have left the area (Beier 1993, Honza *et al.* 1999, Schaefer *et al.* 2006). Later in the season, or in habitat that is suboptimal for Great Reed Warblers, the positions can be reversed and Eurasian Reed Warblers can become dominant because of their high population density, driving individual Great Reed Warbler pairs away by mobbing them (fig. 6.5; Hoi *et al.* 1991). This highlights the fact that it is ecological factors which influence spacing, and dominance can shift from one species to the other, depending on the local situation (fig. 6.4, from Hoi *et al.* 1991).

Singing from a different songsheet

The conflicts between Great and Eurasian Reed Warblers could be seen as having their origins in the marked difference in size alone, but they are not an isolated case. Most examples of aggressive interspecific interactions occur between acrocephalid species of the same size, since competition for resources is probably even more intense (Leyequien *et al.* 2007). Following the novel playback experiments with reed warblers by Clive Catchpole (1977) and Francoise Lemaire (1977) using heterospecific song, the problem of how reed warbler species react to the simulated intrusion into their territory of a singing rival of another species has been tested in other species-pairs and at different sites. Indirect conclusions regarding the intensity of the ecological interactions can be drawn from the differing reactions to the heterospecific song used in the experiment (Catchpole 1978, Catchpole & Leisler 1986, Leisler 1988a,b, Rolando & Palestrini 1989, Ivanitskii 2001, Laußmann & Leisler 2001).

A general pattern has emerged of interspecific aggressive responses, or of interspecific territoriality: reactions to heterospecific song have only been noted in species-pairs with overlapping habitats and breeding seasons, and not in those that are sharply segregated in time and space. The reactions are asymmetrical, since only the males of those species that arrive early in the year respond to the song of later-arriving species, and not vice versa. Thus only Sedge Warblers react to the song of the later migrant Eurasian Reed Warblers, which in turn respond to the song of the even later arrivals, Marsh Warblers, and again not vice versa (Catchpole 1978). In agreement with these findings, Blyth's Reed Warblers began laying on average 10 days earlier in a Russian study area than Marsh Warblers and reacted aggressively to this late-arriving species (Ivanitskii 2001). A response to heterospecific song has only been found where the potential competitor species actually occurs (sympatry, or syntopy at a smaller scale), and not in areas where it is absent (allopatry). This is also true at a more local level: although Eurasian Reed Warblers reacted intensely to playback of Great Reed Warbler song in a reedbed where they occur syntopically, in another reedbed only 5 km away and in which only the former bred, no reaction to the heterospecific playback was recorded (Catchpole & Leisler 1986).

Interspecific aggression always arises in situations where established individuals of a species are frequently exposed to invading individuals of a possible competing species. It does not represent a failure to discriminate between species and has nothing to do with mistaken identity, for in that case an individual of one species would still react to playback in the absence of the competitor in its habitat. The fact is that interspecific aggression confers advantages and arises through learning (Catchpole 1978, Catchpole & Leisler 1986). Its adaptive value consists in the fact that it leads to spacing, to a reduction in the population density of sympatric species, whereby nest predation rates in the area of common occurrence can be reduced. Experiments by Hoi &

Winkler (1994) have demonstrated that in simply structured wetland vegetation a diversification of nest types is less effective in avoiding nest predation than in more complex habitats such as tropical or deciduous forests. As a result, all pairs of different reed-nesting passerine species share the same set of predators and brood parasites and so compete indirectly for enemy-free space (short-term apparent competition).

A convergent, and in many ways comparable interspecific territoriality has developed among the marsh-nesting songbirds of the New World (Orians & Willson 1964; chapter 14). Here, a situation of mutual interspecific hostility between icterids excludes Marsh Wrens from common space, while the wrens in turn attempt to destroy the eggs of other birds, unrelated or even conspecific (Picman 1988).

The benefits of getting rid of the competition

The essence of interspecific competition is that individuals of one species suffer a reduction in fecundity, growth, or survival as a result of resource exploitation or interference by individuals of another species (Begon *et al.* 1990). Until now the precise costs of coexistence for birds have hardly been quantified. Paul & Thomas Martin (2001a,b) measured the consequences of coexistence for fitness-related traits in two congeneric wood warblers, Orange-crowned and Virginia's Warblers, which competitively interact on overlapping breeding territories. As in reed warblers, the slightly larger Orange-crowned Warblers are aggressive towards the later-arriving Virginia's Warblers and respond aggressively to heterospecific playback, whereas Virginia's Warblers avoid actual, as well as playback-simulated interactions with the dominant Orange-crowned Warblers. On two study plots, one or other of the two ground-nesting species was removed (reciprocal removal). The breeding success of the remaining species then increased, when compared with control plots on which both species coexisted, because nest predation rates fell following removal of the competing species. In addition, Virginia's Warblers also used nest sites which otherwise would have been inaccessible to them through preemption or interference by the dominant Orange-crowned Warblers. They were also able to increase their feeding rates. On the control plots, nest predation rates for both species rose as a consequence of nest predators shifting their foraging behaviour to the increased number of ground nests, for example by spending more time searching for this nest type (short-term apparent competition; Martin & Martin 2001a,b). These results support the earlier studies in reed warblers, which used only behavioural playback experiments to infer ecological interactions among coexisting species.

Recognizing other species

As a result of 'cross-species learning' individuals can increase their responses to songs of similar heterospecifics, either related or ecologically competitive species; in other words, they react to heterospecific males just as strongly as to conspecific ones. By studying Blackcaps and Garden Warblers, which are clearly different in appearance, Matyjasiak (2005) recently demonstrated experimentally that songbirds learn to utilize differences in song as well as in plumage in order to identify heterospecific ecological competitors. Both closely related warblers compete strongly for space, with Blackcap males responding to the song of the later-arriving Garden Warblers, defending their territories against them, and excluding them from areas acceptable to both species. Prior to the return of the Garden Warblers, Matyjasiak tested Blackcap males in their territories with playback and mounted model males of both species. When he played Garden Warbler song, almost all Blackcaps attacked the dummy Garden Warblers; when he played their own song, all territory owners behaved aggressively exclusively towards the conspecific model. It was clear that Blackcaps, including many yearling males, had learned to associate Garden Warbler song with plumage and other species differences. In addition, they were also able to remember and recall the association after a time period of eight months.

This ability to associate certain acoustic and visual cues of an opponent and to retain them in memory should be regarded as an evolutionary adaptation, especially in migrants living in cluttered habitats or crowded social environments. It helps birds avoid wasting the time and energy spent on evicting heterospecific intruders from habitats in which they pose no threat (Matyjasiak 2005). In isolated cases (for instance in the absence of conspecific models) misimprinting can sometimes occur through song copied from the 'wrong' heterospecific tutors, which can result in mixed singers and occasionally mixed pairings (e.g. Marsh Warblers; Lemaire 1977, Helb *et al.* 1985; section 2.4). Nevertheless, the results from reed warbler studies suggest that these abilities are only

learned if a competitor species is actually present, and, when taken together with Matyjasiak's findings, there can be little doubt that we can now reject the interpretation reached by Murray (1981), that interspecific aggression is commonly misdirected *intra*specific aggression, or mistaken identity.

The frequency and ubiquity of interspecific territoriality appears to be a particular feature of sylviid warbler breeding systems (Bairlein 2006). It has both ecological causes and far-reaching consequences: species which interact aggressively in habitats with short vegetation or lower productivity can nevertheless coexist in taller vegetation or more productive habitats. The fact that larger dominant species, such as Great Reed Warblers, occupy the more productive habitat segments along ecological gradients leads to the structuring of local guilds, further examples of which can be found among Amazonian birds and also in the *Phylloscopus* warblers of Siberia (Robinson & Terborgh 1995, Orians 2000, Forstmeier *et al.* 2001c).

Competition and shared parasites

Competition between species as a result of shared parasites has also been discussed as a further important force influencing the coexistence and population dynamics of sympatric species. Such interactions are a form of 'apparent competition', whereby the presence of one species decreases the fitness of another through the increased presence of a shared enemy (Holt 1977). This phenomenon is being intensively researched by ecologists using today's improved methods of diagnosing blood parasites with the latest molecular toolkits. In this way, Reullier *et al.* (2006) discovered – in the moving contact zone of Melodious and Icterine Warblers (section 2.4) – that the expanding Melodious Warblers harboured a greater variety of parasites than the receding Icterine Warblers and that this may indicate a better ability to deal with a diverse parasite fauna. In sympatric areas, cross-species parasite transfer was confirmed, and was asymmetric; in other words, the aquisition of parasitic infections of the sister species was commoner in the receding Icterine than in the expanding Melodious Warblers, a fact which presumably contributes to the displacement of the hosts' contact zone. This result supports the idea that parasites can cause incompatibilities between related host populations and thus prevent their secondary contact (Ricklefs 2010).

Summary

Territoriality allows competing individuals to achieve spacing. Acrocephalids and other birds will defend sites of higher quality, thus increasing their chances of attracting females, ensuring paternity, and securing a safe nest site or food supply; each species has its own priorities regarding the function of territory.

In the simply structured habitats occupied by the reed warblers, some of the resources afforded by these sites are unevenly distributed. In some areas the attractiveness of territories can remain remarkably constant over years and can actually be based on 'tradition', when inexperienced birds copy the territorial preference of older ones. Among Great Reed Warblers and Sedge Warblers, owners of the most attractive territories achieve higher fitness (reproductive success) than those with less attractive territories. In many acrocephalids social aggregations are common, in which the song of conspecifics may serve as a cue to others as to the attractiveness of any particular breeding site.

The productivity of reedbeds makes for high population densities; density declines substantially in more open and bushy habitats. Occasionally over 40 active Eurasian Reed Warbler nests can be found in less than one hectare of reedswamp.

Interspecific aggression linked to territoriality is a common phenomenon in wetland-dwelling acrocephalids, as demonstrated by the various degrees of overlap and intense direct interaction between members of species-pairs (e.g. Oriental Reed Warblers and Black-browed Reed Warblers, or Great Reed Warblers and Eurasian Reed Warblers), ranging from total expulsion to partial exclusion or tolerance of a subordinate species. Explanatory factors for such interactions include similar or overlapping resource requirements, limited possibilities for niche segregation in simply structured habitats, and the sheer numbers of all bird species living in high-density populations in these habitats. The phenomenon is possibly intensified by the similarities between acrocephalid species in appearance and song.

In some reedbeds, Eurasian Reed Warblers can only begin their breeding activities once the dominant Great Reed Warblers have ended theirs. Playback experiments with heterospecific song demonstrate that aggressive reactions mostly exist between species with overlapping habitats and breeding seasons, and that it is always the species which arrive the earliest that respond most to the song of the later arrivals.

Reed warblers are an excellent example of how apparent competition leads to interspecific aggression. The presence of a particular species may decrease the fitness of another through the increased presence of a shared enemy, such as nest predators, brood parasites, or parasites. Learning plays the decisive role in the development of this interspecific aggression. The dominance hierarchies created between species can have wider effects, since they also structure local bird guilds.

Photo 6.6 The phenomenon of high population densities in reedswamp also applies to other passerine species; inter-specific aggression is therefore commonplace, even between distantly related taxa such as the Eurasian Reed Warbler and the Reed Bunting shown here. This aggression results from competition for predator-free space (photo Mark Coates).

A song worth warbling

Aquatic Warbler song-flight.

Of all the charms with which birds delight us humans, it is their song that has the greatest emotional impact. Imagine for a moment that we are sitting by an overgrown riverbed in the northeast of the Republic of South Africa. It is a clear morning and lively birdsong can be heard. It emerges from the interior of an impenetrable bush, surrounded by a herb layer of Compositae (Asteraceae) flowers. The singer is a Marsh Warbler, perched well hidden, almost invisible in the bush; his song is of a pleasing beauty, a continuously flowing, warbling chatter of hundreds of motifs, some sweet and silvery, others harsh, interspersed with liquid trills and tremolos, constantly speeding up or slowing down. Its variations of frequency, timbre, and pace are breathtaking. It is easy to see why a Marsh Warbler has been dubbed 'the best rapper in nature' (Constantine & The Sound Approach 2006). Yet even more striking than the astonishing speed and richness of its song is the bewildering blend of perfect imitations of both Palaearctic and African bird species. Again and again our South African bird produces impersonations of familiar European species: the calls of wagtails and Meadow Pipits, song fragments of Great Tits and Dunnocks, continually alternating with calls of African species, even within one and the same motif, such as when calls of Southern Buffbacks or Bleating Warblers are jumbled together with those of Common Chaffinches and European Blackbirds.

Our bird hatched in Lithuania at the end of June of the previous year and found his own way to the African winter quarters where we are now listening to him (chapter 11). Like other young Marsh Warblers he did not learn his imitative motifs from his father, but from a multiplicity of other bird species encountered first of all around his natal nest site, then along his migration route, and finally in his wintering area (Becker & Lütgens 1976). He stored these alien voices initially in his memory over a period of several months, forming a template of species-specific song during the sensory acquisition phase, which is a critical learning (though silent) period. From about January onwards he began to insert them into his own song. The development of his species-specific song begins with simple, non-imitative subsong, in which more and more mimetic elements are incorporated (song crystallization – motor phase). If our Marsh Warbler successfully returns to the northern hemisphere he will defend his first breeding territory, probably not very far from his own birthplace, with practically the same song that so pleases us now. Obviously his song must be good enough to attract a female, and the time between January and June is used only to improve the continuity and exuberance of his vocalization (Dowsett-Lemaire 1981b).

Such astounding mimetic abilities make Marsh Warblers the 'Meistersingers' whose astonishing utterances have captivated generations of nature-lovers and birdwatchers in their European breeding range. New imitations are continually being identified, so the list of mimicked

Palaearctic species in Marsh Warbler song becomes ever longer, and has been estimated at around 212 species (Lemaire 1974, Dowsett-Lemaire 1979b). However, this is not even half the story about our 'mockingbird'! In a series of papers, Francoise Dowsett-Lemaire has shown that, on average, more African than European species are imitated by any individual Marsh Warbler (Lemaire 1974, 1975a,b, Dowsett-Lemaire 1979 a,b, 1981b). This finding is a good example of the 'northern hemisphere bias', whereby the longer research tradition in the north compared with the south has hindered a complete understanding of various phenomena even to the present day.

This chapter will first provide an overview of song diversity in the acrocephalids, before considering the various functions of song, as well as its development in young birds into an extremely complex form of behaviour. We will examine in detail whether ecological and behavioural factors such as habitat, mating systems, or annual cycles have determined differences in several song traits between the species, such as complexity, frequency, and temporal arrangement.

1
Diversity of song
The various species that belong to the Acrocephalidae family differentiate themselves most strikingly in the structure, complexity, and duration of their songs. Since closely related acrocephalids lack strongly-defined or

Photos 7.1 and 7.2 Almost without exception, the song-phrases uttered by reed warblers of the temperate regions are long, elaborate, and complex, while the song itself is delivered almost without pause, especially before pair formation. By contrast, the song-phrases of the sedentary species are shorter and simpler. Generally the resident males sing from inside dense vegetation, like this Clamorous Reed Warbler (right); occasionally they might do so from an exposed perch, like this Saipan Warbler (left) (photos Jens Hering, Jack Jeffries).

diagnostic plumage patterns, for ornithologists their highly species-specific songs are often the most valuable distinguishing characteristic in field identification. Before we start to make comparisons it would be useful to define some of the basic terms we shall be using. 'Element' here is a direct equivalent of 'unit' or 'syllable' elsewhere in our text, and denotes the smallest distinguishable building block of which a song is composed; 'complexity' is mostly concerned with repertoire size, or the total of different unit types that comprise the song; the term 'note' refers to any discrete sound; 'phrase' is used for a distinct passage of elements of the same type, and 'motif' for a connected series of several elements of different types. We shall use 'strophe/song-phrase' and 'song' synonymously for a connected series of elements that is distinguished from a following series by a longer pause. In continuous song the series are longer and the pauses shorter; the reverse is the case in discontinuous song.

Complexity and structure

At one extreme of the song spectrum we find such species as Marsh and Sedge Warblers with their enormously complex and lengthy songs. These species demonstrate a virtually limitless number of possible combinations of different units in a motif or song (Koskimies 1991a, Schulze-Hagen 1991c). At the other extreme, the song of Henderson Island Warblers is such a simple series of three thin longer notes that one is forced to ask if this vocalization is not more a series of calls than a song (Graves 1992; chapter 12). By contrast, in uninterrupted song sections of over 10 minutes duration delivered by the first two species, an astonishing total of more than 3500 single units has been counted (Comtesse 1974, Panov *et al.* 2004). Sedge Warbler males have an average of 68 different units at their disposal (table 7.2), and although these are used repeatedly in the structure of a song, they are always in different combinations. Clive Catchpole discovered that the repetition of syllables, and simple rules for the transition between them, are sufficient for the composition of an infinite variety of different songs, which is why he compared singing Sedge Warblers to 'jazz musicians in performance' (Catchpole 1976, 2000). More soberly, Panov *et al.* (2004) described the composition method of Sedge Warblers as a self-organizing process.

Catchpole's sequence analysis of the units revealed specific rules according to which they are combined to build up an almost endless variety. Each song starts with a long section in which short series of two elements alternate with each other. There is then an abrupt switch to a louder, more rapid, and complex central section in which 5 10 new elements are suddenly introduced in quick succession. In the last part of the song, the patterning is similar to that at the start, except that the two elements used are selected from among those that occurred in the central section. These same two elements are typically those that are employed at the start of the next song (Catchpole & Slater 2008).

Male Paddyfield Warblers have an even larger repertoire of units available to them than Sedge Warblers (including much mimicry) and construct their songs in a similar way (Ivanitzkii *et al.* 2006). By contrast, the song of Dark-capped Yellow Warblers is structured in a completely different fashion: from a few repeated short melodic phrases a song type is built, which is itself repeated a few times (for example three times) before a change to another song type is made (fig. 7.1, showing two song types).

Tempo and syntax

While the various clades do not significantly differ from each other in such temporal aspects as duration of pauses and units (see below; Fiedler 2011), there are nevertheless great differences in other characteristics of temporal structure, such as song-phrase length and patterning (syntax). There is great variety in both the tempo of song delivery over a longer period and how rapidly the notes or units are repeated. We have already met one species (Marsh Warblers) that can quickly change tempo within a song, but the tempo changes encountered in Sedge Warbler song are even more marked. Conversely, the song of Eurasian Reed Warblers displays an extreme of monotony. The male bird utters his grating and squeaky notes at a steady pace, as precisely as a metronome and, to our ears, almost boring in its regular delivery: 'kerr kerr kerr, chirruc chirruc chirruc, ke ke ke, chirr chirr' (fig. 7.2).

On the other hand, species such as Blyth's Reed Warblers, Upcher's Warblers (fig. 7.3), and Eastern Olivaceous Warblers typically repeat longer sections of song made up of complex phrases without pauses in a cyclical pattern. This creates a repetitive, tedious impression, reminiscent of 'a vinyl record with a scratch in it' (Constantine & The Sound Approach 2006).

Songs also vary considerably in tone and timbre, as we can gather from the various words which humans use to describe them: 'squeaky', 'shrill', 'grating', and so on. Species such as Moustached Warblers and Icterine Warblers can combine very different tonal qualities. Song-phrases of the former are often introduced by a series of rising pure-toned whistles, which remind the listener of the crescendo of Common Nightingales (fig. 7.4). The remainder of the song then consists of a warbling segment, reminiscent of Eurasian Reed Warblers or Sedge Warblers, but softer, more varied, and containing mellow notes and trills (for the different functions of these parts see section 7.3).

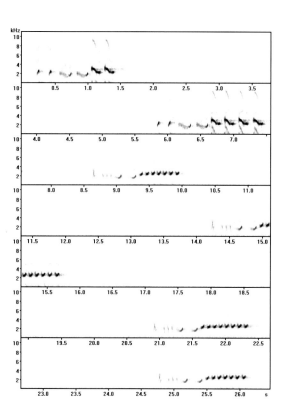

Fig. 7.1 Sonagram of two song types of a Dark-capped Yellow Warbler. The shortish phrases of throaty shrill notes are typical of the species.

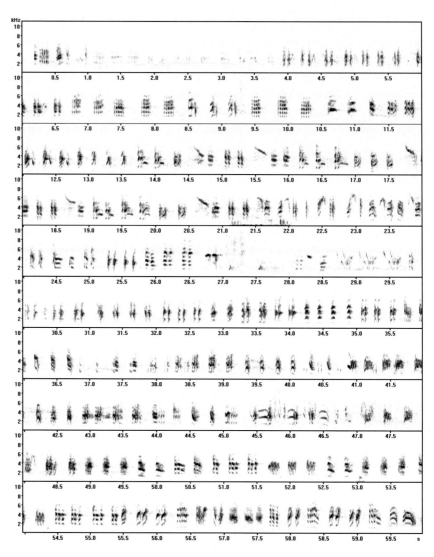

Fig. 7.2 Sonagram of the song of a Eurasian Reed Warbler. Note the typical long phrases and the regular, metronome-like rhythm.

Finally, not only single units but even entire songs can vary in volume, whether under the influence of some intra-specific situation (such as a male using song in territorial defence or directed towards a female) or in a group-specific fashion. For example, the songs of both the large reed war-blers and also Thick-billed Warblers, are louder than, say, Eastern Olivaceous Warblers or other small species.

Thus while some song traits are to be found in quite dif-ferent taxa – for example, cyclicity in representatives of *Iduna* (Eastern Olivaceous Warblers), *Hippolais* (Upcher's Warblers), and *Acrocephalus* (Blyth's Reed Warblers) – other traits, such as frequency and song length (see below) appear to be typical of certain groups. This there-fore begs questions as to whether different clades are dis-tinguished by particular song characters, and, if so, by which, and what influence has phylogeny had on song development? Such issues will be addressed in the next section and also in section 7.5.

Song and systematic relationships

Bärbel Fiedler (2011) has been working on acrocephalid song and looking at the problem of whether the various systematic clusters, or clades, can also be distinguished on the basis of song. From the wealth of song characters she analysed we have extracted those which we think are

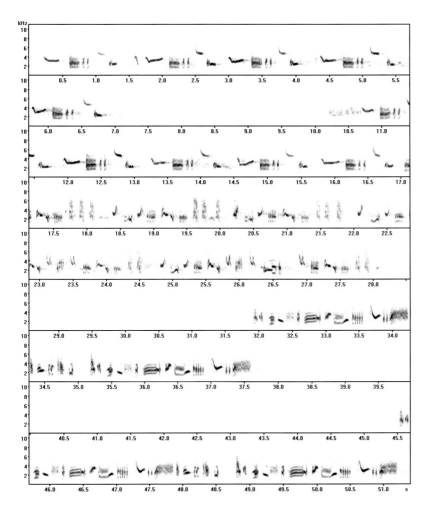

Fig. 7.3 Sonagram of the song of an Upcher's Warbler. Note the repetition of similar motifs.

the most significant, such as tempo, mean frequency, and repertoire size, and have compared them in table 7.1. This shows that Papyrus Yellow Warblers have one of the lowest-pitched songs of all the acrocephalids, consisting of various short sibilant, liquid trills, which are generally repeated several times before the singer moves on to the next one (MacLean *et al.* 2003). Thick-billed Warbler song is a hurried chattering warble of harsh and more musical notes, characterized by longer notes and a flowing delivery (Marova *et al.* 2005). Most species of the *Iduna* group are noted for their very harsh and nasal, even strangulated song. They sing their song-phrases just as quickly as the striped species, but their song flows along rather 'aimlessly' in more or less the same register. As we have seen,

the song of the tropical Dark-capped Yellow Warblers is characterized by its simplicity; they have the smallest repertoire within the family.

The four *Hippolais* species are sharply distinguished from each other, but one feature common to them all is their extremely variation-rich song-phrases, which sound less hard or harsh than those of 'typical acrocephalids' (*Acrocephalus* and *Iduna* songs). The units are among the longest in the entire family, as exemplified by Icterine Warblers.

The group of large reed warblers have deep and loud songs in common, which they mostly deliver in short strophes/phrases (though Basra Reed Warblers are the exception). However the two subgroups *Acrocephalus* and '*Calamocichla*' differ clearly from each other in that the songs of the latter, the tropical swamp warblers, are shorter and simpler, and the units contain more pure tones, lending their song a softer sound than that of the *Acrocephalus* group (Dowsett-Lemaire 1994, Fiedler 2011. Their characteristic loud varied trills are all separated by pauses and calls.

The small plain-coloured acrocephalids all share continuous songs with an extremely complex structure, sounding harsh and scratchy because of their many 'noisy' units. However, Blyth's Reed Warblers are the exception, with their often-repeated motifs consisting of whistles or glissandi. The species of the *agricola* subgroup have the widest repertoires; their very complex songs sound hurried and thin. The striped reed warblers deliver brief units in a

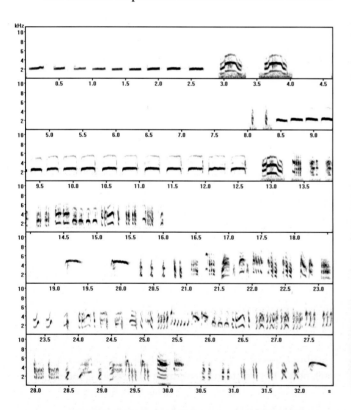

Fig. 7.4 Sonagram of the song of a Moustached Warbler. Note the introductory whistle segments and the second section with its warbling character. Both parts have differing functions, the first being an aggressive 'keep-out' signal to other males, the second aimed at attracting females.

Photo 7.3 Soon after arrival on the breeding grounds, when the fresh vegetation is still low and the dominant colour in the environment is the beige of the old reeds, this singer's bright orange-red mouth lining perhaps functions as a visual signal to accompany the acoustic signal of the song itself (Great Reed Warbler; photo Winfried Wisniewski).

Table 7.1 Characterization of the songs of various acrocephalid taxa based on selected traits. The number of species studied is given in brackets. Numerical values are the species means, or the means of the relevant group derived from these. Tempo refers to the number of elements per second. Repertoire values are the number of elements which make up that repertoire. After data from Fiedler (2011).

Taxon (spp)	Form	Tempo E/s	Mean frequency	Repertoire	Imitations	Impression
C. gracilirostris (1)	distinct phrases	(7.6)*	1 860	16	none	repetitive, sibilant
Ph. aedon (1)	continuous	5.4	3 751	158	many	lively, babbling
Temperate *Iduna* (4)	distinct phrases	6.4	2 428 - 5 192	174 64-292	few	bubbling, hard
Tropical *I. natalensis* (1)	distinct phrases	5.4	2 487	7	none	trilling, melodious
Hippolais (4)	distinct phrases	5.3 3.8-7.3	2 098 - 4 876	133 73-181	many	variable, chattering
Large *A. griseldis* (1)	continuous	2.5	3 024	203	?	hesitant, throaty
Large (4)	distinct phrases	4.1	2 084 - 3 925	102 61-93	few	low, loud
Large *Calamocichla* (5)	distinct phrases	4.9 3.8-6.7	1 605 - 3 116	35 11-75	none	simple, soft
Small *agricola*-group (3)	continuous - distinct phrases	5.3	2 747 - 5 422	216	many	high, jingling
Small plain (5)	continuous	4.6	1 942 - 5 962	155 60-216	many	hard, scratchy
Small striped (4)	distinct phrases - continuous	6.9 5.8-9.3	2 054 - 6 226	68 24-213	variable	high, rapid

* in trills

series, which makes the tempo of their vocalizations the fastest in the acrocephalid warbler family. Their songs are also distinguished by the rapid alternation of noises and pure notes, as well as by their broad frequency range.

The various members of the genus *Nesillas* appear to lack a structured song. Their vocalizations are shrill chatters and simple series of rattling notes. Only Moheli Brush Warblers have a more complex scratchy warbling that can be delivered by both members of a pair in short duets (Louette *et al.* 1988, Kennerley & Pearson 2010).

Dendrogram: a diagram of branches

As with other traits such as morphology and nest characteristics (chapters 5 and 8), we will now try to answer the question of whether the songs of more closely related

species show a closer resemblance to each other than to those of more distant relatives. To this end we shall compare the similarities in measured song characters among the species (see above) with their actual degrees of relatedness derived from molecular phylogenies (note the colour symbols in fig. 7.5). This figure shows how 33 species studied by Fiedler (2011) group themselves according to song. What strikes us immediately is the marked divergence of the songs of Basra Reed Warblers and Icterine Warblers from those of all other acrocephalids, though the songs of these two species do not resemble each other in any way. The remaining species fall into two clearly separate clusters. One (from African to Blyth's Reed Warblers *baeticatus* to *dumetorum*) embraces species that sing in a more flowing fashion and whose songs have a complex structure. The second cluster (from Aquatic War-

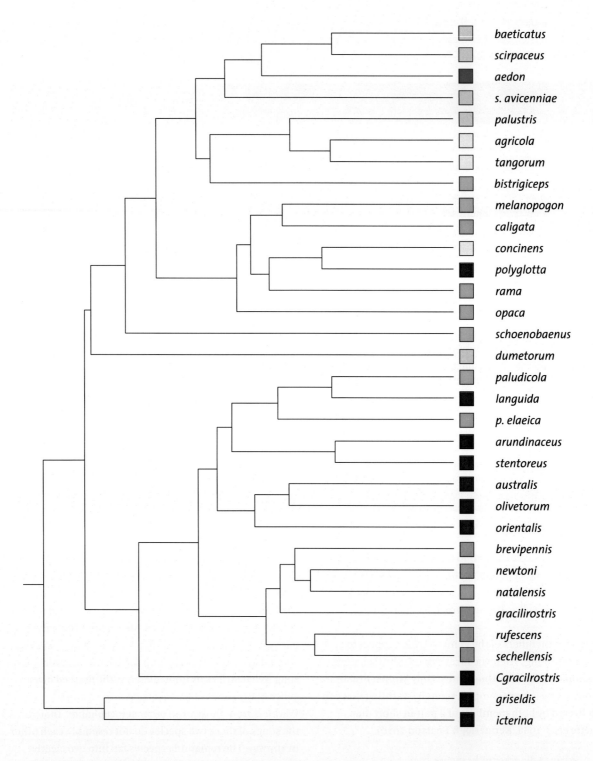

Fig. 7.5 Cluster analysis (dendrogram) of 33 acrocephalid species using song characters. This demonstrates the similarities between species based on measured song traits. Here congruences as well as divergences become apparent when we compare this dendrogram with the phylogenetic tree based on genetic distances in fig. 2.5, recognizable by the corresponding order of the colour symbols in both figures.

Table 7.2 Relationship between song characters and mating system plus parental care in six reed warbler species. After Catchpole (1980), using additional data from Fiedler (2011).

	Song length	Repertoire size	Mating system	Paternal investment
Moustached Warbler	12.3	212	monogamous (cooperative)	very high
Eurasian Reed Warbler	34.0	164	monogamous	very high
Marsh Warbler	16.1	181	monogamous	very high
Sedge Warbler	19.5 – 27.2	68	monogamous (polygynous)	high
Great Reed Warbler	3.2 – 5.1	71	facultative polygynous	reduced
Aquatic Warbler	2.2 – 5.5	24	promiscuous	absent

blers *paludicola* to Papyrus Yellow Warblers *C. gracilirostris*) is made up of species that deliver their simpler songs mostly in short phrases, with a small repertoire and hardly any mimicry. In the first grouping are all the small plain-coloured reed warblers, most of the striped and *Iduna* group, but also Thick-billed and Melodious Warblers. In the second grouping, the systematically isolated Papyrus Yellow Warblers can also be seen as acoustically isolated. The large reed warblers constitute the bulk of the species here, with two subgroups well characterized by similar songs. Two *Hippolais* and *Iduna* species also fall within this cluster.

The three clades whose representatives show the least congruence in the song traits examined, and are therefore spread most radically between the different clusters, are the *Hippolais*, the *Iduna*, and the striped species, thus confirming the diversity of their songs as perceived by the human listener. For instance, Olive-tree Warblers are more akin to large reed warblers in vocalization, while the song of their sister species, Upcher's Warblers, has more in common with that of Aquatic Warblers. Moustached Warblers are in a cluster of species whose songs are very rapid and high-pitched (*Iduna* group, Blunt-winged Warblers, and Melodious Warblers). On the other hand, their close relatives, Black-browed Reed Warblers, find themselves a member of a group of small plain-coloured species. The dendrogram (fig. 7.5) also shows just how strongly the song of Blyth's Reed Warblers differs from that of the other small plain-coloured species.

From these comparisons, we can draw the general conclusion that song is heavily influenced by both phylogeny and lifestyle. In other words, species that are more closely related to each other have songs that are more alike than those of more distantly related species (e.g. the small plain species and the large '*Calamocichla*'). On the other hand all tropical species, regardless of relatedness, form a cluster (Dark-capped Yellow Warblers and the '*Calamocichla*' spp.).

We shall shortly consider which factors can explain this diversity of songs (section 7.5) but first, let us turn to the various functions of acrocephalid songs and their development in young birds.

3

Why do birds sing?

The breeding season song of male birds serves two main functions. As a 'keep-out' signal it helps to proclaim the ownership of a territory and defend it against other males, and as a 'come hither'/advertising signal it attracts and stimulates a conspecific mate. Thus the song has two different audiences, since it is used both for intrasexual and intersexual purposes. The importance of these two main functions of song can nevertheless vary substantially throughout the Acrocephalidae family, depending on the resources being defended and the mating system of the species involved, and thus ultimately on habitat quality (chapter 10). These various influencing factors also determine the length of the song period.

Since Darwin's (1871) insight that the song of Common Nightingales is so 'beautiful' in order to enchant the female, we know that song can represents a vocal 'ornament' that is shaped by sexual selection. We know further that the song of oscine birds ('songbirds') is controlled by a relatively well-defined series of discrete brain nuclei, the 'song system', and that young oscines must learn their song by listening to older birds. Therefore in all songbird species potential singers must first develop, during ontogeny, a neural network of centres in the brain that controls the song-producing muscles.

Reed warbler studies have once more made outstanding contributions to research on all these issues, principally because field studies could be cleverly combined with laboratory experiments.

The case of six reed warbler species

The long-term comparative study of six central European *Acrocephalus* species that we have described in previous chapters can once more be called upon to reveal an important hidden pattern – this time the relationship between the form and function of their songs. Four of the six species (Eurasian Reed Warblers, Moustached, Marsh, and Sedge Warblers) sing lengthy songs (table 7.2), but reduce their length following successful attraction of a female (Catchpole 1973, Dowsett-Lemaire 1981a, Kelsey 1989, Borowiec 1992, Kloubec & Capek 2000). However,

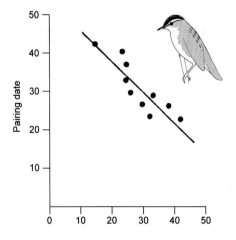

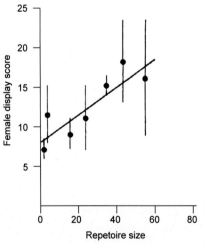

Fig. 7.6 Repertoire size in Sedge Warblers. Top: the relationship between repertoire size and pairing date in wild birds. Males with a larger repertoire pair off sooner. Below: The relationship between male repertoire size and sexual displays in captive females. Females performed higher numbers of wing vibration bouts in reaction to test songs with higher repertoire size (from Catchpole 1980 and Catchpole *et al.* 1984).

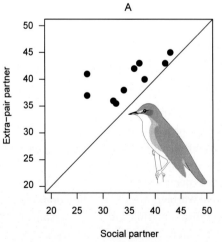

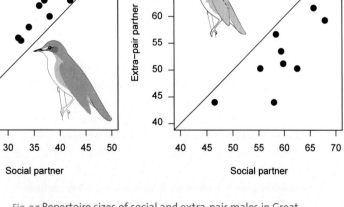

Fig. 7.7 Repertoire sizes of social and extra-pair males in Great Reed Warblers and Sedge Warblers (from Hasselquist *et al.* 1996 and Marshall *et al.* 2007). A: In Great Reed Warblers, females choose males with larger repertoires than their social partners for extra-pair fertilizations. B: In Sedge Warblers by contrast, females choose males with smaller repertoires than their social partners for extra-pair fertilizations.

early arriving and paired male Sedge Warblers can also resume singing in order to try and attract an additional female; that is they become polygynous, as do the closely-related Black-browed Reed Warblers (Hamao 2008, Zajac *et al.* 2008; see below and section 10.5).

Those sections or features of song which contain the male- or female-directed messages are yet to be explained. Only Fessl & Hoi (1996) have looked at this issue in Moustached Warblers, a species in which the male plays an important role in brood care (table 7.2; chapter 10). Using observations and playback experiments in the field, they found that the introductory whis-

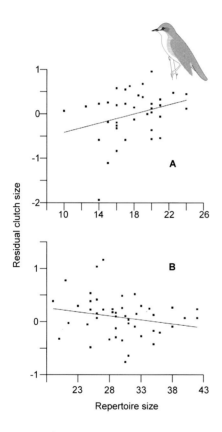

Fig. 7.8 Relationship between repertoire size of male Great Reed Warbler song and clutch size as an indicator of reproductive success. Residual clutch size is adjusted to variation in laying date. A: Period 1982-1993 B: Period 1994-2000 (from Forstmeier & Leisler 2004). The positive relationship between repertoire size and clutch size disappears in years of low population density when all males are able to occupy good territories.

tle segment of the song (fig. 7.4) is a 'keep-out' signal directed at male competitors, while the following complex warbling section is aimed at females.

Great Reed Warblers are different from all other European *Acrocephalus* species in that they are facultatively polygynous and deliver much shorter songs (table 7.2). One part of the male population is always polygynous, meaning that part of the female population must raise their young unassisted and so must base their choice of mate upon territory quality rather than male quality (section 10.5). Great Reed Warbler males drastically shorten their variable song once a pair has formed and sing only short phrases during the fertile phase of their females (Catchpole 1983). Using playback experiments in both field and aviary, Catchpole *et al.* (1986) were able to demonstrate that short songs are more directed at males and longer ones at females. After their primary females have laid eggs, many males resume their full song, and thus attempt to pair with a further female (as do Oriental Reed Warblers; Ezaki 1987). Even briefer than the songs of Great Reed Warbler males are those of Aquatic Warblers, which do not participate at all in brood provisioning. They possess different song types, of which the shorter 'simpler' ones function in male-male aggression, and the longer more complex type in female attraction (Catchpole & Leisler 1989, Schmidt *et al.* 1999). In contrast to all other migratory reed warblers, Aquatic Warbler males maintain their song activity undiminished right through the breeding season (Dyrcz & Zdunek 1993).

In addition to song length, Catchpole (1980) compared a second trait in the male vocalizations of the six central European reed warblers – song repertoire size (table 7.2). On average, the songs of monogamous species contain a greater number of units than those of the non-monogamous. Within a species – in this case, Sedge Warblers – males with larger repertoires mate earlier than rivals with poorer songs (Catchpole 1980; fig. 7.7b). Later experiments (Catchpole *et al.* 1984, 1986) demonstrated that hormone-implanted Sedge Warbler females reacted more to playback of larger repertoires (fig. 7.7b).

On the basis of the clear *interspecific* pattern in table 7.2, which shows that the four monogamous species sing long songs with larger, more complex repertoires, while non-monogamous species have shorter, simpler songs, together with the high *intraspecific* importance of the repertoire

Photos 7.4 and 7.5 The use of air sacs, muscles, and syrinx in song production leads to the throat feathers becoming erect, thus revealing the feather tracts on the skin (Great Reed Warbler and Marsh Warbler; photos Zdenek Tunka, Sergey Yeliseev).

in attracting Sedge Warbler females, Catchpole (1982) suggested that two different types of song structure have evolved in reed warblers and that sexual selection is the most important evolutionary force driving this development. In polygynous species, male-male contests predominate and have produced shorter, simpler songs more suited to the particular problems of territorial defence against rival males (intrasexual selection), since territory quality is of great importance for females. Shorter songs are advantageous because males need time to listen to the reaction of rivals in the pauses. However, within species, females react more to the more complicated songs. In monogamous species, female choice predominates and has produced longer and more complicated songs for sexual attraction (intersexual selection; Catchpole 1982). As early as 1974, Richard Howard observed in American Northern Mockingbirds that both forms of sexual selection make song into a vocal ornament (plume) via the dual function of song (territorial defence *vs* mate attraction). Later Loffredo & Borgia (1986) found that in general the vocalizations involved in female attraction were always more 'beautiful', while those used in male contest were bolder and more striking.

4

Sexual selection and honest signals

Sexual selection results from variation in the mating success of individuals, such as access to different numbers of mates or to mates of differing quality. Males can gain access to females in two ways – by intrasexual competition or by intersexual selection. An evolutionary advantage will be conferred by a song trait that increases the success of an individual in male-male contests – in gaining the breeding territory required by a female, or one that is more attractive to females during courtship. A female preference for such traits of superior song quality will develop if females receive some benefit from mating with the singer of this particular song character and if the character is a reliable indicator of male quality, such as condition or viability. The reliability of signals is vouched for if they entail some costs to the male which other weaker, inferior males cannot afford to pay (Zahavi's handicap principle; Andersson 1994, see also below).

Female benefits can be divided into two categories: direct and indirect. If a song trait correlates with some aspect of a male's phenotype that directly increases the female's reproductive success, such as his ability to defend a good, food-rich, or safe territory or to provision the young, then she receives direct benefits by mating with a *phenotypically* (and mostly also genetically) superior male. The benefits gained by a female are indirect if an indicator trait correlates with some aspect of the *genetic* quality of the male she mates with ('good' or 'compatible' genes for viable or successful offspring). Indirect benefits become effective only in the future, since the female does not profit from them immediately, but only via the reproductive success of her superior offspring.

Studies carried out on reed warblers have helped to reveal the dual function of song variability in gaining territory and in female choice (see above). For an understanding of the latter we must also consider the role played by song, not only in the choice of a social partner but also in the choice of potential extra-pair males. There is a considerable level of extra-pair copulation in many bird species (chapter 10). The benefits gained by a female in extra-pair relationships are overwhelmingly indirect (i.e. genetic), since they arise solely through copulation with an extra-pair male that will play no role in raising the young.

The example of Sedge and Great Reed Warblers

We can regard Sedge Warblers and Great Reed Warblers as model species when it comes to examining which benefits females gain through their choice of a male possessing particular song characters. In general it has been found that male Sedge Warblers with a larger repertoire size are less infected with blood parasites, have higher genetic diversity, occupy larger territories, pair earlier, are mostly polygynous, and give more parental care or provide more food for their offspring than less elaborate singers (Bell *et al.* 1997, Buchanan & Catchpole 1997, 2000a, Buchanan *et al.* 1999, Marshall *et al.* 2003). Furthermore, their song repertoire size increases with age (Nicholson *et al.* 2007). Therefore females that choose males with a larger repertoire as social mates receive not only various direct benefits (a high offspring provisioning rate, for example), but almost certainly indirect benefits as well. More complex song appears to be more effective in both territorial defence and sexual attraction. It is therefore somewhat surprising that males which were able to achieve extra-pair fertilizations actually had smaller repertoires and

smaller territories than their cuckolded rivals (Marshall *et al.* 2007; figs 7.7 and 7.8. Since Sedge Warblers cease singing after pairing, song cues appear to be unavailable for later extra-pair matings and it may be that females switch to other male attributes.

We know from both laboratory and long-term field studies in Sweden and Germany that sexual selection has also influenced repertoire size in Great Reed Warblers (Hasselquist 1998, Forstmeier & Leisler 2004). In experiments, oestrogen-implanted females preferred large-repertoire songs. Females in the wild can expect both direct and indirect benefits if they pair with males with more elaborate songs. Males with a richer repertoire also have greater reproductive success than males that sing simpler songs (Catchpole *et al.* 1986, Hasselquist 1998). However, in contrast to the primarily monogamous Sedge Warblers, sexual selection in the polygynous Great Reed Warblers seems to focus more on the significance of song in territorial occupation. Territory quality is of decisive importance for Great Reed Warbler females, since they often raise their young without male help (resource defence polygyny; section 10.5). When male age and territory quality are taken into account, the established positive correlations between repertoire size and male reproductive success (such as pairing success, clutch size, or annual fledgling production) tend to disappear. This means that the relationship between repertoire size and success is not a direct one, because older males possess both the larger repertoires and the best territories. Females in turn prefer to occupy the best territories (Hasselquist 1998, Forstmeier & Leisler 2004). In the latter study, the positive correlation between song repertoire and clutch size vanishes if the same population is compared at two different times – of lower and higher density. When population density is reduced the intensity of male-male competition also declines, so that practically all males are able to breed in good territories (Forstmeier & Leisler 2004; fig. 7.8).

Although the Swedish and German Great Reed Warbler populations differ in various aspects of sexual selection (Forstmeier *et al.* 2006), characters of song complexity (repertoire size or syllable-switching) in both populations do reflect singer qualities such as viability or longevity, which are clearly attractive for females. In Sweden (but not in Germany) males with a larger song repertoire size sired genetically more viable offspring (Hasselquist

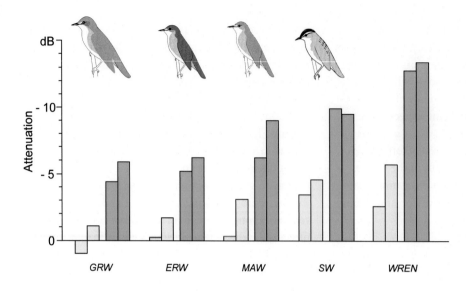

Fig. 7.9 Mean attenuation of songs in 4 *Acrocephalus* warblers and Wren in spring reedbeds (yellow columns) and summer reedbeds (green columns); measuring distances are 4 and 8 m respectively (from Heuwinkel 1990). The lowest attenuation is found in Great Reed Warblers with their low frequency song, the highest in Winter Wrens, which are not native to the habitat.

1998), and in Germany (but not in Sweden) males with longer and more versatile song-phrases lived longer and had a higher mating success (Forstmeier & Leisler 2004). In addition, some males increased their reproductive success by extra-pair copulations, in which only males with larger song repertoires were successful (fig. 7.8b). It is particularly revealing that the resulting offspring survived better, presumably because, by engaging in extra-pair fertilizations with the virtuoso singers, their mothers received 'good' or 'compatible' genes for more viable offspring (Hasselquist *et al.* 1996).

One advantage for a Great Reed Warbler male of having a large repertoire could be that he has different syllables at his disposal with which to compose his song in different years. This allows him to share the syllables that other males are singing and therefore a dialect develops (Fischer *et al.* 1996). The male can thus be recognized as a member of a local, philopatric group that is preferred by females, presumably because such birds know their way around the breeding grounds better than newcomers (Wegrzyn & Leniowski 2010; chapter 8).

Yet it is not only the presence of an especially 'local' element in the repertoire of a male that makes him attractive for females; the quality he exhibits when singing his repertoire is also important. In their Polish study population, Ewa Wegrzyn *et al.* (2010) were able to show that Great Reed Warbler males improved the performance of difficult to sing double-element whistles in subsequent seasons and that the quality determined how successful they were with females. With well-performed whistles a male not

only demonstrates that he has returned to the breeding grounds more often than a poor performer (with regard to philopatry and age) but also his learning ability. To sum up: the studies indicate that the quality of a Great Reed Warbler male, as well as that of his territory, are indicated by his song, and both aspects are taken into account by the females when choosing their partners.

The reliability of the information concerning the quality of a signaller is assured if the signal incurs costs (see above). Birdsong can thus be regarded as a multi-faceted ornament for whose different traits differing costs or constraints must be considered (Gil & Gahr 2002). Potential costs can easily be imagined for some song traits, such as output: a male that sings often and intensively exposes himself to social aggression, raises his risk of being predated, and has to reduce other activities in his time budget. On the other hand, it has recently been found that the metabolic costs of intensive song are actually surprisingly low (Ward *et al.* 2004).

However, it was initially difficult to discern the exact nature of the costs entailed in the production of more complex song and larger repertoire. A satisfactory answer to the question only appeared with the idea that the costs of creating the neural 'song system' are met during the singer's juvenile development. Using Great Reed Warbler nestlings, Nowicki *et al.* (2000) showed that the rate of primary feather growth was positively correlated with the repertoire size of adult males. According to these findings, a large song repertoire in an adult male signalled that he grew up well ('developmental stress' hypothesis;

Nowicki & Searcy 2005). Only young which are of high quality and can meet the extra costs to compensate for any stress (nutritional, parasites, cold, and so on) will have the necessary neural hardware to produce a more complex song. Those of lower quality cannot afford to meet these costs and will signal their inferior quality as adults by producing a less complex song (Catchpole & Slater 2008). That the sizes of important song nuclei in the neural hardware responsible for song production really do correlate with the repertoire size of an individual has been shown in Sedge Warblers (Airey et al. 2000). The size of this region of the brain also develops independently of whether young birds grew up hearing the song of their conspecifics or not. A young Sedge Warbler therefore has the potential to become a 'Meistersinger' even without an example to follow (Leitner et al. 2002).

5 Singing different songs

Song in bird populations is continually moulded, not only by sexual selection, but also by many other ecological or behavioural factors, such as habitat, a sedentary or migratory lifestyle, or body size (Read & Weary 1992).

Habitat and body size

Sound propagation in the environment has specific consequences for signal transmission; the song of a particular species will disseminate better in its native habitat than in non-native surroundings (Chappuis 1971, Jilka & Leisler 1974, Morton 1975). Bird sounds can be altered by the environment in several predictable ways. First, intensity diminishes with increasing distance ('spreading loss'); secondly, in most habitats higher frequencies become more attenuated over distance than lower ones ('frequency-dependent attenuation'); thirdly, sound may become distorted or degraded by reflections from the ground or reverberations caused by the vegetation. A signal must after all impose itself on the background noise. Sound is both diffracted and scattered; wind and temperature gradients divert sound waves and create zones of low intensity (sound shadows). Capek & Kloubec (2002) studied the effect of weather variables on song and found that the song intensity of Great Reed Warblers increased with lower humidity. Through these various influences song usually becomes fainter and imprecise, loses signal content, and becomes 'blurred' with increasing distance between source and receiver (Winkler 2001, Boncoraglio

& Saino 2007). In spite of these negative pressures, there are also particular effects, such as echo creation, that can become positive via the increased range and signalling capability caused by resonance (enhanced transmission, whereby habitat can at times provide a 'concert hall' effect; see below).

In marshes, where there is no open water surface to reflect sound but where the ground substrate absorbs reflections, birds use lower frequencies than in grassland areas, for example (Cosens & Falls 1984). In reedbeds, the habitat of many acrocephalids, attenuation of high frequencies is strong (especially above 5 kHz), but decreases with increasing height of song-posts. By choosing certain preferred heights from which to sing, Eurasian and Great Reed Warblers and Sedge Warblers can lessen any potential dampening of their song (Jilka & Leisler 1974; see also song-flight below). The songs of reed warblers are well adapted to the acoustic properties of their habitats (Heuwinkel 1982, 1990). Hubert Heuwinkel played the songs of four Acrocephalus species in a field of reeds in spring and then again in summer, measuring the attenuation at 4 and 8 m. The songs transmitted especially well through the vernal vegetation, while the song of the 'alien' Winter Wrens in this habitat was most strongly dampened (fig. 7.9). Great Reed Warbler song carried best, because the lower frequencies (particularly below 4 kHz) created resonance effects which contributed to low or even 'negative' attenuation. Heuwinkel was able to demonstrate this effect by showing that certain frequencies in Sedge Warbler song caused the reed stems to resonate.

A means by which a singer can increase the transmission of its song is to orient its body towards the recipient (Brumm et al. 2011). In addition male Sedge and Aquatic Warblers increase the efficiency of their vocalizations by employing song-flights, which are more likely to be directed at females than at rivals. In both species aerial song is actually more complex than song delivered from song-perches (Catchpole 1976, Schmidt et al. 1999). As well as being a visual display, a song-flight helps to broadcast the song in a variety of directions. However the performance of such song-flights is quite different depending on species (Leisler 1975, Schulze-Hagen 1991a). Sedge Warbler males start singing on elevated rigid song-posts before they take off for a 5-second song-flight, ascend, then turn for a brief semicircular flight, followed by a slow gliding descent. The 'towering' song-flight in the absolutely flat and open habitat of Aquatic Warblers is much more stereotyped (introductory drawing). The male starts off silently from a perch, flies steeply upwards with rapid wingbeats to around 10-20 m, begins singing shortly before the peak of the ascent and then glides down again, with spread tail feathers and head thrown back, usually singing right up to the moment of landing.

Low frequencies commonly do best because 'they attenuate less, they are less disrupted by objects and they are the only frequencies that transmit well at ground level' (Catchpole & Slater 2008). So why are such frequencies not more common among songbirds? The probable answer is that body size is responsible. In principle, song frequency is correlated with the body size of a species: smaller species sing higher frequencies than larger. They cannot afford to sing lower because of constraints of size and the length of the vocal tract (Wallschläger 1980, Price 2008). This principle applies to all the acrocephalids studied so far.

However, when Fiedler (2011) analysed which factors have influenced acrocephalid song, she found that the strength of the relationship between body size and frequency became considerably weaker when the phylogenetic relatedness of the species in question was taken into account (table 7.3). The result was that the register in which a species sings was determined not so much by body size but principally by the duration of its stay at the breeding site and its phylogenetic grouping. Sedentary acrocephalids, in other words tropical species, use lower frequencies to communicate with their partner and neighbouring birds in the vegetation. Most of the migratory birds that sing in a higher register than sedentary species do not deliver their song from vegetation but while perched at a point above it, though there are exceptions, such as Icterine, Olive-tree and Thick-billed Warblers, Eurasian and Basra Reed Warblers, which all sing from vegetation. Wiley & Richards (1982) found a similar difference in forest birds, where temperate species tend to sing from higher up, tropical species from lower down. In temperate zone migrants, song output has been selected for a time-limited signalling to attract mates and/or defend a territory, but in tropical birds selection has principally been for ranging, in which song degradation is used to measure the distance to a singer or social partner (Stuchbury & Morton 2001; chapter 12).

Further matches between song and habitat are found mainly in the temporal properties of song: element duration and tempo (table 7.3; Wiley 1991, Badyaev & Leaf

Photo 7.6 In the cool air at sunrise the breath exhaled in singing condenses and becomes visible, like an aircraft 'contrail'. Recent research has shown that birds expend less energy in singing than was previously supposed (Great Reed Warbler; photo Jan Vermeer).

1997, Fiedler 2011. The units in acrocephalid song delivered from low vegetation are shorter than those of species living among taller plants. Species in lower and denser vegetation close to the ground above a wet substrate tend to sing faster, with longer songs and larger repertoires. According to Wallschläger (1985), species in wet, strongly reflective habitat conditions produce a certain amount of information redundancy in that they repeat short units and deliver songs that make a monotonous impression. The situation of the arboreal species mentioned above (i.e. Icterine and Olive-tree Warblers) is completely different, as they sing from inside foliage and encounter more forest-like conditions. Long notes in a narrow frequency band are advantageous here, and they can even be extended in their duration and amplified by reverberations (Slabbekoorn *et al.* 2002, Nemeth *et al.* 2006).

Although two recent meta-analyses by Boncoraglio & Saino (2007) and Brumm & Naguib (2009) confirmed that different song traits represent adaptations to a particular acoustic habitat (acoustic adaptation hypothesis), they also stressed that habitat structures do not fully explain the acoustic properties of birdsong. The most plausible explanation for this can be found in the basic assumption that songs ought to be constructed to maximize transmission and minimize song degradation. However, any resulting potential disadvantages (mainly predators and parasites) were not taken into account, so it could be that a lower locatability by 'hostile' eavesdroppers could be advantageous.

Will they hear me from here?

Are reed warbler songs really loud enough to allow a male to mark his entire territory from a central song-post? Heuwinkel (1982, 1990) studied this problem in several species. First he measured the decline in sound pressure level of the songs of different males over a distance in their natural habitats. Then he determined the range of the songs, that is the distance at which they became masked by the background noise in the corresponding habitat. The effective ranges were then compared with the average radii of the territories for each species. In all reed warblers studied, the volume sufficed to reach every part of an occupied territory from a single central song-post. Only in Aquatic Warblers, with their extensive home ranges, was the effective carrying distance too short for the purpose. To compensate, Aquatic Warbler males are extremely mobile and frequently change their position between widely-spaced song-posts (Schaefer *et al.* 2000). As we have seen, feathered singers can indeed compensate for certain negative acoustic effects of their habitat.

Migration, mimicry and more

Factors such as a sedentary or migratory mode of life and sexual selection seem to exercise greater effects on acrocephalid song than habitat characteristics. There are more correlations (and these correlations are more robust) between such factors and song features, such as duration and complexity, than between habitat parameters and song features (table 7.3; Fiedler 2011.

Table 7.3 Relationship between strength of influencing factors and some song traits in 33 acrocephalid species. The signs (+/−) indicate the correlation values, and bold type represents those correlations that remain statistically significant when the effect of the relatedness of the studied species has been removed. (Based on results of multiple regression with and without phylogenetic correction; Fiedler 2011.)

	Mean frequency	Element duration	Song length	Repertoire size
Body mass	(−)			
Habitat variables		+ vegetation height	(wet ground)	(dry ground)
Residence time on breeding ground	−	(−)	(−)	−
Mating system (incr. promiscuity)			(−)	(−)

The songs of migratory reed warblers are longer with larger repertoires and more mimicry than those of sedentary species (Fiedler 2011). It has been demonstrated for several avian groups at different taxonomic levels that migrants and their songs are under stronger sexual selection than residents and that the repertoire size of various song types is greater in migratory than in non-migratory species (Read & Weary 1992). More complex songs have been reported in those *Vireo* species (Mountjoy & Leger 2001) and in those Greenish Warbler populations (Irwin 2000) which migrate greater distances.

According to Irwin (2000), increasing day length from south to north during the breeding season, and the parallel increase in food supply, allow a greater intensity of sexual selection on song in northern populations or species. Given these prevailing ecological conditions, songs can become longer and more complex. Although acrocephalids with non-monogamous mating systems are associated with shorter songs in *inter*specific comparisons (table 7.3), an *intra*specific comparison shows that polygamous males have more complex songs than monogamous ones (section 7.4).

Vocal mimicry is much commoner among migratory reed warblers (as well as other taxa) than among tropical residents, which begs the question as to whether sexual selection in migratory populations is responsible for greater song complexity through vocal mimicry. In other words, whether mimicry is a mode of repertoire acquisition via female choice or whether it is possible that such mimicry only results from 'imperfect' song learning. Black-browed Reed Warblers are great mimics, but females do not at all prefer those males with larger mimetic repertoires (Hamao & Eda-Fujiwara 2004). A comparison of European songbirds also offers more indications that imperfect learning may be the key factor here (Garamszegi *et al.* 2007). Imitations of other birds are found more often in species that deliver continuous songs and that inhabit a variety of different acoustic environments. Mimicry is rarer in species with longer intervals between songs, and which are limited to a well-defined habitat type. This would all indicate that vocal mimicry can be regarded more as a side-effect of song learning than as an element of complexity or as an adaptation. According to Wiley (2000), in long-range migrants the possibilities for learning have perhaps been exceeded, so birds must develop their songs by improvisation or innovation, or simply by imitation as another form of vocal play. Many species that are good mimics are open-ended learners (e.g. Common Nightingales, Common Whitethroats and European Pied Flycatchers; Eriksen *et al.* 2011).

Oscine songs are of great significance in evolutionary terms because learning plays such a large role in their development and mutual recognition, albeit within genetically determined limits. They are therefore plastic and hence more variable, even over short time periods (through copying errors, for example), than other phylogenetic characters. Conversely, the copying process can ensure that the songs of a species remain similar over large areas and long time periods (Martens 1996; section 2.4). In chapter 12 we will find examples of the opposite phenomenon: rapid changes that have arisen during the colonization of islands.

Summary

The songs of the acrocephalids are highly diverse, with considerable differences in structure, complexity, and duration. They are highly species-specific and often provide the most valuable distinguishing character between species. An analysis of reed warbler songs based on a series of song parameters shows that they are heavily influenced by both phylogeny and lifestyle.

During the breeding season song serves two main functions – as a 'keep-out' signal to other males and as an advertising signal to attract females. This can be observed in several species: the introductory whistles of a Moustached Warbler song-phrase and the short songs of Great Reed Warblers are directed at males, while the warbling section of a Moustached Warbler song-phrase and the long songs of Great Reed Warblers are aimed at females.

Studies of acrocephalids have looked at how both forms of sexual selection – female choice (intersexual selection) and male-male contests (intrasexual selection) – make the song of a bird into a vocal ornament. In monogamous species female choice predominates, resulting in longer and more complex songs for mate attraction; in polygynous reed warblers the emphasis is more on the defence of good territories, resulting in shorter, simpler songs.

Photo 7.7 This Sedge Warbler takes off for a song-flight during which he will deliver a more complex song than when perched (photo Jan van der Greef).

Sedge Warblers and Great Reed Warblers are model species for the investigation of which direct or indirect benefits are gained by females through their choice of males with more elaborate songs. Song complexity reflects a singer's quality (e.g. viability or longevity); males with more complex songs produce more viable offspring. Song complexity is most likely to be an honest signal that can only be afforded by healthy males that have matured without any undue developmental stress.

However, it is not only sexual selection that determines song, but also many other behavioural and ecological factors, such as a migratory or sedentary lifestyle, habitat, and body size. Males of migratory species in the temperate zones sing higher-pitched songs with larger repertoires and more mimicry, and from higher songposts, than their tropical cousins.

The songs of reed warblers are well adapted to the acoustic properties of their habitats. Species in taller vegetation employ longer elements in their songs and make use of resonance effects more than those in lower and denser reedy vegetation. Song can also be effectively targeted by directional singing and by song-flights, such as those employed by Sedge and Aquatic Warblers.

Reed warblers reproduced

Moustached Warbler building its nest.

The reed-dwelling acrocephalids provide an excellent subject for breeding biology studies, partly because their nests are so numerous and are so easily found in their typical habitat. Countless investigations involving thousands of breeding records have been conducted to date, all contributing vital information on the reproductive biology of small passerines. Some reed warbler investigations have analysed data collected over 30 years (e.g. Komdeur 2003a,b, Hansson *et al.* 2004a, Schaefer *et al.* 2006, Halupka *et al.* 2008) or covering a wide geographical area (e.g. Bibby 1978, Schulze-Hagen *et al.* 1996b, Stokke *et al.* 2007). This wealth of breeding biology studies covers the entire spectrum of reproductive biology.

This chapter will therefore explore a very broad field, from nest sites and nest building via timing of breeding and various aspects of reproduction to mortality and adult survival. The life-history traits of acrocephalids in the northern temperate zones and in the tropics have been shaped by their respective environments and are therefore totally different from each other. Comparing them illustrates the contrasts in these 'pace-of-life' strategies which range, for example, from a low reproduction rate in tropical species to a high one in northern temperate species (Ricklefs & Wikelski 2002). The two main factors that determine breeding success are food abundance and predation, both of which will feature prominently throughout this chapter. Since brood parasitism by Common Cuckoos plays a crucially important role in the reproductive success of many acrocephalids, that phenomenon is dealt with in a chapter of its own (chapter 9).

cal branches of coastal bushes, or higher up near the tops of trees, has presumably been important in the colonization of islands. Thibault *et al.* (2002) showed that the nest sites of Pacific reed warblers are less vulnerable to climbing predators such as black rats than those of the sympatric monarch-flycatchers, which are built on horizontal branches lower down (section 12.7).

Photo 8.1 Many reed warblers are excellent nest builders, sometimes suspending their nest on only two vertical stems. In rough weather the robust and neat form ensures that the nest does not slide down the stems, while the deep cup prevents the eggs from rolling out. The Eurasian Reed Warbler nest hidden deep in a sea of reeds is a place of safety (photo Wolfram Scheffler).

1 Masterpieces of construction

Anyone who has ever watched a reed warbler nest can only be impressed by such a masterpiece of nest construction, hanging as it does above the water, often attached to only two reed stems, swaying back and forth in the wind, and yet never sliding downwards or losing its eggs or nestlings. Reed warbler nests frequently survive severe gales completely undamaged, thanks to the fastness of their attachment to the reeds and the curvature and depth of the cup, which ensure that the contents always roll back towards the centre. The birds' adaptations to vertically structured plant communities include the facility to suspend their nests from upright stems. Even the arboreal acrocephalids usually prefer to build on thin, vertically oriented twigs at the ends of branches. Such specialized nest construction is one of the foundations of the considerable ecological success of these birds. The ability to build a well-hidden nest in the verti-

Nest sites

Acrocephalids utilize a wide variety of nest sites. The nests of reed-dwelling species are usually suspended from one to ten *Phragmites* or *Typha* (reedmace) stems; those of the herb-layer species are attached to stinging nettles, meadowsweet, or similar tall perennials; the nests of the wetland-dwellers are often in grass tussocks and those of bush-dwelling acrocephalids and 'tree warblers' are found in low bushes, bamboo, mangroves, and trees of varying heights. Depending on the vegetation lay-er or stratum that the different acrocephalids prefer, their nests are usually placed from a few centimetres to several metres above the ground. In the course of the breeding season the height of a single nest can vary considerably because of the high growth rate of some of the nest-bearing plants; for example the height of Eurasian Reed Warbler nests varied from 50 cm in May to 120 cm in July (Schulze-Hagen 1991c). Great and Oriental Reed Warblers, both reedswamp breeders, differ in their choice of nest height, the former lower than the latter. This possibly results from differing predator pressures: Oriental Reed Warblers lose more nests to ground predators, mainly snakes, while Great Reed Warbler nests are mostly plundered by aerial predators, often harriers (Dyrcz & Nagata 2002). Only Booted and Aquatic Warblers nest close to the ground, the latter obligatorily with their nests in accumulations of old vegetation or leaf litter, sometimes even in burnt hollows or in rodent holes in tussocks (Vergeichik & Kozulin 2006a). The nests of Booted Warblers can be similar to those of the even more ground-loving *Locustella* warblers in that they often have a ramp or run at the entrance (Ernst 2001). The nests of some island reed warblers are the highest suspended nests in the whole family. In general, reed warbler nests are protected against both weather and observation from above, so bush- and tree-nesting species prefer to build them under a covering roof of leaves.

Potential nest sites in the various reed warbler habitats may seem to be abundant when we imagine the millions of *Phragmites* stems in a reedbed. Yet breeding success can be almost wholly dependent on exactly where a female builds her nest. Generally females have to find the best compromise between predation risk, good foraging sites near the nest, and microclimatic requirements; areas with denser vegetation are thus usually preferred. The distance from the edge of the vegetation also plays a role. Some species have developed very narrowly-defined nest site requirements: Great Reed Warblers choose stems five to seven mm thick and always build over water, while Eur-

Photos 8.2 and 8.3 Unlike the tight construction of this Olive-tree Warbler nest, the material in that of the Oriental Reed Warbler is rather loosely put together. The variability of nest types is determined by phylogeny, but also very strongly by nest site and the local ecology (photos Peter Castell, Irina Marowa).

Photos 8.4 to 8.6 Nest building lasts about 5-6 days and demands a great deal of effort from the builder, usually the female. Nest material includes soft stems, reed panicles, plant fibres, and plant wool, which often has to be pulled from the seed head with some force before being carried to the nest (Eurasian Reed Warbler, Great Reed Warbler, Olive-tree Warbler; photos Greg Morgan, Oldrich Mikulica, William Vivarelli).

Photos 8.7 and 8.8 Reed warblers start their nest construction by wrapping adhesive material around vertical stems and connecting it to further stalks. Cross-connections then create a loose platform. Loose material in the cup is packed down using movements of the breast and feet. To create a half-hitch knot, wet material is slung round the support stalk from one side, released, picked up again from the other side, pulled tight, and finally anchored in the nest wall (Great Reed Warbler nest; photo Oldrich Mikulica).

asian Reed Warblers in the same habitat are much more flexible in their preferences (Prokesova & Kocian 2004). The primary females of polygynous male Great Reed Warblers that begin to breed earliest in the best territories build on thicker stems in places with a low stem density than their later-arriving conspecifics (Dyrcz 1988; chapter 10). Together with the selection of a safe nest site, the quality and the size of the nest are among the essential prerequisites for successful reproduction (section 8.4).

Nest building

Reed warblers are skilled nest builders. Thousands of pieces of material must be arranged in such a way as to create a receptacle of the proper size, and they must be firmly interwoven to withstand the pressures that will be exerted on them by the weather, the incubating female, and the weight and movements of the nestlings. The building of the nest is almost always undertaken by the female alone, though there are exceptions where both partners build, among them Millerbirds, Cape Verde Warblers, and Australian Reed Warblers (Morin *et al.* 1997, Higgins *et al.* 2006a, Hering & Fuchs 2009). As part

of the reproductive effort, the construction of the nest represents a substantial investment in time and energy (Hansell 2000) and to such an extent that deserted nests are often used as sources of new nest material.

A comparison of the nest-building process in the four species Great and Eurasian Reed Warblers, Moustached and Icterine Warblers, illustrates clearly how a similarly robust, neat, and solid cup can be created, despite differing locations, attachment, or materials (van Dobben 1949, Kluyver 1955, Leisler 1991a,b, Schulze-Hagen 1991a). Great and Eurasian Reed Warblers suspend their nests from vertical stems or stalks, but Moustached Warblers can build their nests both supported by a base or hung between stems. Similarly, Icterine Warblers are able either to suspend their nest between twigs or place it on top of them. The same construction techniques of interlocking and weaving are used by all species, though to a variable extent. Interlocking materials become entwined with each other through simple construction movements and according to their adhesive properties; in weaving, strands of vegetation intertwine when the bird employs loops and hitches. The adhering components are wet in Great Reed Warblers and Moustached Warblers, dry but sticky in Eurasian Reed Warblers, while Icterine Warblers make particular use of spider silk to hold things together. In other species that hang their nest between vertical structures (e.g. Marsh Warblers, Thick-billed Warblers) the nest wall is built around the stems or bound to them with attachments projecting from the wall ('basket handles').

The three *Acrocephalus* reed-dwellers start their nest construction by wrapping adhesive components around smooth sections of single emergent stems, usually at blocking points such as a stem node or leaf sheath. The attachments are usually then reinforced on two neighbouring stems and connected to further adjacent supporting stalks. Later cross-connections between several support stems create a loose platform or frame, upon which the bird places layers of plant wool and long fibres. Up to this stage the female has been working by holding on to stems from above the construction, but with the next step she begins to shape the thin nest floor with her breast by snuggling down and moulding a depression after every arrival flight with fresh nest material in her bill. This is then laid down, its loose ends looped using the bill and pulled tightly round the support stalks, and finally anchored into the nest wall.

From this stage onwards vigorous breast movements and foot-scrabbling alternate with each other. Coarser material is placed at the bottom of the nest cup, after which the bird lies flat on the nest rim, supported by wing, throat, and tail, or presses herself into the cup. Then the feet scrabble backwards alternately and rapidly, so that the material placed there becomes matted and distributed (introductory drawing). As the wall grows and the cup deepens, material is pulled over from the periphery to the centre in order to reinforce the wall. This action is performed using spider silk, plant down, projecting fibres, or stems. In each individual species the nest wall gets its solidity and stability from the same construction movements, but carried out to varying extents and with differing components. Great Reed Warblers increase the weaving activity, Icterine Warblers use more feet scrabbling and larger numbers of cocoons and spider webs, while Moustached Warblers employ more plant wool, requiring increased stuffing and packing movements of the bill.

The nest is usually completed after just five or six days. With their very short breeding periods, long-distance migrants, such as Marsh and Icterine Warblers, often need only four days, while other species can take up to seven. Tropical or southern hemisphere acrocephalids can take even longer – 10 days for Tahiti Warblers and Dark-capped Yellow Warblers (Bruner 1974, Dean 2005), 10 to 14 days for Millerbirds (Morin *et al.* 1997), and an average of 19 days for Seychelles Warblers (Komdeur, unpubl. data).

While over 90% of the nest material used by the reed-dwellers comes from the *Phragmites* plant itself, and the marsh-nesting congeners most often use dry grass stems and leaves, the bush-nesting and arboreal species employ quantities of bark and bast fibres in nest construction. Other common materials include plant down or wool and fibres, cocoons, algae, water plants, moss, or snake skin (in Basra and Great Reed Warblers; Kennerley & Pearson 2010, Trnka & Prokop 2011), while the tropical species favour coconut and banana fibres, tendrils, and hair. Man-made substances such as plastic, string, thread, wool, or rags are also sometimes found as nest material, so far without any obvious detriment.

Nest characteristics

Most acrocephalid nests are deep cups – their height roughly equal to their external diameter – and conical or cylindrical in form. Some have a thickened lip on the rim, concave towards the inside. Nests of Aquatic and Thick-billed Warblers are comparatively loosely constructed, the latter resembling those of *Sylvia* rather more than *Acrocephalus* warblers.

Some species (Icterine Warblers, Madagascar Swamp Warblers, Saipan Warblers) 'decorate' the outer surface of the nest with light-coloured sticky materials, such as spider cocoons, birch bast, and so on. According to Hansell (2000) this has little to do with reinforcing the nest wall but rather so that the pale spots may have the effect of breaking up the outline and solid shape of the dark nest (disruptive camouflage). The interior consists of fine, soft materials, commonly reed inflorescences from the previous year, removed from the stem with a sharp tug. Several species line the cup with feathers or plant down.

Most reed warblers only build a breeding nest. Exceptions to this rule are Kiritimati Warblers and Australian Reed

Warblers, for example, which also build cock nests (Schreiber 1979, Berg *et al.* 2006). Australian Reed Warbler territorial males frequently construct rudimentary unlined nests suspended from only two support stems. Such nests are not used by the females for egg laying, which distinguishes them from the many multiple-nest-building species, such as Marsh Wrens, which are polygynous and typically continue nest building after acquiring a mate. The best explanation of the adaptive significance of multiple nest building in these acrocephalids is that of mate assessment: individual males might thus be able to signal their current condition or territory quality. Their cock nests may be considered as 'non-bodily' ornaments, and thus as secondary sexual characters (Berg *et al.* 2006; chapter 10).

By comparing the traits of nest site, form, and material, we can attempt to throw light on the question of whether nest construction reflects phylogeny, and therefore whether nest construction characteristics have any taxonomic significance. To this end we compared 18 characters in a matrix (Bochenski & Kusnierczyk 2003; see also www.knnvuitgeverij.nl/EN/appendix-the-reedwarblers). Fig. 8.1 shows the nest similarities in a dendrogram, thus allowing a quick comparison with the actual phylogenetic relationships between the species, symbolized by the colour codes. It can easily be seen that only a slight agreement exists between degree of relatedness and similarity of nest construction. Therefore, on the basis of their nests, the acrocephalids can be better categorized in ecological than in systematic groups.

Three main groupings can be discerned in the dendrogram. The first contains the nests of Aquatic and Thick-billed Warblers, which are not only obviously different from all the others, but also from each other. The second group consists of the marshland species Moustached *A. melanopogon* to Oriental Reed Warblers *A. orientalis*, independent of their systematic position. The third grouping is made up of the island reed warblers together with the genera *Iduna* and *Hippolais*. In the wetland species, the nests of those from deep-water marshes (Moustached *A. melanopogon* to Australian Reed Warblers *A. australis*) resemble each other more than those of species that inhabit reedswamp edges, tall herbage, and bushes (Sedge Warblers *A. schoenobaenus* to Oriental Reed Warblers *A. orientalis*). Remarkably, with the exception of Millerbirds, the nests of all island species resemble each

other, independent of origin (*Acrocephalus* or *Calamocichla*) and of their geographical range, whether in the Atlantic, Indian, or Pacific Ocean. The nests of the striped reed warblers differ from each other most strongly. By contrast, in the members of *Hippolais* and *Iduna* the nests of several sister species are remarkably similar (e.g. Melodious and Icterine Warblers, Dark-capped and Mountain Yellow Warblers).

Distinctive innovations in nest construction have occurred in the course of the evolutionary history of several bird families. For instance a technological change to using mud as a construction material took place in the swallows (Hirundinidae), and there were changes to new materials and building techniques in the weavers (Ploceidae). Whereas similarities in nest construction also reflect relationships within swallows and weavers, such a connection cannot be made in other groups such as swiftlets (Apodinae), for example (Lee *et al.* 1996, Hansell 2000). Similarly, when considered in the striped species alone, acrocephalid nest construction does not seem to be very strongly influenced by phylogeny, as they not only use very different nest sites (high up between vertical stems or on the ground) but can also build differing nest cups using a variety of materials. It is most likely that evolutionary changes in the reed warblers led first to changes in the nest site, with resulting new challenges in nest attachment. It remains unclear whether these developments were accompanied by innovations involving changing materials and building techniques, or were actually facilitated by them (as a preadaptation). To judge by the variety of building techniques in the family, quantitative changes in the use of already existing construction methods alone could have been sufficient to solve any problems.

2

Getting the timing right

Most members of the Acrocephalidae family have one distinct breeding season within a year. In general, the timing of reproduction maximally matches the peak of optimal food availability (Gill 2007). The length and timing of this peak can be distributed over the year quite differently in different climatic zones, and, on a finer scale, may even differ between different habitats. Almost all insectivorous passerines of strongly seasonal high-latitude regions are migratory, whereas those in the tropics are, by contrast, sedentary. Accordingly, while breeding

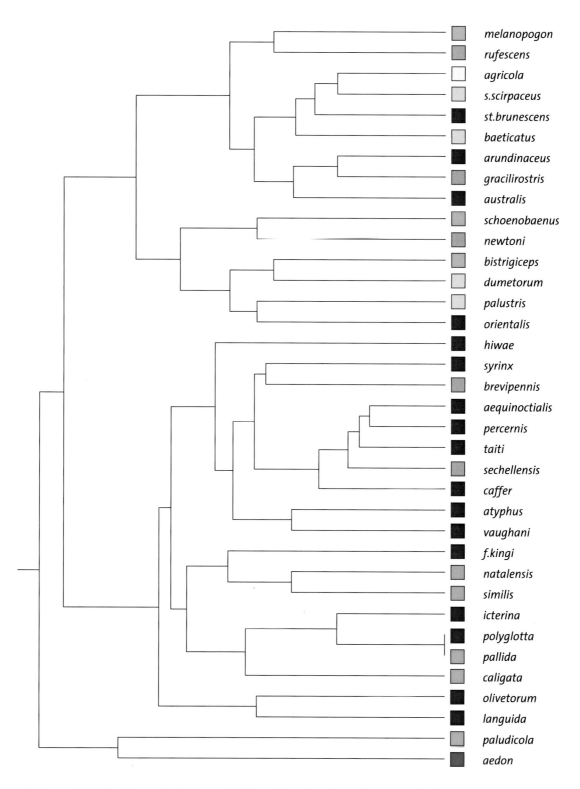

Fig. 8.1 Cluster analysis (dendrogram) of 36 acrocephalid species using 18 nest characters. This demonstrates the similarities between species based on measured nest traits. Here congruences as well as divergences become apparent when we compare this dendrogram with the phylogenetic tree based on genetic distances in fig. 2.5, recognizable by the corresponding order of the colour symbols in both figures.

seasons tend to be longer and less synchronized in the tropics, high-latitude species usually breed seasonally and in a highly synchronized fashion. Their annual breeding and migration cycles are generally controlled by rigid endogenous rhythms (section 11.1).

The idea of internal 'calendars' that would allow migratory birds to time their seasonal activities was postulated long ago on theoretical grounds (e.g. Rowan 1926, Aschoff 1955). Researchers reasoned that migrants would be unable to receive information about suitable conditions for reproduction while they were far away from their breeding grounds. They would therefore need to consult an inner clock to determine the right time to leave their winter quarters. Later, this idea was confirmed experimentally by Ebo Gwinner and colleagues, who kept migratory songbirds under captive conditions (Gwinner 1967, 1986). Although the birds were kept ignorant of any temporal information, they showed recurring activities at roughly the correct time each year. However, over a period of years these activities drifted away from the calendar year, revealing their endogenous nature as 'circannual' (from *circa* = about and *annus* = year). Such timing programs interact with environmental timing cues – predominantly photoperiod (the length of the daylight part of each day) – to help birds keep track of time. Species for which internal programs have been confirmed include the acrocephalid Marsh Warblers (Berthold & Leisler 1980).

Decades of subsequent research revealed that such endogenous programs were not only relevant for migration, but also for the timing of moult and reproduction. Most birds regress their reproductive system dramatically once the breeding season is over, presumably in order to save energy and minimize body mass. Prior to breeding, the reproductive axis has to be reactivated, a process which takes approximately one month. Many species, including tropical taxa, rely on endogenous clocks to restart their reproductive system in advance of the time they expect to be in breeding condition (Gwinner 2005). However species differ greatly in the way they use environmental information to time these activities. Those in habitats with little seasonality, or with unpredictable seasonal conditions, rely much more on direct environmental cues, such as food availability, rainfall, or temperature change (Stutchbury & Morton 2001, Hau *et al.* 2008) than on internal clocks. Currently Seychelles Warblers, a tropical acrocephalid species, are noteworthy for their direct responsiveness to environmental cues, which allow them to switch between annual and continuous breeding seasons (Gwinner 2005, Komdeur & Daan 2005). Thus, once again, acrocephalids can serve as an instructive example of natural diversity in behaviour and adaptations to cope with specific environments.

Members of the Acrocephalidae family display an astonishing variation in their reproductive phenologies (in Eurasia between April and August, in the temperate zones of

Photos 8.9 to 8.11 Many species hang their nests on vertical stems, but others build in trees (a Cape Verde Warbler nest) or on the ground (an Aquatic Warbler nest). Occasionally the latter species places its nests in a cushion of moss in a hollow or in a hole excavated by a rodent in the ground (photos Jens Hering, Alexei Kozulin).

Photos 8.12 and 8.13 Sometimes incubating Eurasian Reed Warblers press themselves deep into the nest cup so that they are bare-ly visible from the outside. The species uses plenty of spiders' webs to strengthen the rim of the nest. The nest cup of this Dark-capped Yellow Warbler is not so deep (photos Garry Prescott, Warwick Tarboton).

the southern hemisphere from September to January) and both the timing and duration of their breeding seasons are related to their migratory behaviour. The breeding season of long-distance migrants often starts later and is shorter than that of short-distance migrants or residents (even at the same latitude). It is the species-specific habitat type, especially the type of food exploited during breeding, that determines breeding phenology, which in turn can influence the migration strategy of a species (Tökölyi & Barta 2011).

In the northern temperate zone, reedswamps are less seasonal than a range of bushy and arboreal habitats. Accordingly, reed-dwellers are able to utilize an extended food supply, which in turn means that their breeding time-windows are relatively broad (chapters 3 and 13). They include species that migrate over a variety of distances, even long-distance migrants, though the extreme long-distance travellers are not among them. By contrast, in the herb-, bush-, and tree-dwelling *Iduna*, *Hippolais*, and *Acrocephalus* species, the breeding time-windows are narrower, since these taxa exploit a very late and temporally limited food supply, often in conditions of summer drought in semi-arid continental or Mediterranean regions. For instance, there is an unusually late spring arrival of Western Olivaceous and Olive-tree Warblers, which are confined to the southernmost Mediterranean zone; they appear on their breeding sites only in early May. This is probably explained by the late sprouting of new leaves on the evergreen and deciduous trees and bushes in their habitats, so that their prey becomes avail-

able in large numbers only from June onwards (Blondel & Aronson 1999). Three members of this group – Icterine, Olive-tree, and Marsh Warblers – have the shortest breeding periods of all the acrocephalids and can consequently afford to make the longest journeys.

The most extreme example can be found in Marsh Warblers, which winter in southern Africa (section 11.2) and arrive latest of all on their European breeding grounds. Marsh Warblers' habitats are characterized by vegetation consisting of fresh growth of late, comparatively short-lived herbs on their breeding grounds, along their migration routes, and in their winter quarters (Schulze-Hagen *et al.* 1996b). The entire laying period of central European Marsh Warbler populations lasts only around 50 days, in stark contrast to their closest relative in reedbeds, Eurasian Reed Warblers, which have a laying period of 120 days (Dowsett-Lemaire 1978, Schulze-Hagen 1991c).

In the tropics there is considerable climatic seasonality, despite the absence of pronounced changes in day length, so avian breeding varies temporally and spatially. Here local conditions, and hence breeding periods, can change rapidly within only a few kilometres (Hau *et al.* 2008). Seychelles Warblers are the best-studied example of a tropical reed warbler and have an extremely restricted range on four tiny Seychelles islands (section 12.6). The core population is found on the 29-ha island of Cousin, which is under the influence of both a strong south-easterly monsoon and also of a weaker northwesterly monsoon, to a different extent on each coast. These slight

Photo 8.14 For reasons of economy smaller prey items are bundled together in the bill when brought to the nest as food for the young (Great Reed Warbler; photo Oldrich Mikulica).

differences in rainfall are directly reflected in the breeding seasons. Most Seychelles Warblers on the island breed during the southeasterly monsoon, the time of greatest food richness. The start of both monsoon and breeding season varies between years by four months (May to August; Komdeur 2006). While the birds occupying the better territories have only one breeding season and apparently find that sufficient, many pairs on the much food-poorer southeast coast breed a second time during the northwesterly monsoon (December to February), which brings a further if smaller peak in arthropod numbers. Individual pairs can therefore breed at six-monthly, or semiannual, intervals (Komdeur & Daan 2005).

Despite the variability of the rains and the resulting increase in food supply, almost all pairs start their breeding cycle around two months before the start of the peak in insect numbers. This period corresponds exactly to the time required for courtship, nest building, incubation, and hatching. When the birds started earlier or later, their reproductive output was much lower (Komdeur 1996a, Komdeur 2006). The enormous immediate importance of climate and food availability for breeding was tested by Komdeur (1996a) in transfer experiments, taking birds from Cousin to the neighbouring islands of Aride and Cousine. They both lie in close proximity and have a similar climate and vegetation, but food is more abundant than on Cousin. On Aride, with its good year-round food supply, the same pairs that had bred only once a year on Cousin now breed continuously year-round. On Cousine, with seasonally changing and more limited food resources than Aride, the transferred pairs increased the number of broods per breeding season. In another insular species, Henderson Island Warblers, the breeding period is also dependent on maximum food supply (Brooke & Hartley 1995). Two main breeding seasons per year are also known from Nightingale and Tahiti Warblers and Cape Verde Warblers, while Nauru, Tuamotu, and Marquesan Warblers can breed year-round if their food supply is constantly high.

3 Reproductive effort

A brief excursion into life history

Life histories are sets of evolved traits or attributes that interact with environmental variables to determine the fitness of an individual member of a population. Fecundity and adult survival or mortality (sections 8.5 and 8.6) are the primary life-history attributes. The trade-offs between the two are the traditional focus of life-history analyses. The lifetime reproductive success of an individual, that is the number of young raised during its lifespan, is then the decisive measure of fitness. As small passerines, acrocephalids characteristically have the following life-history patterns when compared with other avian forms such as seabirds or raptors: low survival before breeding (*ca* 15%), early maturation (at one year), moderate fecundity (three young per annum), and high adult mortality (50%; Gill 2007). These parameters also vary within the passerines, and in a particularly striking fashion between species from different latitudes.

According to Ricklefs & Wikelski (2002) and Robinson *et al.* (2010), in tropical and southern hemisphere species most of the variation in life-history traits falls on a slow–fast continuum, in which a longer average life span is associated with reduced investment in single reproductive events (slow reproduction rate, slow development of the young), while the opposite traits are typical of northern temperate species. In northern latitudes, highly seasonal environments can cause increased mortality because insectivorous birds have to migrate, thus incurring higher risks and costs. On the other hand, higher food availability during breeding favours greater contributions to individual reproductive events, characterized by larger clutch sizes, faster development, and shorter duration of parental care.

In the following sections we will continually return to the pronounced differences in these contrasting, 'pace-of-life' strategies of the acrocephalids in the northern and southern hemispheres. They reflect the differing demands made by the environment on northern migrants and tropical residents. Since all of them belong to one phylogenetically homogeneous group, the widespread acrocephalids are once more well suited for comparative studies.

As well as the number of broods per season, among the most important of all components of reproductive effort are clutch size and the extent and duration of parental care that is necessary for the incubation, nestling, and post-fledging periods. Some temperate acrocephalids, such as Sedge and Aquatic Warblers and Eurasian Reed Warblers, have a relatively broad breeding time-window that allows them to raise two consecutive broods, but the majority of long-distance migrants can raise only one brood and undertake lengthy migrations.

Clutch size

In general, clutch size increases with latitude (from south to north, e.g. Russell *et al.* 2004), and is therefore generally smaller in altricial non-migrants than in migrants (Jetz *et al.* 2008). It is subject to considerable fluctuations in the various climatic zones, ranging from one egg in Seychelles Warblers and Cook Islands Warblers to up to seven at the northern edge of the distributional range of Sedge and Booted Warblers (Butyev *et al.* 2007, Kennerley & Pearson 2010; table 8.1). Species in higher latitudes are able to make use of the excessive abundance of arthropods in the northern summer to feed their broods, while the surplus food supply available to the tropical and southern hemisphere species is considerably less (fig 8.2).

In many temperate zone acrocephalids the mean clutch size decreases with increasing date through the breeding season, a phenomenon known as the 'calendar effect'. This seasonal decline in clutch size has been studied in many single-brooded altricial bird species. In long-distance migrants, the reduction is *ca* 20% of the clutch size maximum. This effect is caused by the declining 'value of an egg' and has various ultimate causes, such as the decline in food supply or the reduction of the chances of

Table 8.1 Key variables in the breeding biology of migratory and resident acrocephalids. Note the marked differences. n = number of species

		Clutch size	Incubation period (in days)	Nestling period (in days)	Time to independence (in days)
migrant	$\bar{x}$	4.7	12.9	12.6	25.4
> 43°	range	4-7	11-16	7-16	
	n	15	13	12	5
migrant	$\bar{x}$	3.6	13.0	13.0	24
43 - 23°	range	3-5	12-14	12-15	
	n	10	7	6	1
tropical & southern hemisphere resident	$\bar{x}$	2.3	13.8	14.5	73
< 23°	range	1-4	11-18	12-21	30-120
	n	19	9	8	4

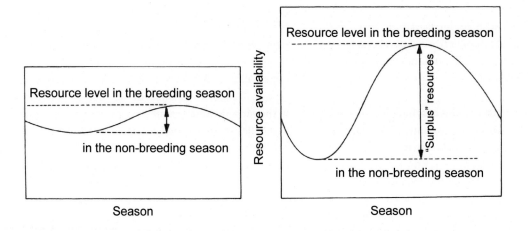

Fig. 8.2 The seasonality model explaining clutch size variation between tropical and temperate regions. In the seasonally productive temperate latitudes (right) there is more 'surplus' food available for raising the young than in less seasonal tropical environments (left), where competition for surplus resources is higher, owing to populations being near carrying capacity on a year-round basis (from Bairlein 1996; after Perrins & Birkhead 1983 and Ricklefs 1980).

juveniles to survive or to acquire experience as the season progresses (Daan *et al.* 1988, Price *et al.* 1988, Møller 1994). The immediate 'decision' by the female to lay a particular number of eggs is evolutionarily 'predetermined' in that the maximum number is laid exactly within the time-window of the best rearing conditions. As early as 1982, Lars von Haartmann discussed the fact that the calendar effect in long-distance migrants is differently regulated from that in residents or short-distance migrants, but it has recently become a lively topic once again. In Great Reed Warblers (as in other long-distance migrants such as Eurasian Reed Warblers and Pied Flycatchers) seasonal clutch size decline in a population results from each individual female 'adjusting' her clutch size according to the calendar ('absolute clutch size determination'). This was impressively demonstrated in a long-term study of colour-ringed Great Reed Warbler females. In different years individual females – that were described as 'ready at different times' – laid clutches whose size corresponded exactly to the clutch-size reduction in the population as a whole (Schaefer *et al.* 2006). This form of clutch-size variation means that clutches become larger as the start of the egg-laying period gets earlier. In Eurasian Reed Warbler populations, whose breeding season is starting earlier because of global warming, early clutches are now increasing in size to the exact extent predicted by the extrapolation of the line of regression of clutch size and laying date (fig. 8.3; Schaefer *et al.* 2006).

Seeing it through from incubation to post-fledging

Following the laying of the last egg the incubation period lasts between 10 and 18 days. The period lengthens slightly from the northern temperate zone to the tropics in line with the fast–slow continuum, with tropical eggs taking *ca* 9% longer to hatch (Ricklefs 1969, Geffen & Yom-Tov 2000, Martin *et al.* 2007; table 8.1). Runts are often found in acrocephalid nests as a result of asynchronous hatching, and their frequency increases with greater food supply and more favourable weather. Using an experimentally increased food supply, Eikenaar *et al.* (2003) were able to demonstrate that this situation increases incubation attendance, leading to greater hatching asynchrony. Better food availability was therefore shown to be a proximate factor in modifying the extent of incubation and hatching asynchrony.

In the cooperatively breeding Seychelles Warblers and Henderson Island Warblers, two females occasionally lay their eggs at different times in the same nest. In extreme cases this can mean that nestlings contribute to the incubation of later eggs (Brooke & Hartley 1995).

Young acrocephalids leave the nest at 10-15 days old (table 8.1), but, if they are threatened, reed warbler nestlings can jump from the nest when only eight or even seven days old (section 8.5). Given the high predation rates on marsh-nesting birds, adaptations that reduce the

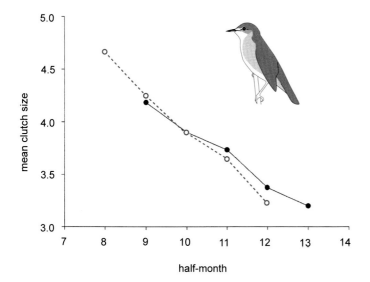

Fig. 8.3 Comparison of seasonal decrease in Eurasian Reed Warbler clutch size between the periods 1973-1987 (closed circles, black line) and 1988-2003 (open circles, red line; after Schaefer *et al.* 2006). At a certain date in the season clutch size remained constant over the years, but mean clutch size increased due to the earlier initiation of egg laying in the population caused by global warming.

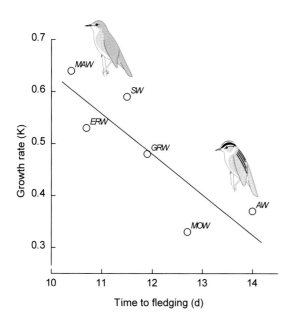

Fig. 8.4 Negative relationship between duration of nestling period (d) and growth rate (K) in six central European reed warblers. Growth rate is fastest and nestling period shortest in Marsh Warblers.

length of the vulnerable nestling growth period are of crucial importance. Nestling growth rate is modified by three key factors: predation, food abundance, and parental investment (Ricklefs 1969, Kleindorfer *et al.* 1997). In six European *Acrocephalus* species the growth rates of nestlings correlate negatively with the nestling period (fig. 8.4). Their growth rate constants vary substantially between 0.64 (fast) and 0.33 (slow), reflecting complex species-specific interactions between levels of predation, food supply, and nestling provisioning. On the one hand Moustached Warbler development is slow, reflecting food limitation, low predation, and opportunities for multiple broods, while on the other, Marsh Warblers have the fastest growth rate and the shortest nestling

period, reflecting a better food supply and higher predation for a single breeding attempt. The long nestling period of Aquatic Warblers stands out particularly, and is presumably accounted for by uniparental care. In this species, characterized by high food supply, a faster growth rate would be expected, yet it is *de facto* slower because of the low feeding efficiency of female uniparental care. In addition, it is likely that their relatively secure nest sites, resulting in low predation rates, also play a role here (Dyrcz *et al.* 1994, Kleindorfer *et al.* 1997). The nestling period of the socially monogamous acrocephalids is also longer if one of the parents is experimentally removed (Eurasian Reed Warbler; Duckworth 1992; see also section 10.4).

Photos 8.15 to 8.17 Shortly before leaving the nest the young are fed almost non-stop. When they leave the nest, sometimes at an age of 7 or 8 days if disturbed, they are still fragile and their feathers are not fully developed. They remain for a time in the neighbourhood of the nest, drawing attention to themselves by loud begging calls (Eurasian Reed Warbler, Great Reed Warbler, Clamorous Reed Warbler; photos Alfred Limbrunner, Oldrich Mikulica, Jens Hering).

In species on tropical islands the young tend to remain a little longer in the nest: for example, 16 days in Millerbirds, 18-21 days in Henderson Island Warblers and Seychelles Warblers, though only 14 days in the latter on the food-rich island of Aride (Komdeur 2006, Kennerley & Pearson 2010). Although the nestlings of tropical songbirds generally appear to grow more slowly than those in northern latitudes, they do not consistently stay longer in the nest, which means that they fledge at a relatively smaller size, probably as a reaction to the high nest predation pressure in the tropics (Ricklefs 1969, Robinson *et al.* 2010).

Little is known about parental care for the young after fledging. The post-fledging period, or time to independence, lasts a further 10 to 17 days, but is strikingly extended in most tropical species. In Henderson Island Warblers it lasts at least six weeks, while in Kiritimati Warblers the young can remain in the parental territory for several months and in Seychelles Warblers even for as long as six months (Milder & Schreiber 1989, Brooke & Hartley 1995, Komdeur & Richardson 2007; table 8.1; section 8.5). Since the examples given above are of island warblers, the longer time spent with parents could also be explained by the limited dispersal opportunities for the young.

Brood losses

The number of surviving offspring is a key component of life-history evolution. Marshes offer their avian inhabitants high prey abundance and frequent insect emergence, but they are also characterized by high nest predation rates (Orians 1961). Predation is the most important cause of breeding failure, accounting on average for 60-75% of all losses (Ricklefs 1969, Schulze-Hagen et al. 1996b), which could be a consequence of the structural simplicity of reedbed architecture and the relatively high nest density (Picman et al. 1988, Hoi & Winkler 1994). Furthermore, open marshes are particularly vulnerable to extreme weather events such as heavy rains, flooding, storms, or drought, which dramatically reduce breeding success in some years, or even prevent entire populations from breeding at all, as in the case of Aquatic Warblers (Post 1988, Vergeichik & Kozulin 2006b).

The sometimes remarkably high density of nests in reedbeds is also an important factor in determining breeding success (Hoi et al. 1991; chapters 6 and 14). Reedswamp predators do not discriminate between passerine nest types, so above a particular threshold of the total nest density of all species, predation pressure increases disproportionately, as searching for nests becomes more profitable for predators (Hoi & Winkler 1994). Since all reedbed breeders are exposed to the same predators, this leads to apparent competition between them, with nest spacing and interspecific aggression as a consequence (sections 6.2 and 14.2). Under such conditions 'enemy-free nesting space' is a limited resource.

Predators

Chief among the predators on acrocephalids are other birds, mammals, and reptiles. Over a wide area Common Cuckoos are not only a major nest parasite but also the commonest nest predator, responsible for up to 68% of all losses (chapter 9). Eggs and young in reedbeds are also taken by rails (Rallidae) and Eurasian Bitterns; some Little Bitterns also specialize on reed warbler nests, plundering them systematically (Hocke 1903, Dyrcz 1981). In cold weather Montagu's Harriers patrol over sedge fens and react to the increased begging calls of Aquatic Warbler nestlings (Halupka 1999). Western Marsh Harriers have a better chance of finding Eurasian and Great Reed Warbler nests when Phragmites stems are blown apart by the wind (Dyrcz 1981).

Away from reedswamps, Eurasian Magpies and Red-backed Shrikes take eggs and often young; at woodland edges Eurasian Jays are common nest robbers (Catchpole 1974, Bussmann 1979, Duckworth 1991, Ferry & Faivre 1991a). On islands, frequent predators of acrocephalid eggs are the endemic bird species, such as Seychelles Fodies or the Telespiza finches for Millerbirds on Nihoa (Komdeur & Daan 2005, Morin et al. 1997).

Presumably for reasons of apparent competition, reed warblers such as Great and Eurasian Reed Warblers will destroy and possibly eat the eggs of other congeners, such losses being substantial at times (Hoi et al. 1991, Honza et al. 1999, B. Stokke, pers. comm.). Recent observations throw suspicion on the smaller Eurasian Reed Warblers of robbing the nests of the larger birds (Trnka et al. 2010). In addition, egg destruction has been recorded intraspecifically in the polygynous Great Reed Warblers, where secondary females destroy the broods of primary females in order to become the assisted individuals (Bensch & Hasselquist 1994, Trnka et al. 2010; sections 6.2 and 10.5).

Mammalian predators include red foxes, raccoon dogs, stoats, weasels, polecats, or domestic cats. In Belarus, shrews can drastically reduce the breeding success of whole Aquatic Warbler populations in some years (Vergeichik & Kozulin 2006a). High losses can also be attributed to rodents. In a German study, 55% of all Sedge Warbler eggs were eaten by water voles. In drier habitats frequent nest robbers include rats, wood mice, yellow-necked mice, harvest mice, and common dormice. On the Pacific islands, black rats and Pacific rats are sometimes responsible for extremely high losses (Rogge, typescript 1959, Schulze-Hagen 1991b,c, Thibault et al. 2002). In southern Europe, but chiefly in Asia, Africa, and on a few Pacific islands, snakes are very important predators. In central Honshu, Japan, no fewer than four species of nest-robbing snakes, including Elaphe and Natrix species, have been recorded simultaneously in one study area (Bussmann 1979, Hamao 2005). Two skink species have been known to plunder nests in the Seychelles (Komdeur & Daan 2005).

Breeding losses can vary greatly in different years, caused mainly by unfavourable weather or fluctuations in predator density, such as plagues of small mammals. There is also a distinct seasonal variation in losses to predators.

There are also enormous differences between species, even sympatric ones (Schaefer *et al.* 2006). As a rule of thumb, losses are at a maximum during the peak of the breeding season (Bibby 1978, Hoi & Winkler 1988).

Nest defence and anti-predator strategies

Because nest predation is a prime determinant of reproductive success in passerines, strategies that minimize its consequences should be strongly favoured by natural selection (Martin 1993). The more carefully nests are hidden, the lower will be the losses (Batáry & Baldi 2005). The most important determining factor of nest site quality appears to be the density of the vegetation and hence protection from view as far as possible on all sides. When Marsh Warblers and Eurasian Reed Warblers nest adjacently to each other, losses due to predation in the former are significantly less than in the latter, because Marsh Warbler nests are better concealed by the stems and leaves of the herbaceous vegetation than those of the Eurasian Reed Warblers, which are higher up in the reeds and often easily visible from above. In Namibia, nests of African Reed Warblers were found in the tallest and densest patches in the territory (Schulze-Hagen 1984, Schulze-Hagen *et al.* 1992, Ille & Hoi 1995, Eising *et al.* 2001).

Yet the quality of the nest site does not always protect against predation. Generally speaking, breeding at the edges of vegetation or in smaller patches exposes birds to higher risk. Eurasian Reed Warbler nests in the vicinity of trees suffer greater losses than nests in homogeneous *Phragmites* stands, because trees serve as vantage points for Common Cuckoos and Eurasian Jays. Successful nests are closer to the open water than unsuccessful ones, while nests in deep-water marshes are generally more secure than those in drier wetlands. Icterine Warbler nests suspended high up in trees have a high breeding success (Ferry & Faivre 1991a, Leisler 1991b, Picman *et al.* 1993, Procházka 2000, Baldi & Batáry 2005). In Japanese Black-browed Reed Warblers, a grassland species, nests attached to thick and stable reed stems are more often robbed because snakes find it easier to climb them than thinner stalks (Hamao 2005). Even the size of a nest can influence the probability of predation: the smaller and denser nests of Eastern Olivaceous Warblers were less predated than the larger-sized nests (Antonov 2004).

On the whole, vegetation cover and food abundance show a negative relationship (Hoi & Ille 1996, Ille *et al.* 1996;

chapters 3 and 4). This is why reed warbler parents often burden themselves with long flights to food-rich areas comparatively far from their broods in the nest, frequently using distant communal feeding sites away from the reedbed (Duckworth 1991, Ezaki 1992, Ille & Hoi 1995, Poulin *et al.* 2000, Eikenaar *et al.* 2003). However, as soon as young Moustached Warblers leave the nest, they are led by the adults to places of high food abundance within the reedbed (Taylor 1993).

There are also clear interspecific differences in the active nest defence techniques used by acrocephalid parents. On seeing a predator, Eurasian Reed Warblers give alarm calls earlier and more intensively than Marsh Warblers, whose nests are, as we have seen, better camouflaged. While it is mostly only the female Marsh Warblers which sound the alarm, both Eurasian Reed Warbler partners will cooperate in nest defence. In fact, active nest defence in this species seems to be more important than nest concealment, especially since many neighbours can join in communal alarm-calling and predator mobbing. By contrast, Marsh Warblers have to depend more on their own efforts because of their lower population density. Given their physical size and strength, Great Reed Warblers can drive predators off more successfully than their smaller relatives, and perhaps this is why losses through predation in this species are substantially lower than those suffered by Eurasian Reed Warblers in the same reedswamp. A positive correlation between level of nest defence and fledging success was also demonstrated in Moustached Warblers (Dyrcz 1981, Schulze-Hagen & Sennert 1990, Ille *et al.* 1996, Hansson *et al.* 2000, Kleindorfer *et al.* 2005, Schaefer *et al.* 2006.)

In general, the participation of males in nest defence appears to be higher in those acrocephalid species whose males contribute more to incubation and feeding assistance to their females than those with less paternal care (Halupka 2003; chapter 10). Male nest defence is greater in Eurasian Reed Warblers than in Sedge Warblers, which are in turn more active than Aquatic Warblers. This relationship can also apply intraspecifically: polygynous Great Reed Warbler males with two simultaneously active nests defend the nests of primary females less vigorously than monogamous males (Trnka & Prokop 2010). While male nest attendance aimed at deterring predators is rather reduced in tropical species, increased nest guarding by the male Seychelles Warblers contributes to a clear improvement in breeding success, since predation by

Photo 8.18 The fledged young, almost tailless, look rather 'unfinished' compared with the adult birds, such as this Sedge Warbler, but in contrast to the development of their wing and tail feathers, their climbing and grasping abilities are already very impressive. Early fledging contributes to reducing the high nest-predation risk (photo Huib Tom).

skinks or fodies occurs in the incubation or brooding breaks when the female is searching for food (Komdeur & Kats 1999). Anti-predator reactions differ according to predator and vegetation type: ground-nesting Aquatic Warblers will mob a mounted polecat more vigorously, more loudly, and more closely than a mounted harrier. Nests in low vegetation are defended more intensively against ground predators, higher nests more against aerial predators (Halupka 1999, Kleindorfer *et al*. 2003, Kleindorfer *et al*. 2005).

Photo 8.19 Predation of eggs and nestlings is the main cause of losses at the nest stage, ranging on average between 50 and 60%. In Belarus predation on Aquatic Warblers by shrews can be heavy in some years. The shrews mostly just nibble at the nestlings (photo Alexej Kozulin).

Photo 8.20 Some Little Bitterns in reedbeds specialize in plundering reed warbler nests (photo Oldrich Mikulica).

There is a close correlation between both growth rate and nestling development and the deployment of anti-predator responses such as crouching or jumping from the nest. Nestling anti-predator responses also correlate with parental alarm-calling (Kleindorfer *et al*. 1997). When the nestlings have attained around 50% of their fledging weight they are capable of jumping from the nest when the adults utter alarm calls or a predator is detected near the nest. Nestlings of Marsh Warblers can do this much earlier than those of Moustached or Aquatic Warblers. The very early ability of reedswamp-dwelling acrocephalid young to climb, and hence to leave the nest early, helps to avoid total losses in the late nestling period (Impekoven 1962, Kleindorfer *et al*. 1997, Halupka 1998). A brood of Aquatic Warblers can jump instantly in all directions, a form of behaviour that probably counteracts attack by shrews. Other predation-avoidance strategies are helpers at the nest or splitting the fledged brood into two groups cared for separately by male and female parents, a common technique in Marsh Warblers and Eurasian Reed Warblers (Kleindorfer & Hoi 1997).

5 Reproductive success – fitness matters

As indicated above, the interactions between life-history attributes and environmental factors are complex. An important element in this system is annual fecundity, which is positively correlated with adult mortality (high extrinsic mortality – high fecundity). Recent studies appear to show a general 'rule' that species in higher latitudes invest more in offspring *quantity* than those in the tropics, which in turn invest more in the *quality* of fewer offspring (Geffen & Yom-Tov 2000, Ricklefs & Wikelski 2002, Russell 2000, Russell *et al*. 2004, Brawn 2006, Robinson *et al*. 2010).

Fecundity in migratory birds is modulated by four factors, which, when taken together, almost completely explained the annual variance in fecundity in a long-term study of Black-throated Blue Warblers (food abundance 35%, predation 29%, double-brooding 19%, and renesting 15%; Nagy & Holmes 2004). Annual reproductive output can also vary strongly in reed warblers, as the translocation experiment with Seychelles Warblers showed, where it rose 23-fold under especially favourable conditions (Komdeur 1996a). Reproductive output can even be very different in closely related species living near to each other, as a

Photo 8.21 Harriers are also important predators of marshland birds. While Marsh Harriers prefer to rob reed warbler nests made visible by wind parting the reeds, in cold, wet weather Montagu's Harriers concentrate their efforts in wet grasslands by listening out for the begging calls of the hungry nestlings (photo Klaus Nigge).

meta-analysis of thousands of nests illustrated (Schulze-Hagen *et al.* 1996b). While single-brooded Marsh Warblers produce an annual average of 3.3 fledglings per female, Eurasian Reed Warblers produce 3.8 fledglings per female (despite substantially higher predation losses), thanks to their longer laying period, up to five replacement clutches, and some second broods. Unfortunately we have only limited information on adult mortality in both species (the expectation being that it is higher in Eurasian Reed Warblers than in Marsh Warblers).

In addition to the calendar effect discussed in section 8.3 above, other factors can cause the decline of various fitness components over the breeding season. According to life-history theory, an increase in reproductive effort negatively affects future reproduction; we can thus expect reduced reproductive success in replacement broods for reasons other than the course of time *per se*. To examine the possible effects of previous reproductive effort on different fitness components, Hansson *et al.* (2002a,b) analysed in Swedish Great Reed Warblers the frequency and success of replacement broods in relation to the time of the season and the length of the previous breeding attempt. These two parameters represent a measure of reproductive effort. After controlling for laying date they found that both re-laying frequency and size of replacement clutches were independent of both the duration of the previous breeding attempt and the timing of its completion. On the other hand, in replacement broods, both the number of fledglings and proportion of recruits declined according to the duration of the previous breeding episode. Evidently the workload during the first attempt had reduced female performance, negatively affecting the development of the eggs. In other words,

Photo 8.22 In Asia, snakes play an important role as predators. In Japanese marshlands four nest-plundering colubrid species(*Natrix*, *Elaphe*) occur sympatrically. Here, a Japanese Rat Snake (*E. climacophora*) slithers through the vegetation near a Black-browed Reed Warbler's nest (photo Shoji Hamao).

hatching success and chick survival were reduced because of poorer incubation and feeding performance, though some females coped with the burden better than others.

The great significance of individual differences between females in their fertilization and incubation abilities has also been confirmed in recent analyses of hatching success in Great Reed Warblers. A few females clearly had poorer hatching success than the majority, independent of the negative effect of parental relatedness (Knape *et al.* 2008). The proportion of adults with no offspring during a breeding season fluctuates between 20 and 40%, and some individuals do not reproduce at all in their lifetime (Great Reed Warblers, Hasselquist 1995; Seychelles Warblers, Komdeur 1996b). Breeding performance correlates with age, and is at its best within the middle age ranges in Great Reed Warblers and Seychelles Warblers (Leisler *et al.* 1995, Komdeur 1996b, Komdeur & Richardson 2007).

Inbreeding and dispersal

If the genetic similarity between parents is high the hatchability of their eggs can be negatively affected. This 'inbreeding depression' has been recorded in both Sey-

chelles Warblers and Swedish Great Reed Warblers, and is attributable to the small island population of the former and to the fact that the birds of the Swedish population descend from a few founder individuals. In Great Reed Warbler pairs whose partners were closely related to each other, hatching success declined. This was particularly true for the 4% of breeding pairs that were especially closely related. Other fitness components such as clutch size or proportion of fledglings and recruits were, by contrast, unaffected (Bensch *et al.* 1994, Hansson *et al.* 2004a). Nevertheless, Hansson *et al.* (2001, 2004b) were able to show that surviving and non-surviving siblings differed to varying degrees in genetic diversity (heterozygosity) and that those that were genetically more diverse had higher survival rates. The Swedish Great Reed Warbler population represents one of the few examples in avian science where the fitness–heterozygosity correlations could be studied. In Seychelles Warblers, an elegant cross-fostering experiment demonstrated that the young of more heterozygous mothers had better survival rates (though not those of more genetically diverse fathers), although only in unfavourable seasons when survival chances were lower (Richardson *et al.* 2004, Brouwer *et al.* 2007).

The effect of the dispersal of first-time breeders on their reproductive success is substantial. In Sweden, philopatric Great Reed Warbler males produced more fledglings and recruits throughout their lives, and survived better than immigrants (Bensch *et al.* 1998). Conversely, those immigrant males categorized as long-distance dispersers had the lowest lifetime reproductive success. Their song did not match the local dialect, their repertoires were correspondingly smaller, and their mating success lower than in philopatric birds (section 7.4). In females, short-distance dispersers recruited more offspring per year than both philopatric individuals and long-distance dispersers. These results suggest that long-distance dispersers were of poor phenotypic quality and less adapted to local social conditions. The lower lifetime reproductive success of philopatric females is obviously related to the costs of inbreeding in the studied population (Hansson *et al.* 2004c).

Life span and adult survival

The many comparisons between acrocephalids of very different backgrounds and lifestyles have repeatedly confirmed the crucially important role played by latitude in their life-history traits. Another vital regulating factor (alongside fecundity - sections 8.3 and 8.5) in life history is adult mortality or life expectancy, which also differs in characteristic ways. Annual adult survival is on average *ca* 10% higher in tropical and southern temperate than in northern temperate regions (Robinson *et al.* 2010). Thus, six long-distance migratory acrocephalids have a mean adult life expectancy of 1.4 years, but in African insectivores this is more than twice as long at 3.1 years (Peach *et al.* 2001). This pattern can be well observed in the superspecies *A. scirpaceus-baeticatus*: in Eurasian Reed Warblers the annual survival rate of first-calendar-year birds is 22%, but in African Reed Warblers this value is 76.5% (Long 1975, Peach *et al.* 2001, Thaxter *et al.* 2006). In Seychelles Warblers adult annual survivorship can even reach 84% (Komdeur 1996b).

Despite their generally low life expectancy, both migrant and tropical resident acrocephalid warblers hold quite a few age records: several Sedge and Marsh Warblers are known to have reached at least 9 years of age, Great Reed Warblers and Icterine Warblers have attained at least 10 years, the oldest Eurasian Reed Warblers were at least 12 and 14 years old when retrapped, while an Aquatic and an Icterine Warbler in captivity reached 9 and 14 years respectively (Long 1975, Redfern 1978; own unpublished data). Perhaps unsurprisingly, the absolute record-holder among the reed warblers is a tropical species: a Seychelles Warbler on Cousin reached the ripe old age of 21 years (Komdeur & Daan 2005).

Reed warbler mortality in temperate zones is above all else determined by events and conditions during migration and in their winter quarters; for example, the mortality of Sedge Warblers (though not Eurasian Reed Warblers) correlates with the amount of rainfall in the Sahel region, and is clearly higher in times of drought (Peach *et al.* 1991, Thaxter *et al.* 2006; chapter 13).

Summary

Reed warblers are skilled nest-builders; their ability to suspend their nests between vertical stems was probably an important precursor of their evolutionary success. During their evolutionary history this ability would have facilitated the frequent switches between scrub and reedbed or settlement on remote islands. The form of their various nest types is determined by phylogeny as well as by local ecological conditions and nest site.

Although the breeding strategies of acrocephalids in the northern temperate regions are fundamentally different from those in the tropics, in both zones the warblers rely on an endogenous clock to time their breeding. However, translocation experiments with Seychelles Warblers have shown that tropical species are influenced more by external cues, such as food availability. In optimal conditions they can change from a narrowly limited season to year-round breeding.

In the temperate zone, several species that nest in the herb and shrub layer of drier habitats exploit a specific, very late and temporally limited food supply, and consequently have extremely short breeding seasons. The species with the longest migration routes, such as Marsh Warblers, belong in this group. By contrast, in temperate wetlands the food supply is less seasonal, and hence acrocephalids there have longer breeding seasons, but also have to cope with higher rates of nest loss. In order to reduce these, many reedswamp-dwellers breed in denser, but food-poorer patches of reed and so are forced to make long flights to forage in more productive areas.

As countermeasures against frequent nest predation, reed warblers have developed a range of adult nest defence strategies and nestling anti-predator behaviour patterns, which may even affect the growth rate of the young. Such strategies have been shaped by the hunting techniques employed by their main predators, by nest position and concealment, and by the robustness of parental nest defence and the level of parental care.

Reed warblers breeding in higher latitudes invest more in offspring quantity, which corresponds to higher adult mortality, while those in the tropics invest in fewer offspring (though their time to independence is strikingly extended) and have longer life spans. Clutch size varies with the seasonality of the habitat, and is highest of all among long-distance migrants, whose young develop rapidly. Clutch sizes also decrease over the breeding season in the temperate zone – the so-called 'calendar effect'.

9

Coping with the Cuckoo's trickery

Male Great Reed Warbler mobbing a Cuckoo.

Silently and secretively the female Common Cuckoo glides down from her tree perch to the Eurasian Reed Warbler's nest in the reeds below, removes one host egg, pauses for a few seconds during which she lays her own egg, swallows the stolen one and then disappears as quickly and unobtrusively as she came. Year after year, large numbers of reed warblers are confronted with an alien egg in their nest. From the viewpoint of the brood parasite, many acrocephalid species are ideal hosts because of their high local breeding densities and the accessibility of their nests. In many parts of Europe and Asia they are the commonest hosts of cuckoos (Moksnes & Røskaft 1995), whose breeding success is significantly higher with some acrocephalid species than with the majority of other hosts (Moskát & Honza 2002, Kleven *et al.* 2004, Antonov *et al.* 2007a). From the viewpoint of the reed warblers, parasitism by cuckoos, with its accompanying nest-robbery, is in many cases the commonest cause of breeding failure, considerably reducing the breeding success of populations on a local scale (Schulze Hagen 1992, Barabás *et al.* 2004; section 8.4). Ground-breaking discoveries concerning cuckoo brood parasitism and the defence mechanisms of hosts have been made as a result of studies on acrocephalid warblers. Eurasian Reed Warblers are the equivalent of laboratory white mice for avian brood parasitism researchers and have been the ideal study species for classic and widely-known experiments and fieldwork (Davies 2000, Stokke *et al.* 1999, 2007a, Krüger 2007).

1
Brood parasitism by Common Cuckoos

Common Cuckoos (hereafter called Cuckoos) are distributed throughout almost the entire Palaearctic Region. Suitable hosts for these obligate brood parasites are insectivorous songbirds, which are mostly considerably smaller than the Cuckoos themselves. Most hosts have open nests, can attain high breeding densities, and are commonly distributed over wide areas; in Europe the principal hosts constitute a group of approximately 25 species. Genetically distinct Cuckoo strains (also called gentes; singular gens) can be distinguished, each one indicating the specialization of many Cuckoo females on a single host species, and whose eggs are generally good mimics of the host's own (Stokke *et al.* 1999, Davies 2000, Aviles *et al.* 2006, Soler *et al.* 2009, Fossøy *et al.* 2011). Different gentes can occur in sympatry (Antonov *et al.* 2010, Fossøy *et al.* 2011). In addition to host specialization, a possible alternative mechanism for finding host nests is 'habitat imprinting' (Teuschl *et al.* 1998). In habitats with a high host density, several Cuckoo males and females can be present (fig 9.1); their often overlapping territories

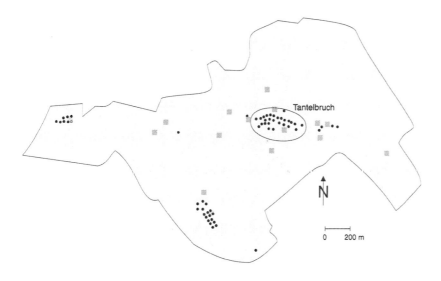

Fig. 9.1 Distribution of Common Cuckoos (blue squares) and of their local hosts Eurasian Reed Warblers (black dots = territories) on a study site in Germany (from Schulze-Hagen *et al.* 1996a).

Photo 9.1 Trees are used by female Cuckoos as vantage posts to observe the nest activities of potential hosts. The parasitism rate in reedbeds correlates positively with the number of bushes and trees (Hungary; photo Csaba Moskát).

adaptations and counter-adaptations exhibited by hosts and parasite are textbook examples of co-evolution, a process that can be regarded as a continuous 'arms race' consisting of several stages (Davies & Brooke 1989a,b, Stokke *et al.* 2005, Fossøy *et al.* 2011). Host defence is constructed in successive lines: first, the selection of a safe nest site and the mobbing of the parasite prior to egg deposition (Welbergen & Davies 2009); secondly, some form of rejection of this parasitic egg (Davies & Brooke 1988, Stokke *et al.* 1999); thirdly, countermeasures directed against the Cuckoo chick (Grim 2007, Kilner 2010). However, once the young parasite is hatched, host defence is almost non-existent (sections 9.3-9.5). Among the parasite's adaptations are a relatively small egg size, accelerated embryonic development (starting when the egg is still in the oviduct and known as 'internal incubation'), egg mimicry, unusually thick egg shell, and eviction behaviour of the chick (Davies 2000, 2011, Birkhead *et al.* 2011).

Photo 9.2 This Eurasian Reed Warbler is alarm-calling loudly in front of its nest (just visible behind the bird) towards the female Cuckoo approaching the nest in order to deposit her egg there. Male Eurasian Reed Warblers regularly guard the nest during the egg-laying phase (photo Thomas Krause).

are on average 60-70 ha in area (Nakamura & Miyazawa 1997). Although their breeding sites are usually reedbeds and bushy areas, Cuckoos will fly up to 20 km to forage in woodland, where they find their main food of hairy caterpillars. The females spend most of the day on their breeding territories, and in the course of the relatively long breeding season of some of their hosts, individual Cuckoo females can lay up to 25 eggs. Eggs are laid every second day, and the breeding season of Cuckoos may last about 1-1.5 months.

Cuckoos avoid the high costs of parental care, which is an enormous burden for the host birds. In contrast to nest predation, which often merely compels a victim to nest again, for a reed warbler, a young Cuckoo in the nest means not only complete breeding failure for the entire season, but also a reduction of its own survival chances because of the increased time and energy expended in caring for the Cuckoo chick. Furthermore, many host species are short-lived and have only a few opportunities to reproduce throughout their lives. Brood parasitism is therefore a strong selective force which creates several types of anti-parasite defence, through reciprocal evolutionary changes in interacting species. The often subtle

2

Acrocephalid hosts and parasitism rates

A number of acrocephalids are common or very common hosts of Common Cuckoos, for example, Eurasian, Great, Oriental, Black-browed and Blyth's Reed Warblers, Marsh, Paddyfield, Thick-billed, Sedge, Booted, and Olivaceous Warblers. They all have high local breeding densities and each species has its own Cuckoo gens. By contrast, even though Cuckoos are not uncommon in their habitats, other species such as Moustached Warblers are only very rarely parasitized. Possible explanations for this low parasitism rate are low breeding densities and habitats lacking the necessary trees from which Cuckoos can survey the nests of potential hosts (Røskaft *et al.* 2002a; section 9.3).

Greater Swamp Warblers in central Africa and African Reed Warblers in east Africa are known to be hosts of Klaas's Cuckoos; Madagascar Swamp Warblers are parasitized by Madagascar Cuckoos (Urban *et al.* 1997); Thick-billed Warblers are parasitized by both Common and Lesser Cuckoos (Y. Shibnev, pers. comm.). Other cuckoo species play a less important role as brood parasites of acrocephalids.

The average parasitism rate of the other more prominent Cuckoo hosts in western Europe (including species such as Dunnocks, Pied Wagtails, and Meadow Pipits), is relatively low and ranges from about 1 to 3%. In *Acrocephalus* warblers the rate is often many times higher. A meta-analysis of over 18 000 nests in 52 studies from western and central Europe yielded a mean parasitism rate for Eurasian Reed Warblers of 8.3% and for Marsh Warblers of 6.2% (Schulze-Hagen 1992). However, the level of parasitism can vary substantially, both locally and between years (section 9.7), and under increasingly favourable conditions it can rise, for instance from 0% to 82% within a period of eight years in Eurasian Reed Warblers (Schulze-Hagen *et al.* 1996a). The proportions can also vary considerably on a larger geographic scale. The most important predictor of parasitism is host density, as shown by a comparison of 16 Eurasian Reed Warbler populations throughout Europe (Stokke *et al.* 2007a; fig. 9.2).

Parasitism in Marsh Warblers can reach rates of 44% and 28% in east-central and southeast Europe respectively (Kleven *et al.* 2004, Antonov *et al.* 2006b). Extreme rates of *ca* 65% have been described in the Great Reed Warblers of the Hungarian Plain, where 58% of all Cuckoo eggs were

found in multiply-parasitized clutches containing two, three, or even four Cuckoo eggs (Moskát & Honza 2002, Moskát *et al.* 2009 b). This high rate of 65% has been constant over many decades and is one of the highest rates of Common Cuckoo parasitism known anywhere. On the other hand, Great Reed Warblers in Sweden and Greece are not parasitized at all (Moskát *et al.* 2002; see also section 9.7). While only 0.5% of Sedge Warbler nests in Britain contained a Cuckoo egg, the figure in Moravia (Czech Republic) was 26% (Bibby 1978, Kleven *et al.* 2004). In northwest Bulgaria, Eastern Olivaceous Warblers are counted among the principal Cuckoo hosts of the region, with a parasitism rate of 27%, (Antonov *et al.* 2007a).

3

On the battlefront: before the egg is laid

Female Cuckoos search for host nests by hiding among treetops and observing the host's nest-building activity (Gärtner 1982). Nests situated closer to trees are more likely to be parasitized than those further away. Moreover, the higher the tree, the higher the risk of parasitism (Øien *et al.* 1996, Moskát & Honza 2000, Antonov *et al.* 2007a). At Wicken Fen in Cambridgeshire, UK, all parasitized

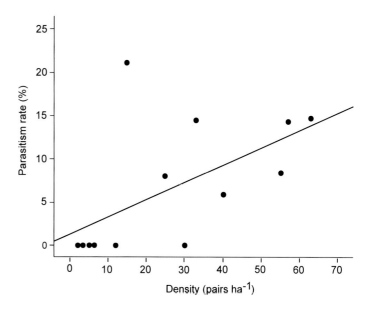

Fig. 9.2 Positive relationship between parasitism rate and host density among 16 Eurasian Reed Warbler populations. The fitted line is the linear regression line (from Stokke *et al.* 2007). The greater the host density, the higher the parasitism rate.

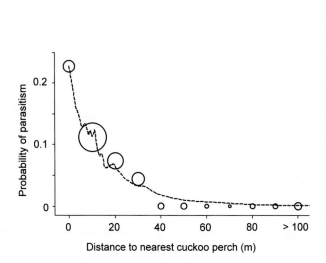

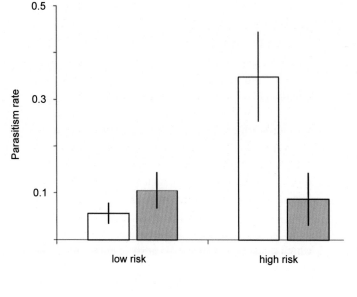

Fig. 9.3 Probability of parasitism in the Wicken Fen area, UK, related to the distance (m) to the nearest Common Cuckoo perch (from Welbergen & Davies 2009). The parasitism risk decreases with increasing distance from perch.

Fig. 9.4 Nests with owners that mob the Cuckoo are parasitized less than expected. The figure shows actual parasitism according to whether reed warblers mobbed (dark columns) or did not mob (light columns) a model Cuckoo placed at nests that had a parasitism risk lower or higher than average (from Welbergen & Davies 2009).

nests were within 28 m of the nearest observation perch (Welbergen & Davies 2009; fig 9.3). In a meta-analysis of 15 Eurasian Reed Warbler breeding biology studies, the parasitism rate reached 31.6% in fragmented reedbeds interspersed with bushes and trees, 13.3% in narrow reed belts with an adjoining bushy zone, but only 1.2% in extensive, mainly bush-free reedbeds (Schulze-Hagen, unpubl. 1990). Nest exposure is also an important factor, since parasitized nests tend to be in stands of lower vegetation. A decrease in parasitism probability with season again reflects a decrease in nest conspicuousness as the seasonal growth of vegetation increases (Antonov *et al.* 2007a, Stokke *et al.* 2008, Welbergen & Davies 2009).

One host adaptation would therefore be to avoid brood parasitism at the nest site selection stage. Eurasian Reed Warblers and Marsh Warblers should avoid building their nests near trees, and in fact Eurasian Reed Warbler population density is actually higher in reedbeds that are mostly free of bushes and trees (Øien *et al.* 1996). Additionally, at higher breeding densities a dilution effect and a more effective warning call system contribute to a reduction of the individual parasitism risk (Welbergen & Davies 2009). In Eurasian Reed Warblers, a positive correlation between egg volume and distance of the nest

from the nearest tree indicates that the better nest sites are taken by older and more experienced birds and that there is competition for sites further away from trees (Øien *et al.* 1996). In Japan, Oriental Reed Warblers were also subjected to a higher level of parasitism in suboptimal habitats, which were occupied in the main by young, naïve breeders. These findings indicate that parasitism is non-random with regard to both nest site and host age and experience (Lotem *et al.* 1992, Stokke *et al.* 2005, 2008; section 9.7).

Having selected a nest site that is as safe as possible, another front line of defence against brood parasitism is aggressive mobbing prior to egg deposition by the parasite. Mobbing is an effective anti-parasite measure, well studied in Eurasian Reed Warblers (fig. 9.4; Welbergen & Davies 2009, Campobello & Sealy 2011), though one counter-adaptation against mobbing is the intimidating hawk mimicry of the Cuckoos themselves (Welbergen & Davies 2011). Anyone who has witnessed up to 12 excited and loudly scolding reed warblers pursuing a female Cuckoo through a reedbed will not easily forget the experience. Just hearing the call of the Cuckoo female, and especially seeing her flying over, is enough to trigger a wave of alarm-calling throughout the reedbed. In Eurasian Reed

Warblers mobbing is a stereotyped audible (bill snaps, rasp calls) and visual display, including direct physical attacks towards the parasite. Eurasian Reed Warblers that had observed conspecifics mobbing a Cuckoo increased the intensity of their nest defence, indicating that social cues are more important than personal experience, with the brood parasite as a basis for a behavioural change. Once they have young in their nests and the parasite is no longer a danger, the birds stop reacting to Cuckoos, whereas alarms directed at Eurasian Jays or Eurasian Sparrowhawks continue, as both are potential predators of the fledged young. (Duckworth 1991, Welbergen & Davies 2009, Campobello & Sealy 2011).

In nests where Eurasian Reed Warblers had contact with Cuckoos before the first egg was even laid, 40 per cent were abandoned. Once laying began, Cuckoo appearances did not cause increased desertions, but enhanced egg-rejection strategies (Davies *et al.* 2003). In the Czech Republic, video recordings and experiments using a Cuckoo mount demonstrated that if the Eurasian Reed Warbler hosts were absent while the Cuckoo was laying, then none of them rejected the parasitic egg, whereas 67% of those which were present rejected it (Moksnes *et al.* 2000; section 9.7).

At the end of nest building and during the laying phase of Eurasian Reed Warblers, the bird sitting in the nest was almost always the male. Females did not increase nest guarding, perhaps because of the increased time needed for foraging during egg laying. Male nest attendance was at its most intensive when one or two eggs had been laid. However, increased male nest attendance at the one-two-egg stage was not at the expense of mate guarding, because this declined anyway when laying began, and did not lead to increased paternity loss compared with controls. Sitting on the nest may also dissuade the parasite from laying, since it can see that the hosts are in attendance and would be alerted to reject the alien egg (Davies *et al.* 2003). Great Reed Warbler males, which do not incu-

Photo 9.3 Nest guarding and mobbing are effective front lines of defence against brood parasitism. This Great Reed Warbler fiercely mobs a model Cuckoo, seemingly undaunted by the fact that the wooden model is being held in the hand of a researcher (photo Oldrich Mikulica).

Photo 9.4 Before the Cuckoo lays her egg in the hosts' nest she eats one of their eggs (Great Reed Warbler egg; photo Oldrich Mikulica).

bate, played the key roles in Cuckoo-mobbing and nest guarding, while females were responsible for egg rejection (Pozgayová *et al.* 2009). Nests of polygynous Great Reed Warblers were more often successfully parasitized than nests of monogamous pairs, because polygynous males assisted less in nest guarding during the egg laying period (Trnka & Prokop 2011; section 8.4).

In Eurasian Reed Warblers the bulk of parasitism take place on the second and third days of laying (Davies *et al.* 2003). The female Cuckoo lays her egg in the host nest only in the course of the afternoon and not, like her host, in the early morning. During the average 41 seconds that she spends on the nest she removes a mean number of 1.5 eggs (Moksnes *et al.* 2000).

4 On the battlefront: an alien egg in the nest

Once the parasitic egg is in the nest, efficient egg recognition becomes of crucial importance to the host in reducing the costs of parasitism. If a Cuckoo's egg is recognized as alien then the reed warblers will attempt to reject it. The alien egg can be ejected by grasping it in the mandibles or by pecking at it long enough to make a hole, in which case the hosts then drink the contents and carry the shell out of the nest. If the hosts are unable to remove the unusual egg they can either reject it by desert-

ing their clutch, or accept it (Antonov *et al.* 2009). The fewer eggs that remain in the nest after parasitism (e.g. in situations where the Cuckoo removes more than one egg) the higher the likelihood that reed warblers will desert that nest. It has been shown that nest desertion by Great Reed Warblers after the successful ejection of the alien egg could be due to reduced clutch size and not to any other anti-parasite defence mechanism (Moksnes *et al.* 2000, Antonov *et al.* 2006b, Moskát *et al.* 2011). The message for Cuckoos is clear: do not lay too early – before host eggs are laid – nor remove too many host eggs from their nest!

From time to time strikingly large reed warbler nests are discovered, which closer inspection reveals to be double nests. Here, the rejection has taken the form of the host bird building over the parasitized clutch and using the new nest cup for the replacement clutch. In Oriental Reed Warblers only the females incubate and therefore only females reject, whereas in Eurasian Reed Warblers both sexes incubate and both reject (Davies & Brooke 1988, Lotem *et al.* 1992). The many differences in the defence strategies of the various acrocephalids are one of the reasons why not only the mode but also the scale of egg rejection varies so greatly (sections 9.6 and 9.7).

In general, the better the egg mimicry of the parasite, the poorer the rejection ability of the hosts (Stokke *et al.* 1999, Antonov *et al.* 2006b). Around one-third of Cuckoo eggs in Great Reed Warbler nests are such an excellent match that even experienced observers find it difficult to distinguish them from the similarly-sized eggs of this large reed warbler (Moskát & Honza 2002). Those in nests of Marsh, Thick-billed, and Olivaceous Warblers can also perfectly resemble the respective host eggs. Specialized egg morph evolution is correlated with larger tracts of homogeneous habitat, where one host species is predominant, such as the Hungarian reedbeds with their large populations of Great Reed Warblers (Soler *et al.* 2009, Fossøy *et al.* 2011). However, the majority of Cuckoo eggs in reed warbler nests show only an intermediate mimicry and are thus easily recognized (sections 9.6 and 9.7). Intermediate egg types occur typically at the junctions of different habitats, where different Cuckoo gentes live in sympatry and interbreeding is likely. Recent research indicates that egg colour and egg volume are not inherited exclusively maternally but also via paternal genes (Aviles *et al.* 2011, Fossøy *et al.* 2011).

In southern Moravia, no difference could be found in mimetic quality between Cuckoo eggs rejected and those accepted by Eurasian Reed Warblers. The acceptance of poorly matching eggs indicates either an early stage of the coevolutionary arms race between parasite and hosts, or that there is gene flow from unparasitized populations (Stokke *et al.* 1999, section 9.7). Good egg mimicry with a particular host might be disadvantageous where host density is declining, whereas laying a more generalist Cuckoo egg, such as the Garden Warbler type, would more easily allow the parasite to switch host species (Edvardsen *et al.* 2001, Grim 2002). Egg mimicry can be improved by the host specificity of the female Cuckoo in concert with selection by the hosts against poorly match-ing eggs. Moreover, improved egg mimicry may also result from Cuckoos actively seeking host nests with eggs that are closer in appearance to their own. Support for

this 'cuckoo preference hypothesis' has been provided by recent studies (Aviles *et al.* 2006, Cherry *et al.* 2007a,b; for pattern mimicry of host eggs by Cuckoos, see also Stoddard & Stevens 2010). Using spectrophotometry and photography, in a Eurasian Reed Warbler population in Denmark only recently parasitized by Cuckoos, Aviles *et al.* (2006) found that the level of matching between Cuckoo and Reed Warbler eggs increased rapidly over a few decades, and that Cuckoos selected nests in which to lay rather than laying at random.

The ability to reject eggs requires that the hosts recognize their own eggs. There are some indications that reed war-blers learn what their eggs look like when they breed for the first time, and first time breeders are more likely to accept Cuckoo eggs than experienced breeders (Lotem *et al.* 1992). Distinguishing an unusual (Cuckoo) egg

Photos 9.5 to 9.7 In many cases the Cuckoo egg perfectly matches that of the host species (for example, Marsh Warblers in Bulgaria; four randomly-chosen clutches containing host and Common Cuckoo eggs, the parasite egg being the first one in each case, left and top right); in other cases the match is not per-fect (Eurasian Reed Warblers in the Czech Republic, below. Specialized egg morph evolution is correlated with larger tracts of homogeneous habitat where one host species is predominant, while intermediate egg types tend to occur at the junctions of different habitats, where different Cuckoo gentes live in sym-patry and interbreeding is likely (photos Bård Stokke, Anton Antonov, Oldrich Mikulica).

Photo 9.8 Other cuckoo species also lay eggs that perfectly mimic those of their acrocephalid hosts; shown here is a Lesser Cuckoo egg with a Thick-billed Warbler clutch in the Russian Far East (photo Yuri Shibnev, Nature PL).

involves recognizing it as different and deciding to reject it. It should be easier for hosts to detect the parasite's 'forgery' if their own clutch has a distinctive signature, for instance if the variation among their own eggs is reduced and all eggs of the clutch look very similar (Stokke *et al.* 2007b, Davies 2011). Indeed Eurasian Reed Warblers that rejected an artificial non-mimetic egg had a significantly lower intraclutch variation in egg appearance than those that accepted it. The higher *inter*clutch variation that results from lower *intra*clutch variation in a reed warbler population makes egg mimicry even more difficult for the parasite (Stokke *et al.* 1999).

Two costs are incurred by the host in egg rejection: recognition errors, arising when the host accidentally removes one of its own eggs, and rejection costs, when the host damages its own eggs due to the increased pecking efforts required on the remarkably thick shell of the Cuckoo egg (Røskaft *et al.* 2002b, Stokke *et al.* 2002). These costs can be relatively high, and have been recorded in nearly 50% of Cuckoo egg-ejection events by the small *Acrocephalus* and *Iduna* species (Davies 2000, Stokke *et al.* 2011; section 9.6). Hosts that damage their

own eggs during ejection are likely to desert their clutches. Eurasian Reed Warblers lost on average 1.2 of their own eggs per rejection event. Even when an individual nest remains unparasitized, if the parasitism rate is high, and if the warblers see a Cuckoo close to their nest, they may throw out their own eggs (Davies 2000). The more mimetic the parasite's egg, the more rejection is delayed, usually by one or two days. This can be explained by the initial failure to recognize the Cuckoo's egg which results in latency in the release of rejection behaviour. An additional component of the delay can also be apparently inefficient host rejection behaviour, which probably reflects a lack of previous experience (Antonov *et al.* 2008b). In Hungary, where Cuckoo eggs are a good match, Great Reed Warblers made three times more recognition errors than did Oriental Reed Warblers in Japan, where Cuckoo eggs are less mimetic (Stokke *et al.* 2007b). Possibly because of this high rate of recognition errors Hungarian Great Reed Warblers accepted the alien egg more often than Japanese Oriental Reed Warblers (Davies 2000, Moskát & Honza 2002).

5

How to deal with this strange new chick

Soon after hatching, the parasitic chick starts to evict all eggs or nestlings from the nest. During this procedure it is strange to observe that the parent warbler does not even interrupt its brooding as the young Cuckoo underneath it pushes egg after egg up the nest wall. In contrast to their highly developed egg rejection behaviour, it appears puzzling that the host parents blindly accept the entirely different-looking young Common Cuckoo. Exchange experiments with nestlings of different species have shown that most passerine hosts simply fail to discriminate at the chick stage and will feed any begging mouth in their nest. Chick recognition appears to be maladaptive, because it would be deleterious if mis-imprinting at a first breeding attempt led to the subsequent rejection of all own young in future unparasitized attempts (Lotem 1993, Davies 2000) Intriguingly, the eggs of almost all parasitic cuckoos hatch in advance of host young. Their shorter incubation periods have possibly been selected to prevent the evolution of learnt chick recognition in their hosts (Kilner 2010, Birkhead *et al.* 2011). Nevertheless, there is increasing evidence that some Australian hosts can recognize and reject the parasitic chicks (three species of bronze-cuckoos). Consequently, a reciprocal selection for visual mimicry of host young by the bronze-cuckoos concerned has evolved (Kilner 2010, Sato *et al.* 2010, Tokue & Ueda 2010, Langmore *et al.* 2011).

A young Cuckoo is fed at about the same rate as a brood of four Eurasian Reed Warbler nestlings, but despite this it is dependent on its foster parents for at least 10 days longer than their own young. Neither its larger size nor its orange-red gape alone is sufficient to stimulate increased feeding activity by reed warbler hosts. Instead, the key is the continuous, rapid, and loud begging call, which sounds like a whole brood of hungry chicks (fig 9.5; Davies *et al.* 1998 Schulze-Hagen *et al.* 2009). Reed warblers can adjust their provisioning rate to the number of chicks, and are even capable of adequately feeding a brood of eight young under experimental conditions. This workload exceeds that required by a Cuckoo chick, so the Cuckoo does not in fact work the hosts to their maximum capacity. The visual stimulus of its single open gape is not supernormal, but in fact subnormal when compared with that of a brood of four reed warblers. As it gets older it has to intensify its begging calls to compensate for its increasingly subnormal gape stimulus (Kilner & Davies 1999, Kilner *et al.* 1999). Young Cuckoos are perfectly attuned to the chick-provisioning strategies of their hosts and have developed host- or gens-specific begging displays and – at least in the gens specialized on Eurasian Reed Warblers – an innate reaction to the host's distinctive alarm call. In exchange experiments it has been demonstrated that reed warbler-Cuckoos fall silent at the warning call of their foster parents, while Dunnock-Cuckoos show no reaction when placed into reed warbler nests (Davies *et al.* 2006, Madden & Davies 2006, Davies 2011).

However, acrocephalids might also be able to discriminate against parasitic nestlings to some degree, not during the early chick stage but rather near fledging, when they use the duration and amount of care as cues. This phenomenon is called 'chick discrimination without

Fig. 9.5 Intense begging calls by a nestling Common Cuckoo. Sonograms (2.5 s) of the begging calls of a single reed warbler chick (top), a brood of four reed warblers (middle), and a single Cuckoo chick (bottom; from Davies *et al.* 1998).

recognition' (Grim *et al.* 2003, Grim 2007). The critical time for chick discrimination seems to be around days 9 to 13, when Eurasian Reed Warbler parents actively stimulate the fledging of their own young by reduced feeding. If this happens, then a young Cuckoo, still immature and not yet able to fledge, may starve. Approximately 15% of the Cuckoo chicks in the southeast Czech Republic were deserted by the host Eurasian Reed Warblers (Grim *et al.* 2003). In one case the Reed Warbler parents dismantled the nest under the hungry Cuckoo chick to build the next one a few metres away (O. Mikulica, pers. comm.).

As well as factors such as the size of the host, the higher breeding success of Cuckoos in Great Reed Warbler nests might perhaps be explained by the longer fledging period of this species (on average 3 days longer than Eurasian Reed Warblers; Leisler 1991b, Schulze-Hagen 1991c; see also below).

The breeding success of Common Cuckoos

Despite foster parents struggling for 5-6 weeks to raise the young parasite, which is nearly twice as long as the rearing of their own brood demands, Common Cuckoo breeding success is not high. In England 31.9% of all Cuckoo eggs laid into Eurasian Reed Warbler nests produced a fledgling, while this rate was only 26.5% in the nests of Dunnocks, the Cuckoo's second commonest host

Table 9.1 Successive lines of reed warbler defences in relation to Cuckoo trickery (modified after Davies 2011).

Stage	Reed warbler defences	Cuckoo trickery
Nest	Nest further from cuckoo vantage perches	Monitor hosts to find nests Time parasitism effectively
	Cryptic nest	
	Secretive behaviour	
	Mob female cuckoo	Secretive behaviour
		Hawk mimicry
Egg	Desert nest or increase egg rejection if cuckoo seen at nest	Secretive and rapid laying
	Reject foreign eggs	Host egg mimicry
		Stronger egg shells
	Egg signatures	Mimic signatures
Chick	Reject/discriminate foreign chick	Mimic signatures Manipulative signals to exploit hosts
All stages	Vary defences in relation to parasitism risk	Secretive behaviour
	Rely on indirect/direct cues to parasitism	

(Davies 2000). For three sympatric acrocephalids in southern Moravia the success rate was 30.4% with Great Reed Warblers, 16.4% with Eurasian Reed Warblers, and 4.3% with Marsh Warblers. The larger size of the Great Reed Warbler relative to its congeners makes this species a more successful host; Cuckoo chicks grow faster and become significantly larger at fledging in the nests of Great Reed Warblers than in those of Eurasian Reed Warblers or Marsh Warblers (Kleven *et al.* 1999). They may also survive better after fledging. The reason for the Cuckoo's low breeding success with Marsh Warbler hosts lies in the latter's very high rejection rate of Cuckoo eggs (Brooke & Davies 1987, Kleven *et al.* 1999, Kleven *et al.* 2004, Antonov *et al.* 2006b).

6 Mounting a varied counter-attack

The acrocephalids represent the only example in which we can compare the interactions between Cuckoos and their hosts in a whole group of closely related species. An interesting question arises as to whether there is a specific Cuckoo gens for each reed warbler species ('host preference hypothesis'; see Aviles *et al.* 2006), or whether Cuckoo females simply do not differentiate between the very similar species and lay their eggs at random in the nests of this family ('nest site hypothesis'; see Edvardsen *et al.* 2001). Radiotelemetry studies indicate a strict host preference by female Cuckoos in southern Moravia, where four *Acrocephalus* species (Sedge and Marsh Warblers, Great and Eurasian Reed Warblers) coexist in the same area. However at this study site, despite well-developed host specialization, the Cuckoo egg morphs were rather generalized in appearance (Edvardsen *et al.* 2001, Aviles *et al.* 2006; see also section 9.4). By contrast, in a Bulgarian area of sympatry, host-specificity seemed to be less strict (Fossøy *et al.* 2011), while in western Germany adjacent nests of Marsh Warblers and Eurasian Reed Warblers were parasitized by the same female Cuckoos (Schulze-Hagen *et al.* 1992).

The different reed warbler species show considerable variation in their levels of parasitism and in the mode of their host defence (Kleven *et al.* 2004, Stokke *et al.* 2005, Dyrcz & Halupka 2006; table 9.2). The type of defence depends on several species-specific traits, among them body weight and bill size. Larger hosts such as Oriental and Great Reed Warblers are capable of physically attacking Cuckoos. Sometimes egg-laying Cuckoo females are so vigorously attacked that they leave many feathers

Photo 9.9 In heavily parasitized Great Reed Warbler populations of the Hungarian Plain, with parasitization rates of 50-60%, cases of multiple parasitism are not unusual, where several Cuckoo females lay their eggs in the same host nest. This Great Reed Warbler nest contains four Cuckoo eggs, all excellent mimics in colour and size (photo Csaba Moskát).

around the nest. Molnar (1944) reported five cases in which a Cuckoo female drowned following desperate struggles with hosts. On the other hand, the smaller reed warblers are not able to drive Cuckoos away successfully. In one area of sympatry, while 40% of all Great Reed Warbler pairs directly attacked a Cuckoo mount at the nest, this behaviour was not seen at all in Eurasian Reed Warblers (Dyrcz & Halupka 2006).

The most important defence mechanism by small acrocephalids is egg rejection and methods vary depending on host bill size. While the larger-sized and strong-billed

Great and Oriental Reed Warblers can easily eject a Cuckoo egg by grasping it between their mandibles, this is impossible for the smaller acrocephalids. Instead they reject it by puncturing it or even by desertion. Whereas large reed warblers eject the Cuckoo egg without any damage to their own eggs, rejection costs in the smaller species are relatively high (section 9.4). As predicted from their finer bill, these costs were higher in Eurasian Reed Warblers than in Marsh Warblers (Antonov et al. 2006a,b). With their even finer bills, Olivaceous Warblers were mostly unable to puncture the thick-shelled Cuckoo eggs. Despite the fact that they had discriminated and strongly pecked all Cuckoo eggs, 47% of them were finally accepted. It seems puzzling why desertion does not always follow unsuccessful puncture or ejection attempts in Olivaceous Warblers (Antonov et al. 2009, but see Davies et al. 1996).

In sympatrically occurring populations of Marsh and Eurasian Reed Warblers parasitized by the same Cuckoo females, Marsh Warblers were strong but Eurasian Reed

Table 9.2 Different scale and type of egg rejection in various acrocephalid species and populations parasitized by Common Cuckoos [Gärtner 1982 (Marsh Warblers), Davies & Brooke 1988 (Eurasian Reed Warblers), Lotem et al. 1992 (Oriental Reed Warblers), Øien et al. 1998 (Eurasian Reed Warblers), Moskát & Honza 2002 (Great Reed Warblers), Kleven et al. 2004 (Marsh Warblers, Eurasian Reed Warblers, Great Reed Warblers), Antonov et al. 2006b (Marsh Warblers), Antonov et al. 2006c, 2009 (Olivaceous Warblers)]. *Rejection rate and acceptance rate (not shown) are complementary and their sum is 100%.

Species	Parasitism rate %	Rejection rate* %	of which: (in %)	desertion	ejection	built over
Marsh Warbler	28-45	50-87		10-30	35-90	0-5
Eurasian Reed Warbler	11-21	18-38		50-79	21-50	0-5
Great Reed Warbler	34-65	27-34		59	35	6
Oriental Reed Warbler	20-21	60		21	74	5
Olivaceous Warbler	27	50		88	12	0

Warblers were weak rejectors of the poorly mimetic eggs, with rejection rates of 85% *vs* 15% respectively (Schulze-Hagen *et al.* 1992, Kleven *et al.* 2004). This substantial variation can be explained by the 'spatial habitat structure' hypothesis, whereby Marsh Warblers – given the structure of their specific habitat – are more exposed to parasitism and have therefore evolved a more efficient

egg rejection behaviour than have Eurasian Reed Warblers (Røskaft *et al.* 2002b, 2006; section 9.7). What is more, the narrow breeding time-window of Marsh Warblers may exert a stronger selective pressure to breed successfully at the first attempt, compared with Eurasian Reed Warblers' wider breeding period window (Schulze-Hagen *et al.* 1996b; chapter 8).

Cuckoo-related breeding losses by reed warblers

Female Cuckoos represent a double threat of both parasitism and predation to acrocephalid warblers. They may parasitize the nest but they may also depredate clutches to force the hosts to lay a replacement clutch, thus creating a potential parasitism reserve. One to three host eggs are removed from many nests during the laying period without the Cuckoos laying any of their own (Davies 2000). Following such partial losses, the clutch is often deserted. Host eggs are thus clearly an important source of food for Cuckoos.

The higher the parasitism rate, the higher the rate of predation by Common Cuckoos. Predation rates on Eurasian Reed Warbler clutches might be 2-4 times higher than those attributable to pure parasitism (fig. 9.6). Breeding success in unparasitized populations is therefore clearly higher than in those that are parasitized. Up to 68% of all breeding losses in Eurasian Reed Warblers, and up to 45% in Marsh Warblers, can be attributed to Cuckoos (Schulze-Hagen 1992). Unparasitized Eurasian Reed Warbler nests fledged 0.48 young per egg laid. Among the parasitized populations, ejectors had a breeding success of 0.29 young/egg, while the corresponding value for acceptors was 0.03. Great Reed Warblers in a heavily parasitized Hungarian population had an average fledging success of 0.21 young/egg, which is too low for a self-sustaining population and is compensated for by immigration from outside the area (Øien *et al.* 1998, Moskát & Honza 2000; section 9.7).

Photo 9.10 Generally, the better the egg mimicry the poorer the rejection ability of hosts. This Oriental Reed Warbler in Japan is just removing the foreign egg from its nest (photo Toshiyuki Yoshino).

Flexibility in a world of change

Despite the evolution of egg recognition abilities and rejection behaviour, significant numbers of Cuckoo eggs are still accepted by hosts. In many acrocephalid populations individuals show only intermediate host defences against a Cuckoo egg, since both rejection and acceptance occur. There is considerable variation between individuals of the same population as well as between adjacent populations (Davies *et al.* 1996, Davies 2000, Stokke *et al.* 2008). How can such variability be explained? There are indications that both gene flow, which weakens the genetic response in a parasitized population, and other individual reactions (e.g. phenotypic plasticity) are involved. Habitat structure also plays an important role in explaining the variability of these reactions (Lotem *et al.* 1992, Brooke *et al.* 1998, Stokke *et al.* 2005, Røskaft *et al.* 2006). The situation is complex, but consideration of the following factors will help to shed some light.

Gene flow, immigration, and host tolerance to multiple parasitism

In Hungary, a unique situation has developed between Great Reed Warblers and Cuckoos. An unparalleled high parasitism rate of *ca* 65% has remained almost constant for a period of at least 70 years, and has been combined with multiple parasitism (i.e. the laying of more than one – and in some cases up to four – Cuckoo eggs in a single Great Reed Warbler nest; Moskát & Honza 2002, Moskát *et al.* 2009a, Takasu & Moskát 2011). This raises two questions: How can such a high parasitism frequency be maintained over long periods without evolving efficient countermeasures? Why does such a heavily parasitized host population not go extinct?

There are two possible explanations. First, when the nest contained a single Cuckoo egg, the Great Reed Warblers rejected 40% of the foreign eggs, but this fell to only 12% when the nest contained more than one Cuckoo egg. Multiple parasitism makes for increased host tolerance, because an increased number of mimetic eggs in the nest simply overrides a host's recognition abilities by decreasing the chance that any one individual egg will be perceived as foreign. In its turn, increased host tolerance is likely to contribute to increased parasite density, thus adding to the persistent heavy parasitism rate (Moskát *et al.* 2009a, Takasu & Moskát 2011).

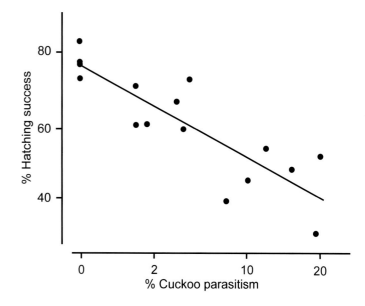

Fig. 9.6 Negative relationship between hatching success and intensity of Cuckoo parasitism in 16 studies of Eurasian Reed Warbler breeding biology (from Schulze-Hagen 1992). The marked decrease in hatching success with increasing parasitism rate can be explained by the high level of egg predation associated with parasitism by Cuckoos.

Secondly, such a heavily parasitized population is a 'sink', which can only be sustained if connected – through the immigration of naïve breeders – to spatially separated, less parasitized 'source' subpopulations of a common metapopulation. This exact pattern is predicted by the spatial habitat structure hypothesis, whose parameters include the patchiness of parasitized and unparasitized subpopulations, their size, and their rate of dispersal and hence gene flow (Barabás *et al.* 2004, Røskaft *et al.* 2006). In fact, with a low level of philopatry and no genetic differentiation in the region, a comparison between three Hungarian populations, located 40-130 km from each other and suffering from varying levels of parasitism (1-68%), could not find any difference in the responses to experimental parasitism, thus demonstrating a strong gene flow between populations.

It is the *combination* of gene flow with increased host tolerance which accounts for the fact that heavy parasitism rates can be stably maintained over such long periods of time (Moskát *et al.* 2008, Takasu & Moskát 2011).

Photos 9.11 and 9.12 Sometimes strikingly large reed warbler nests are found, which, on closer inspection, are revealed to be double nests – an uncommon defence against the parasite. In the vertical section through this Eurasian Reed Warbler nest the old clutch can be seen in the lower storey, while in the upper the visible parts of feather sheaths indicate that the replacement brood has successfully fledged. In another case the replacement nest has been built a little above the old nest, but it contains yet another Cuckoo egg (photos A. Ritter, Oldrich Mikulica).

Habitat structure

In some Marsh Warbler populations there is a high parasitism rate (up to 45%) combined with very low Cuckoo breeding success, ranging around 4% (Gärtner 1982, Schulze-Hagen *et al.* 1992, Antonov *et al.* 2006b). Among all European *Acrocephalus* species, the habitat of Marsh Warblers is almost certainly the most evenly distributed within a matrix, and also contains the highest frequency of vantage points. This tends to result in a similar parasitism pressure throughout the whole population and, according to the spatial habitat structure hypothesis, a more effective rejection behaviour may have thus evolved faster in Marsh Warbler populations than in populations of their congeners (Røskaft *et al.* 2006; section 9.6). The fact that the Central European *Sylvia* species occupying the same habitat as Marsh Warblers all show strong Cuckoo egg rejection behaviour fits this hypothesis well. Given their similar habitat requirements, we might expect that the situation would be the same in Blyth's Reed Warblers.

Phenotypic plasticity

In populations with low parasitism, acceptor hosts will probably do best because they do not incur the cost of recognition errors (see above). Calculation has shown that egg rejection in Eurasian Reed Warblers only pays off above a parasitism threshold of 19% (Davies *et al.* 1996). In Britain, defence in these reed warblers increased with a rise in the rate of parasitism, only to fall again within a

few years when the rate declined. Moreover, the rate of egg rejection varied quickly depending on seasonal and local levels of parasite density (Brooke *et al.* 1998, Lindholm 1999). Reed warblers are able to distinguish Cuckoos from other nest predators and specifically adjust Cuckoo-mobbing to local parasitism risk. Mobbing by neighbours on adjacent territories enhances a pair's own nest defence. Obviously, social learning provides a mechanism by which hosts rapidly increase their defences against brood parasites. Both enemy-specific social transmission and individual experience, enable hosts to track fine-scale spatiotemporal variation in parasitism, and may even influence the coevolutionary trajectories and population dynamics of brood parasites and hosts. Rapid changes in egg rejection and other host defences represent the outcome of adaptive phenotypic flexibility, which makes good sense in a varying environment (Davies & Welbergen 2009, Welbergen & Davies 2009, Campobello & Sealy 2011).

Combination is the key

The nature and mode of host defences, their causes, intensity and the time periods involved, can vary enormously between the different acrocephalid species, depending on variation in parasitism frequency and environmental conditions. We now know that a *combination* of factors involving gene flow and related issues, habitat structure, and phenotypic plasticity are all essential elements in our understanding of this fascinating topic.

Photos 9.13 - 9.15 A further adaptation against brood parasitism is a low intraclutch variation in egg appearance, ensuring that the hosts can detect the alien egg more easily. The low *intra*clutch variation of individual Eurasian Reed Warbler females has the additional consequence of a higher *inter*clutch variation in egg appearance in the population as a whole, making egg mimicry even more difficult for the parasite (from left to right: high, medium, and low intraclutch variation in three different Eurasian Reed Warbler clutches; photo Bård Stokke).

Photos 9.16 and 9.17 In contrast to their highly developed egg rejection behaviour, hosts of the Common Cuckoo blindly accept the entirely different-looking young parasite. After hatching, the young Cuckoo evicts all eggs and nestlings from the nest, while the parent warbler does not even interrupt its brooding during the entire process (Marsh Warblers; photos P. H. Olsen, Karsten Gärtner).

At present the coevolutionary arms race between Common Cuckoos and their warbler hosts has no winner. However, we should not forget that in many populations only about 5% of reed warblers will ever be confronted with a young Cuckoo in their nest. Therefore selection is always acting more strongly on Cuckoos, because they will always need to find a host while hosts will only rarely have to face a 'Cuckoo in the nest'.

We know that local Cuckoo populations undergo cycles of increase, decline, and extinction (Lindholm 1999, Krüger et al. 2009). On the other hand, acrocephalid warblers can reach a surprising age of up to 12 years, and can thus return to breed at a particular site where they may encounter variable parasitism rates during their lifetime. In addition, Common Cuckoo numbers have declined severely in the past 30 years, especially in central and western Europe, one of the causes being a mismatch in egg-laying phenology between some hosts and their brood parasite because of recent climate changes (Brooke et al. 1998, Møller et al. 2011). However, the substantial increase in the populations of Eurasian Reed Warblers, also as a result of climate change (section 13.4), is currently favouring Cuckoos. Man-made climatic and environmental changes will further contribute to the unpredictability of the interactions between warblers and Cuckoos. Flexible behaviour by both hosts and parasite confers great advantages in their ever-changing world, though a great deal of further study will be necessary before the complex Cuckoo-host system is completely understood.

Summary

Given their high local breeding densities and the accessibility of their nests, reed warblers are the most frequent hosts of the Common Cuckoo over much of Eurasia and have thus facilitated ground-breaking studies on brood parasitism by ornithologists. Several species have their own Cuckoo gens, a host-specialized female line.

The average parasitism rate in Eurasian Reed Warblers is around 8%, though the level of parasitism can vary spatially and temporally, reaching 65% in some Great Reed Warbler populations in Hungary. Parasitized nests are closer to trees used by Cuckoos as vantage points than non-parasitized nests. Safer nest sites are preferentially taken by older birds; parasitism is therefore not random, but influenced by host experience. An effective front line of defence prior to egg deposition is aggressive mobbing, the observation of which by neighbours can enhance a pair's own nest defence. Social learning plays a decisive role here, and such phenotypic flexibility makes good sense in a varying environment.

The better the Cuckoo egg mimicry, the poorer the rejection ability of hosts. Specialized egg morphs produced by Cuckoos are correlated with large tracts of homogeneous habitat where one host species is predominant, but the egg morphs are more variable in habitats with greater variation. A young Cuckoo is fed at about the same rate as a brood of reed warbler nestlings, and its rapid, loud begging call does indeed sound like a whole brood of hungry chicks. Eurasian Reed Warbler nestlings can leave the nest as early as 9-12 days old, but young Cuckoos only fledge after 16-22 days. This can therefore be a critical time for Cuckoo survival, with around 15% of reed warblers not feeding the nestling parasite for long enough and so they often starve – a phenomenon known as 'chick discrimination without recognition'.

Different acrocephalids react differently to brood parasitism, depending on factors such as their body weight and bill size. While the larger species can easily eject the alien egg, the smaller ones reject it by puncturing or deserting it. Despite the evolution of anti-parasite measures, significant numbers of Cuckoo eggs are still accepted. For instance, the parasitism frequency in a Hungarian Great Reed Warbler population has been stable at a high rate for decades, possibly because the evolution of an effective host defence has been slowed down by the immigration of naïve breeders and the occurrence of multiple parasitism. On the other hand, in Eurasian Reed Warbler populations the rate of egg rejection can change rapidly, depending on seasonal and local levels of parasite density.

Photos 9.18 and 9.19 The Cuckoo chick remains in the host nest for up to three weeks. Some Eurasian Reed Warblers are able to discriminate against parasitic nestlings in that they cease feeding the young Cuckoo after 9-13 days, the time when their own young normally fledge. Duration and amount of care are obviously used as cues. This phenomenon is called 'chick discrimination without recognition' (Black-browed Reed Warbler in Japan (below) and Eurasian Reed Warbler in the Czech Republic feeding a Cuckoo; photos Toshiyuki Yoshino, Oldrich Mikulica).

10

A battle of the sexes

Mate-guarding is very pronounced in the socially monogamous Eurasian Reed
Warblers. After pair formation, the male follows the female's every move.

Today it is difficult to believe that it has taken so long to discover what we now know about the sexual behaviour of birds. The widespread prudishness of an earlier age delayed the study of avian mating systems for over a century (Birkhead 2008). An analytical viewpoint was automatically distorted by this prudery so that obvious interpretations did not even enter people's minds.

We ourselves have also been made only too aware of such effects. When we studied Aquatic Warblers in the Hortobágy puszta in Hungary for the first time in June 1973, in the lowest shelf of a mist net we found an individual male that had been killed by a weasel. When the bird was dissected we were struck by the enormous size of his testes, which almost filled the entire abdominal cavity, and by the large, extensively coiled seminal glomera in the cloacal protuberance. Such peculiarities were unknown to us from other acrocephalids, or even from other passerines. We subsequently combed the literature for references to this phenomenon, but found only remarks that such a male protuberance is useful for the storage of sperm outside the body cavity and for sex determination (by humans) in the breeding season; we could find nothing on the possible reasons for a variation between species (Mason 1938, Wolfson 1954). It was not until 20 years later that the pieces of the jigsaw began to fall into place, thanks to advances in the observation of behaviour, DNA fingerprinting, and laparoscopic measurement of testes size in various reed warblers. These oversized organs of sperm production and storage are naturally connected to high levels of multiple mating in both sexes and sperm competition (Birkhead et al. 1993), which in the acrocephalids are most pronounced in Aquatic Warblers (Schulze-Hagen et al. 1995). Yet even here time has not stood still; today the tendency to promiscuity in female birds is regarded as a mating behaviour trait among individuals ('correlated infidelity'), whose heritability and genetic architecture is currently the subject of research in large-scale study series (Forstmeier et al. 2011).

Studies of acrocephalid mating systems date back to the middle of the last century, when Kluyver (1955) in Europe and Kikuchi et al. (1957) in Japan discovered that Great Reed Warbler and Oriental Reed Warbler males are polygynous. The majority of the research projects which followed developed in two directions that positively complemented each other: detailed single-species studies, which also took individual mating strategies into account, and comparative studies involving several species. The groundwork for these was laid in the detailed species accounts given in the handbooks of Glutz & Bauer (1991) and Cramp (1992). The starting-point and stimulus for such comparative studies can be found in the pioneering work of Crook (1964) on the relationship between social systems and resources in weaverbirds, as well as the discovery of the relationship between polygyny and

Photo 10.1 Many resident acrocephalids in the tropics and southern hemisphere have a strong pair bond and are socially monogamous, such as these Lesser Swamp Warblers in South Africa. Their habitats are not subject to substantial change between the seasons and some species hold year-round territories, defended by both partners (photo Rob Wicnand).

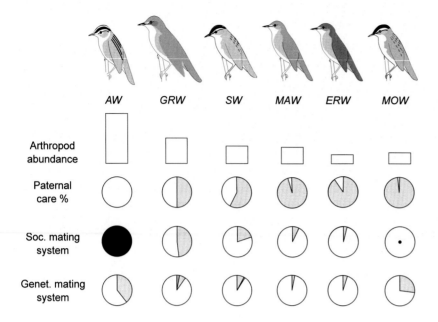

Fig. 10.1 Relationship between food supply, paternal care, social mating system, and genetic mating system in six European reed warblers. Food supply is represented as number of arthropods in the territories of the species (data from Schulze-Hagen & Flinks 1989, Schulze-Hagen et al. 1989, Hoi et al. 1995). Parental care (incubation and feeding of young): grey = percentage paternal share. Social mating system: white = percentage monogamy, small black filled circle = cooperative breeding, grey = percentage polygyny, black = no pair bond, 'promiscuity'. Genetic mating system: grey = percentage extra-pair young. Social mating system and paternal care are from data summarized in Leisler et al. 2002, genetic mating system from data in Hasselquist et al. 1996, Langefors et al. 1998, Buchanan & Catchpole 2000, Leisler & Wink 2000, Dyrcz et al. 2002, Davies et al. 2003, and Blomqvist et al. 2005.

marsh-grassland habitats in North American passerines by Verner & Willson (1966). In 1977, in their work on the significance of the quantity and distribution of the most important resources for breeding, Stephen Emlen and Lewis Oring (1977) finally came up with an ecological framework for understanding and predicting animal mating systems. International workshops at Neusiedler See in 1990 and in Osaka in 1994 (see Urano et al. 1995) brought together reed warbler researchers from various countries, fostering teamwork in the study of breeding systems, by which we mean parental investment and mating systems, which substantially determine both the form of the pair bond and of the overt social system of a species.

The preceding two chapters dealt principally with factors influencing or resulting from breeding success. Chapter 10 is concerned with mating systems, and specifically with the various forms of cooperation between partners when each wants to pass on their own genes but often has conflicting interests with the other. We shall look first at

Photo 10.2 In the temperate regions, with their high seasonality, many wetland habitats have an abundant food supply. This, as well as differing territory quality, are the preconditions for the development of social polygyny, which is common among Oriental Reed Warblers (photo Irina Marova).

the participation of the sexes in caring for the brood (parental investment) in acrocephalids and try to answer the questions that arise as to whether this has coevolved with the social system and how both might relate to habitat quality. Thereafter we shall consider aspects of mate choice, both of a social partner and of an extra-pair partner, and finally the question of what might be the group's ancestral breeding system.

Breeding systems and resources

The case of six marsh-nesting species.....

Acrocephalid breeding systems show more variation than in almost any other passerine group. Variability and diversity in these systems are grounded in ecology, as a comparison of just six central European *Acrocephalus* species has shown (Leisler 1985), in which breeding systems are similar to those of other songbirds that feed their altricial young with small arthropods. The majority of the six species are socially monogamous, that is to say that one male and one female are paired with each other during a breeding season and the care of the offspring is divided more or less equally between both parents (fig. 10.1). Social monogamy and care by both parent birds are found above all in those species that live in relatively food-poor habitats. Their typical prey species are mostly small and carried to the nest by both parents in bundles (multiple prey loading; section 4.3). In species that inhabit more productive habitats and are able to capture larger prey items, there is an increase in the proportion of polygynous males (i.e. those paired with more than one female). In polygynous situations, males only participate in caring for the offspring of their primary female, while the secondary female, and any further females (up to four can be paired with one male Great Reed Warbler) must raise their brood without male help. Thus the number of females that provision the young unassisted increases while the level of paternal effort declines, so that ultimately, in the most productive habitats, paternal care is less valuable and the bond between the sexes can finally dissolve altogether, as happens in Aquatic Warblers (Leisler & Catchpole 1992, Hoi *et al.* 1995).

Fig. 10.1 depicts the general pattern of the relationship between social mating systems and parental investment, as well as the interrelationship between them and prey

Photos 10.3 and 10.4 Both parental care and social mating systems are determined by the food supply in the habitat. Blyth's Reed Warbler is an example of a socially monogamous species living in a comparatively food-poor habitat. Both parent birds bring mostly small prey to the nest (near Moscow, Russia; photos Sergei Eliseev).

biomass/habitat richness. In the socially monogamous Moustached Warblers, females are unable to raise their offspring without male support (Leisler 1991a, Kleindorfer *et al.* 1995). However, in addition to social monogamy, it has been discovered that there are intruding male Moustached Warblers which help with incubation as well as provisioning and defence of the young (a form of cooperative breeding, Fessl *et al.* 1996; see also section 10.4). Another species in which social monogamy and biparental care is the rule can be found in Eurasian Reed Warblers. Although many European populations have been studied intensively, opportunistic polygyny has only been reported very occasionally (see Leisler & Catchpole 1992). In these two species the paternal share of incubation and feeding the young almost equals that of the females. There is a high need for biparental care, and consequently an experimental widowing of Eurasian Reed Warbler females led to a reduced breeding success (Duckworth 1992). The sexes of both species co-ordinate their incubation sessions with each other, or vary them according to the prevailing environmental conditions (e.g. ambient temperature, Kleindorfer *et al.* 1995).

Marsh Warblers are also monogamous, although some males attempt polygyny by becoming polyterritorial. As in the other two congenerics, male Marsh Warblers incubate without a prominent incubation patch; however their paternal care is, on the whole, more variable than in the other two species (Hoi *et al.* 1995), which suggests that the rate of male dispensability may be a little higher. Sedge Warblers have been classified as a socially monogamous species, though polygyny has been found in more populations than in the previous three species, and mostly at a higher rate, with a range of 0-19% (Zajac & Solarz 2004). Male care in this species is essentially restricted to provisioning the young and the paternal contribution to incubation is only marginal (Borowiec & Lontkowski

Photos 10.5 and 10.6 Wet, low grasslands, the habitat of Aquatic Warblers, are particularly rich in food. Here females, which rear their brood alone (uniparental), bring larger prey items to the nestlings. This wealth of food allows pair bonds to be dispensed with and the evolution of promiscuity to take place (photos Franzika Tanneberger, Alexei Kozulin).

1988). Great Reed Warbler males do not incubate at all, which is also the case in other species of the great reed warbler superspecies complex. In different years and different areas between 10 and 38% of males have been regularly recorded as being polygynous (facultative polygyny; Dyrcz 1995, Bensch 1993, Hasselquist 1994). While it is true that males defend the brood of their secondary female, they feed only the young in the first-hatched brood, which are almost always the offspring of the primary female. In Aquatic Warblers there is no pair bond and females always raise the brood alone in a system of uniparental care, consisting of incubation, brooding, feeding, and nest defence, which was only discovered by Günter Heise in 1970. Until then the false assertion that both sexes participated in brood care was stubbornly repeated in the literature (Niethammer 1937, Schulze-Hagen & Leisler 2012, *in press*).

Both parental care and social mating systems are significantly correlated with the food supply in the corresponding habitat of the six species (fig. 10.1). Species gradually move away from social monogamy and biparental care as the richness of the habitat increases. In species whose females frequently care for their nestlings unsupported, or even uniparentally as in Aquatic Warblers, the brood can be supplied with larger prey items than in species in which biparental care is the rule. Moreover, the two species with 'emancipated' females (Great Reed Warblers, and Aquatic Warblers) have stronger bills, possibly as an adaptation for catching larger prey (Leisler & Catchpole 1992). Fig. 10.2 shows the relationship between bill size and variation in the mating system: species with relatively deep bills breed in habitats with high arthropod abundance and show a low level of male contribution to offspring care. Even when allowing for phylogenetic effects, this correlation remains strong (Forstmeier & Leisler 2004).

......and the case of other acrocephalids

Today we have an abundance of breeding biology information at our disposal and we are now in a position to correlate habitat quality, mating system, and paternal care for a further 11 species from the total acrocephalid breeding range, including representatives of all clades such as *Iduna* yellow warblers and *Hippolais* tree warblers. Leisler and colleagues (2002) reconfirmed for all 17 species that – independent of habitat demands – a significant association between mating system and food supply/

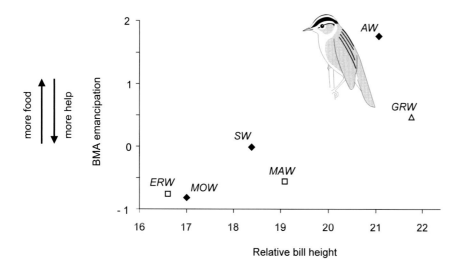

Fig. 10.2 Relationship between variation in emancipation from biparental care and bill morphology in six European reed warblers. BMA (bivariate major axis) is a combination of food supply data and paternal help; relative bill depth is expressed as percentage of bill length. Species with relatively stronger bills breed in habitats with a rich food supply and show a low male contribution to parental care. Different symbols denote the phylogenetic relationships (after Forstmeier & Leisler 2004, poster).

habitat type exists, with monogamy occurring in poor habitats and promiscuity in good habitats.

The great majority of the entire group breed in poor environments, and in such areas males take a greater share in raising the young than in more productive habitats. In each of the major clades (i.e. large, small unicoloured, and striped reed warblers) there are species that breed in more productive habitats and which tend to be polygynous: for example, Aquatic Warblers, Great Reed Warblers, and Sedge Warblers in central Europe; Oriental Reed Warblers and Black-browed Reed Warblers in East Asia. In four species which inhabit poor habitats, cooperative breeding has also been described in addition to social monogamy (Leisler *et al.* 2002; section 10.4).

The strength of the pair bond between male and female in the course of a breeding attempt is determined not only by the indispensability of paternal assistance in rearing the young, but also by the male's role in territory defence or anti-predator measures (e.g. Hoi & Winkler 1988). In the insular reed warblers, an interesting trend can be discerned towards longer-lasting pairbonds and a more intensive participation by the male in caring for the brood (chapter 12).

2

Social versus genetic mating systems

For a better understanding of reed warbler social and genetic mating systems we must first look at some of the basic concepts. By inferring mating systems from social behaviour, Lack (1968) found that more than 90% of all passerine subfamilies were monogamous, that is, they form some kind of pair bond during breeding, and he proposed the necessity for biparental care as an explanation. Slightly later, Trivers (1972) argued that males and females should pursue differing reproductive strategies to maximize their own individual reproductive fitness, often at the expense of their partner (sexual conflict). He made two important general predictions about the mating behaviour of each sex: by producing small but very many gametes, males would behave in a way that increases their opportunity for additional matings (i.e. go for quantity), while females, which invest far more energy into each offspring by laying large but few eggs, would try to increase the genetic fitness of their offspring (i.e. go for quality). As a result, Trivers also predicted widespread infidelity across species.

At the time this idea was not pursued because it was widely considered that in monogamous pairings members of a given pair mated only with their nesting partner, males would have few opportunities for additional copulations, and female behaviour seemed to be directed more towards a mate that would help raise the young than towards mating with a top-quality sire. New behavioural observations and the application of molecular genetic techniques for proof of paternity over the last 25 years finally demonstrated that the young of even socially monogamous species are often not the offspring of the social male, and the whole concept of extra-pair paternity (EPP) came to prominence (Birkhead 1987).

We now know that true genetic monogamy occurs in only about 14% of passerine species and that in the remaining 86% of the species studied at least some of the females lay clutches with mixed parentage. In socially monogamous species an average of 19% of broods contain young that are not the offspring of the male of the pair. Thus a considerable proportion of female birds have been revealed to be sexually polyandrous, having multiple male mates (Griffith *et al.* 2002). The new molecular tools have totally revolutionized our view of avian mating systems and virtually turned Lack's early, 'idealistic' picture of monogamy on its head, making it necessary to distinguish between 'social' and 'genetic' mating systems (sections 10.3 - 10.6).

The widespread 'infidelity' of female birds means that sperm from different males often compete to fertilize the eggs in their reproductive tracts. Sperm competition has become a central element in sexual selection theory (post-mating sexual selection; Birkhead & Møller 1992) and has often resulted in reproductive success having to be measured anew (Bensch 1996, Komdeur & Richardson 2007). The significance of paternity guards has also been re-evaluated in recent times; this is the concept of mate-guarding, where males shield their females during the fertile period from competitors and frequent copulations. (Birkhead & Møller 1992).

The wide distribution of this behaviour has also led to a new interpretation of territoriality in male birds (Møller 1987; chapter 6). It was always fairly clear that some apparently socially monogamous males could father more offspring than others by extra-pair copulations (EPCs), which increased their reproductive success, and for a long time research into EPP was carried out only from a male perspective. It only became clear much later that females can also play an active role in initiating promiscuity, an insight which came as a result of much diligent research by two women (Smith 1988, Gowaty 1996).

Why should females be promiscuous?

Females usually choose their social partners and extra-pair partners according to different criteria. Most desirable as a social partner is a good parent that can directly contribute to the reproductive success of the female. However, genetically inferior males may announce that they will be good parents by singing their 'special' song and manipulate females into mating with them by helping ('helpfully coercive males'; Gowaty 1996). In addition, the first choice of a social partner often takes place under pressure of time, especially in migratory songbirds, and other circumstances can also limit the range of available mates (e.g. Sullivan 1994, Janetos 1980).

Although our knowledge of the ecological and social factors influencing the extent of extra-pair young (EPY) across species remains poor (Neudorf 2004), several meta-analyses have identified differences in male parental care as a robust predictor of variation in extra-pair mating behaviour in many species – EPP frequency declines with increasing need for paternal care. (Møller 2000, Arnold & Owens 2002, Griffith *et al.* 2002, Westneat & Stewart 2003, Brommer *et al.* 2010). In extreme situations females might rarely get the chance to be promiscuous, if male help is essential and males retaliate by reducing it. However, females of higher quality, or in better environmental conditions, can bear any possible costs involved in EPCs better than females of lower quality (Gowaty 1996, 1999). At the same time, males can also be restricted in their pursuit of EPCs by temporal or physiological (hormonal) factors arising from their paternal commitments.

A further recently discovered trend which we must take into account when studying our Acrocephalidae family is that species breeding in closed habitats with reduced visibility have been shown to have significantly higher EPP rates than those breeding in more open habitats (Blomqvist *et al.* 2006).

The proportions of mixed paternity so far established in our six wetland *Acrocephalus* species with varying paternal investment are shown in the lowest row of fig. 10.1 as 'genetic mating system'. We will discuss this in more detail against the background of the corresponding social mating system, monogamy, polygyny, and so on in sections 10.3 - 10.6.

Whether females gain selective advantages by mating with multiple males (polyandry), and why there is such marked variation in EPP both within and between species, is still much studied and controversially debated (Arnqvist & Kirkpatrick 2005, Akcay & Roughgarden 2007, Griffith 2007, Brommer *et al.* 2010). Adaptive explanations for female polyandry include direct benefits (e.g. when the EPC partner is a neighbouring male that permits the female to forage for food in his territory or to obtain other additional resources; Gray 1997) and indirect benefits (e.g. obtaining genes that increase offspring fitness – viability or sexual attractiveness (see also section 7.4). Extra pair mating behaviour may take place even if females obtain no obvious benefit from it at all. Recent work on Zebra Finches has shown that positive selection on males to sire EPY may also lead to increased 'promiscuous' behaviour in the females that are related to them as a correlated evolutionary response. In other words 'Casanova-fathers' will produce 'philandering daughters' simply because the promiscuous behaviour is heritable and positively genetically correlated (Forstmeier *et al.* 2011).

In several species, a higher genetic diversity (heterozygosity) was found in the EP-offspring than in the maternal half-siblings (Kempenaers 2007, Stapleton *et al.* 2007). As a likely link between mate choice and genetic inheritance of viability in offspring, the major histocompatibility complex (MHC) has come under increased scrutiny. This highly variable set of functional genes plays a role in triggering adaptive immune responses. For instance, Westerdahl *et al.* (2005) showed that Great Reed Warbler individuals with a higher MHC diversity, or certain MHC alleles, were better protected against malaria infections than birds with a simpler MHC, while in Seychelles Warblers, juveniles with 'good' MHC genes survived much better than those with a lower diversity or without a specific gene variant (Brouwer *et al.* 2010; section 10.4).

However, recent findings indicate that in some circumstances a better performance by EPY might also be accounted for by non-genetic effects, in that the EPY are placed earlier in the laying order ('mothers give interloper's offspring a head start in life'; Magrath *et al.* 2009).

3

Monogamy

It is important for females of socially monogamous species to be paired with a male that is prepared to provide a high level of parental effort whilst at the same time possessing good genetic quality. We know that EPP can be the norm even in socially monogamous passerines, so an explanation is required as to why a socially monogamous species can also, to a large extent, be genetically monogamous, that is, have a very low EPP rate of less than *ca* 5%. So far we do not have paternity analyses for most of the many socially monogamous acrocephalids, in particular for those which do not live in wetlands, such as the *Hippolais* and *Iduna* species.

In Eurasian Reed Warblers and Marsh Warblers, the rate of extra-pair offspring is low at 6% and 3% respectively (Leisler & Wink 2000, Davies *et al.* 2003; fig. 10.1). The poor habitats of both species seem to have trapped them into biparental care, and intense mate-guarding by the males apparently allows few opportunities for infidelity, despite the dense vegetation. We shall consider the third monogamous wetland species, Moustached Warblers, in more detail in the section on cooperative breeding (section 10.4).

In contrast to Eurasian Reed Warblers and Marsh Warblers, the cues for mate choice in Sedge Warblers are very well studied. While Sedge Warblers are primarily socially monogamous, a varying proportion of male birds are polygynous (section 10.5). Females apparently mate socially with phenotypically attractive and healthy males to gain direct benefits, namely a high offspring provisioning rate (Buchanan & Catchpole 1997). Among several male and territory characteristics, only territory size, song repertoire size, and song-flight activity had a significant influence on the pairing date, which is regarded as a measure of female choice (Buchanan & Catchpole 1997). If genetic characteristics are taken into account, as well as individual blood parasite burdens, then it can be seen that song repertoire size indicates not only a male's individual genetic diversity or quality, but evidently also his potential parental effort (section 6.1). Non-infected males had a larger song repertoire and their provisioning rate as a measure of parental effort was higher than in parasitized males (Marshall *et al.* 2003; section 7.4). By contrast, territory size and song-flight activity – a measure of current health and energy rather than long-term fitness – were not correlated with parental effort (Buchanan *et al.*

Photo 10.7 Cooperative breeding evolved in a small number of sedentary acrocephalid species, mainly living on islands. In this social system, individuals help care for young that are not their own (though sometimes related), at the expense of their own reproduction. The facultative cooperative breeding system of Seychelles Warblers has become one of the best-studied helper systems. In this species a maximum of up to seven birds can live together in a territory; here the breeding pair, two helpers, and two already independent young (photo David Richardson).

1999, Buchanan & Catchpole 2000a). Thus female Sedge Warblers select males using multiple cues that reflect different aspects of the qualities of both the males themselves and their territories (section 7.4).

The EPP rate in Sedge Warblers was studied in two different areas, Sweden and Britain, and was found to be similar at *ca* 8% of offspring in both (fig. 10.1; Hasselquist & Langefors 1998, Buchanan & Catchpole 2000b). In Sweden, males that sired EPY tended to arrive earlier and to have a higher social pairing success than other males (cuckolded, genetically monogamous, and unpaired). In addition, they were close neighbours of the cuckolded males. Thus Sedge Warbler females appear to choose attractive and/or superior male neighbours as EPC partners. In contrast, and somewhat surprisingly, the British

females engaged in EPCs with males that had a smaller song repertoire and smaller territories than their social mates (Marshall *et al.* 2007), thus indicating a significant negative relationship between traits of social and genetic fathers (section 7.4).

4 Choosing to breed cooperatively

In a small number of species, sexually mature young birds forgo the opportunity to breed themselves, either temporarily or for life, and instead become part of a reproductive community by assisting conspecifics with their brood care. Various, not necessarily mutually exclusive, hypotheses have been advanced to explain such cooperative brood care systems (Hatchwell & Komdeur

2000). For example, a shortage of territories or partners can deter young birds from breeding ('ecological constraints'), or it can be more advantageous for them to remain in a good natal territory and help the breeding pair raise future siblings ('benefits of philopatry'). Helpers are frequently related to the nestlings via their parents (primary helpers) but this need not necessarily be the case (secondary helpers). Although the existence of helpers has been known since the 1930s, the paradox in evolutionary biology of apparently altruistic helping was only satisfactorily resolved by the introduction of the theory of genetic evolution of social behaviour in 1964 (Hamilton 1964). According to the concept of inclusive fitness, the lifetime reproduction of an individual is the sum of its own reproductive success (direct fitness) and the genetic benefit gained from supporting kin, or relatives (indirect fitness). Indirect fitness consists of the reproductive success of all assisted individuals weighted according to their degree of relatedness. Hence two forms of selection are at work here: direct selection acting on variation in individual reproductive success, and indirect selection via the influence that an individual has on the reproductive success of its relatives.

The case of Seychelles Warblers

It was a lucky coincidence that led to the cooperative brood care system of Seychelles Warblers becoming one of the most intensively studied helper systems of all: a programme aimed at saving the species combined with a carefully thought out scientific action plan. In 1968 the birds were confined to the 29-ha Cousin Island, having disappeared from all other islands of the archipelago. Once Cousin was designated a nature reserve and the original vegetation restored, the small population recovered and grew steadily. When the saturation level of the island had been reached, a number of surplus birds were translocated to two neighbouring islands, in 1988 to Aride and in 1990 to Cousine. These two populations soon also reached their habitats' carrying capacity, and in 2004 some of the warblers were taken to another island, Denis (Komdeur 2006). Since 1985 a working group led by the Dutch scientist Jan Komdeur has been responsible for the protection of the species, the documentation of its population development, and the study of its reproduction, in which every individual bird is identified. The translocation of birds from the saturated founder population to uncolonized islands supplied material for the elegant field experiments that provided evolutionary

biologists with sensational results. With the help of molecular techniques for genetically determining the parentage of young birds and the relationships between all individuals in the entire population, the complex social system of the species could be analysed more precisely, whereas this had previously only been partially successful using family trees (Komdeur & Richardson 2007).

Paired birds occupy and defend a territory for as long as nine years (Komdeur 1994a). As a rule the clutch consists of only one egg, though if females have access to plenty of food they can lay more. The young fledge at 18-20 days but remain dependent on their parents until they can feed themselves at around three months (section 8.3). The breeding success of the warblers is low – just under a quarter of the offspring survive their first year – but by contrast the life expectancy of adults is high (Komdeur 1991). Sexual maturity is reached in the first year, but some individuals remain in the parental territory and delay the time of their own reproduction, or even forgo it altogether. Whether cooperative breeding occurs depends crucially on territory quality, which in turn is determined by food supply. The way in which variation in territorial quality affects demographic characteristics is summarized in table 10.1 (Komdeur 1992, 2003a).

A territory can be occupied by birds with the following status: by the territory-holding pair and by female or male subordinates which are also sexually mature. Subordinates can be non-helpers, non-reproductive helpers, or reproductive helpers. Hence, under certain circumstances, it is not only the breeding pair in the group that can breed, but also low-status females. A maximum of up to seven birds can live together in a territory (the breeding pair, three helpers, and two young born there). Additionally there are birds that roam around, known as 'floaters'.

With the help of field experiments many questions were answered over the long duration of the study (Komdeur & Richardson 2007). How long would it take for the areas on Cousin that were vacated following the translocation of individuals to be reoccupied and which birds would take them over? What would the young birds born on the newly colonized islands do? Would they remain in their natal territories and help their parents, or would they move away to unoccupied parts of the island and establish new territories?

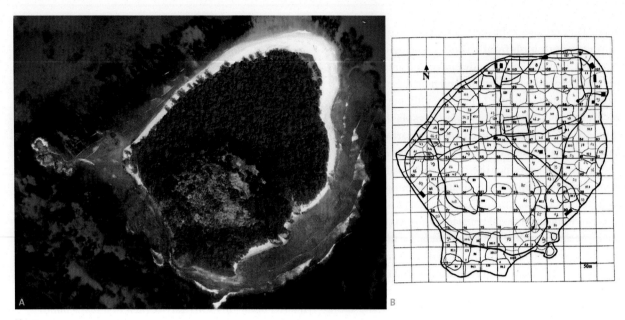

Fig. 10.12

A   Aerial view of Cousin Island, Seychelles. Since the 1980s every Seychelles Warbler territory has been known to the researchers studying them. Many territories have remained constant over several years (a, photo Google Earth). Suitable habitat for
the warblers on the island is saturated and the quality of the territories varies; both factors promote cooperative breeding.

B   The map shows the territories in 2008. High-quality territories are located in the upper centre, low-quality ones in the coastal areas, medium- quality in between (courtesy of J. van de Crommenacker).

The results were clear cut. The newly unoccupied territories on Cousin were immediately taken over, the good
areas by birds that were themselves hatched in good territories, the less good areas by birds from less good territories, and the poor ones almost exclusively by birds from
poor natal territories. Therefore, subordinates from good
or less good territories never moved into areas which
were worse; in other words, the decision to disperse is
influenced by the quality of both the natal territory and of
the newly vacated one.

During the initial growth phase of the small original population on Cousin, there was plenty of room for new territories. There were no helpers in the 1970s. Only when all
good territories were occupied did the young begin to
remain in their natal territories as helpers instead of
founding new territories in areas of lower quality. The offspring of the birds released on Aride and Cousine behaved
exactly as the founder population on Cousin before its
saturation in the 1970s. When there was no space for further territories they delayed their dispersal. Their decision
to stay is determined by habitat saturation, by the variation in habitat quality, and by the presence of parents (see

below). In order to answer the question of why some individuals help, it was at first necessary to determine who
benefits from the help given, and the answer was that both
helper and recipient gain an advantage.

Extra-pair paternity and the benefits of helping

Contrary to the situation in most avian helper systems, in
Seychelles Warblers it is the female offspring that are
more likely to remain in their natal territories to help
with brood care. Male young more commonly disperse
and only 25% of young males become subordinates in
non-natal territories. These subordinate birds can occasionally act as secondary (non-related) helpers in such
territories. If male individuals do assist they do so only to
a small extent and independent of their degree of relatedness to the nestlings (see below). Female subordinates,
on the other hand, preferentially support those nestlings
that are their closest relatives. By helping, subordinates
gain direct fitness benefits in different ways: first, by their
own genetic parenthood, that is, via paternity or laying an
egg in the nest of the breeding pair; secondly, by gaining
brood care experience that will improve their own breeding performance; thirdly, by gaining advantageous experi-

	Low quality territory	Medium quality territory	High quality territory
Number of yearlings produced per pair per annum			
Pairs without helpers	0.19	0.51	0.85
Pairs with one helper	0.22	0.85	1.62
Percentage survival per annum			
Up to one year	30%	67%	86%
Adult birds	76%	88%	91%
Percentage of 1-year-olds that stay with parents	29%	69%	93%

ences for when they occupy their own territory. Initial genetic studies revealed that 44-47% of low-status females successfully laid an egg in the nest of the dominant female (usually their own mother). This means that in territories with more than one female almost half of the offspring are born to female helpers, but since the majority of territories are held by breeding pairs without helpers, only a total of 15% of all offspring in all territories are produced by female helpers (Richardson *et al.* 2001). Furthermore, only 15% of subordinate males were able to ensure paternity, and this was never *outside* their own group (Richardson *et al.* 2002). Hence low-status individuals can increase their direct fitness benefits via parenthood only *within* their own breeding group.

In the translocations to new islands, birds with varying breeding experience were used – males and females with their own breeding experience, with experience as helpers, and also completely inexperienced birds. It was therefore possible to test whether previous activity as a helper improved an individual bird's breeding performance. This was indeed found to be the case; birds of both sexes with helper experience had better skills than inexperienced individuals. However for females, which build the nest and incubate the clutch alone, this advantage was considerably greater than for males.

In cooperative breeding species that live in habitats with a lack of available territories, helping can usually facilitate the later occupation of a territory. However, this is not the case for subordinate male Seychelles Warblers,

where only non-helpers, known as 'budders', have such an opportunity. The reconstruction by Komdeur & Edelaar (2001a,b) of the life histories of many subordinate males that initially had no chance themselves of breeding in high-quality territories led to the surprising result that the male offspring behaved either as helpers or as budders. Since helpers never became budders and budders never became helpers, there are therefore two alternative strategies. Before a budding, or splitting off, had even occurred, the future budders were more aggressive towards neighbours and helped more in territorial defence, whereby territories could increase in area. However, their annual breeding success was low at first and did not differ from the indirect increases in fitness of helpers in high-quality territories. Fig. 10.3 shows the fission and gradual enlargement of territories by two males that remained in their natal sites and that later succeeded in acquiring females and breeding. At an average age of 3.7 years, the long-lived budders had left these originally budded territories and filled high-quality breeding vacancies, either on adjacent territories (66.7%) or on their natal territory (33.3%). Young birds never managed to attain their own territories by evicting the dominant male, but only by expansion or inheritance. At the same time, vacant breeding territories were never inherited or taken over by helpers. Overall, the costs for such subordinate birds of helping to defend a territory appear to be less than those arising from brood care (alloparenting).

Before the indirect fitness benefits for any individual helper can be estimated, its degree of relatedness to the nestlings profiting from its help must be determined. When this was done by genetic analysis, the result came as another big surprise. Forty per cent of the young (of both high-status and low-status females) are not the off-spring of the dominant breeding male of their own group. These young are the result of copulations with males outside their own group which are occupiers of both good and poor-quality territories. Such an EPP rate is unusually high, both for a cooperatively breeding species and for island-dwellers (Griffith 2000). Given this, both the direct and indirect reproductive success of each subordinate was calculated over a three-year period and compared with breeding successes in territories of a similar quality without helpers. The low degree of relatedness between helpers and the young that were not their offspring (which of course resulted from the high extra-group paternity rate) was taken into account. The average degree of relatedness to the young was considerably lower in male than in female helpers. As fig. 10.4 illustrates, indirect fitness benefits (expressed as offspring) are substantially less than direct benefits for females and males. Direct fitness benefits – in female helpers six times greater than indirect – are therefore the driving force in the helper system of Seychelles Warblers and are responsible for the fact that more females than males remain in their natal territory (Richardson *et al.* 2002).

Despite the greater importance of direct fitness benefits in the cooperative system of this species, subordinate individuals should also maximize their indirect benefits by preferentially feeding their closest relatives. How can a subordinate determine its relationship to nestlings in the group if one partner in the high-status pair is replaced by an outsider and if the females still show a strong tendency to promiscuity? To answer this conundrum, Richardson *et al.* (2003) analysed the behaviour of female subordinates against the background of their relatedness, selecting only those females in the group that had not laid any eggs themselves. Female helpers that breed always provision nestlings. An exact computation of the degree of relatedness between attending subordinates and the new nestling generation showed that subordinate females actually only assisted in brood care if they were closely related to the young. It also showed that subordinates fed nestlings significantly more often in the presence of their mothers (through which they were unequivocally related to the nestlings) than when the dominant breeding female had changed. By contrast, the

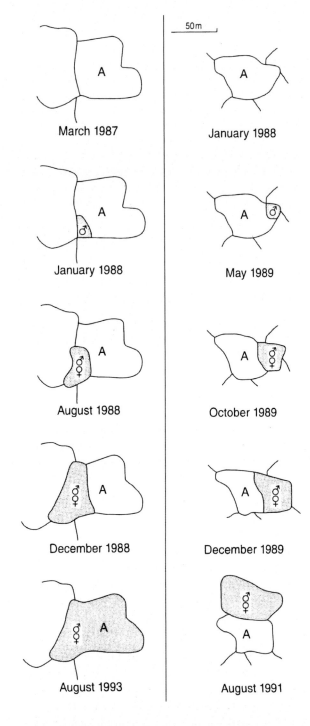

Fig. 10.3 The process of territory fission and inheritance as shown by two Seychelles Warbler males, hatched in March 1987 and January 1988. Shaded area = land gained by the son of the original pair that occupied territory A. The territory boundaries of the original breeding pair and of their son are shown over several years. After about one year the son budded off part of the natal territory; subsequently a female was acquired and the territory expanded. Finally the budding males inherited their natal territories through the deaths of their fathers (from Komdeur & Edelaar 2001b).

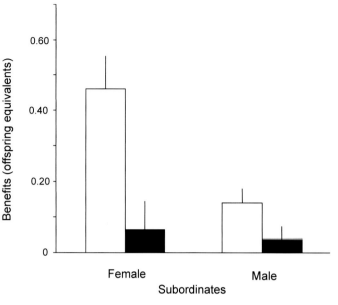

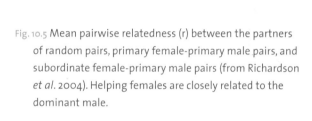

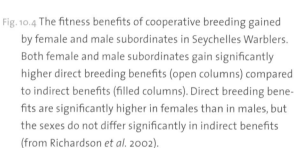

Fig. 10.4 The fitness benefits of cooperative breeding gained by female and male subordinates in Seychelles Warblers. Both female and male subordinates gain significantly higher direct breeding benefits (open columns) compared to indirect benefits (filled columns). Direct breeding benefits are significantly higher in females than in males, but the sexes do not differ significantly in indirect benefits (from Richardson *et al.* 2002).

Fig. 10.5 Mean pairwise relatedness (r) between the partners of random pairs, primary female-primary male pairs, and subordinate female-primary male pairs (from Richardson *et al.* 2004). Helping females are closely related to the dominant male.

presence or absence of the putative father did not influence the feeding frequency of the helpers. This means that subordinates define their feeding effort, not via a direct assessment of their relationship to the nestlings, but indirectly via the continual presence of their mother, the only sex to which they are reliably related.

In Seychelles Warblers, mating between closely related individuals is not completely avoided (5% incestuous matings; Richardson *et al.* 2004). Inbreeding manifests itself in this species by reduced heterozygosity of the young, meaning reduced genetic diversity in the offspring. This has no direct effect on the survival chances of the young bird itself, though it does have a disadvantageous influence on subsequent generations in the female line (section 8.5). On average, secondary females are very closely related to the dominant male, and are often his daughters (fig. 10.5). They reduce the proportion of incestuous matings by being more promiscuous than the dominant females, hence producing fewer inbred young – their extra-pair young are more heterozygous than within-pair young.

On the other hand dominant females prefer to be fertilized by extra-pair males more often when the diversity of disease-detecting genes (of the MHC complex) of their social mates is lower than average (Richardson *et al.* 2005). The researchers have now monitored the fate of the genotyped juveniles for 10 years and have found a positive association between MHC traits and juvenile survival. By not being faithful to a pair-male with low genetic quality, females ensure that their offspring do not end up with below average levels of MHC diversity and therefore lower survival chances (Brouwer *et al.* 2010). Such strategies are especially important in small bottlenecked populations with reduced genetic diversity (sections 8.5 and 13.6).

Benefits for dominant birds and offspring sex ratio

What benefit does the dominant breeding female really receive from helpers? Theoretically she should only share her reproduction effort with another female, therefore accept female helpers, if her total fitness is thereby greater than without subordinates. Richardson and colleagues

(2002) were able to show that breeding success in a territory with a helper increased significantly, by 0.18 additional offspring per helper; the offspring of the subordinates themselves were not included in the calculation. In addition, the dominant pair gains indirect reproductive benefits through the reproductive success of the low-status birds to which they are related. However, too many female helpers will directly reduce the breeding success of a dominant pair by food competition as well as by egg breakage during simultaneous incubation. Furthermore, the indirect reproductive benefits to the primary female via her related helpers decrease when competition among subordinate females for local breeding opportunities becomes more intense. Since helpers bring advantages for the dominant pair only in good territories – in poor ones they are a disadvantage as food competitors – then in those good territories females ought to be preferentially tolerated, or even produced, while in poor territories sons that disperse ought to be tolerated. In fact this does appear to be the case; in good territories pairs without helpers produced 87% daughters, in poor territories 77% sons (Komdeur *et al.* 1997).

How can offspring sex ratio be manipulated to suit ecological conditions? This question is dealt with in sex-allocation theory, according to which the costs entailed in reproduction and in raising young should be in a favourable ratio to the potentially attainable fitness benefits to the parents. The participation of their own offspring in future reproduction scenarios can be optimized by the skewing of the primary sex ratio. This is to be expected when the fitness benefits attached to rearing sons or daughters in a population vary in a predictable fashion, as is the case in Seychelles Warblers. If habitat resources decline in quality the birds should invest in the sex that disperses from the parental territory, in other words, in sons.

By simple experiments (e.g. removing helpers by capture, or translocation of breeding pairs from poor to good territories) Komdeur and colleagues (1997) were able to demonstrate that Seychelles Warblers are capable of 'choosing' the sex of their young, depending on environmental conditions. Pairs from suboptimal territories which had previously produced sons immediately switched to the production of daughters following translocation to good territories. Pairs in high-quality territories with two or more helpers which had produced mainly sons began to produce daughters 'in order to' top up the

helpers as soon as one had been experimentally removed. In birds, females are the heterogametic sex, meaning that they have different sex chromosomes (zw), which allow them to determine, in different ways, the primary sex of their offspring at various stages in the reproductive cycle. During the translocation process of bringing the warblers from Cousin to other islands, Komdeur and his group were able to answer another much discussed question. When do females determine the sex of an egg, before or after ovulation? Birds that had been moved to Aride, the island with excellent environmental resources – the food supply here is three times richer than on Cousin – laid clutches with two eggs. In addition, the females adjusted the sex ratio to the particularly good quality of the habitat by producing in total 86.6% daughters. The decisive point came when they had determined not only the sex of the first egg to favour daughters, but also, and to an even greater degree, that of the second, which was laid 24 hours after the first, as is the norm in passerines. Other mechanisms of gender determination after ovulation (e.g. absorption of zygotes of the unwanted sex) could be excluded, since this would have required an extension of the time between eggs to more than 24 hours. Sex determination in Seychelles Warblers is therefore pre-ovulatory and occurs just before meiosis, and certainly at least before the point in time at which the egg leaves the follicle (Komdeur *et al.* 2002). Recent experiments with other species have confirmed that female birds can selectively absorb ovarian follicles of the 'unwanted' sex and that the stress hormone corticosterone plays a role in this process (Pike 2005, Pike & Petrie 2006).

Formation of families and grandparent helpers

Analyses of datasets for Seychelles Warblers extending over 25 years provided further important and far-reaching results. Eikenaar *et al.* (2007) found that the replacement of a parent plays an important role in the decision of a subordinate to disperse. By remaining with their parents in a territory, subordinates gain fitness advantages as helpers or through nepotistic behaviour on the part of the parents. However, these can be reduced or cease altogether if a parent is replaced by an unrelated dominant bird. It was indeed found that subordinates were more likely to disperse when a parent was naturally or experimentally removed – and subsequently replaced by a step-parent – than when both parents remained on the natal territory. This supports the notion that prolonged parental care promotes the formation of family groups.

Richardson *et al.* (2007) also documented that in the course of this long study 14% of dominant females were ousted from their position and replaced by related females (a daughter, sister, or niece), even when they were not too old to continue breeding successfully. On the other hand, if the dominant female died, then the new female would be more distantly related to her predecessor. Although some of the deposed females left the group, the majority (68%) remained in the territory and became subordinates, caring for the young of the next generation, usually those of their own daughters. An average of around 10% of all subordinates in any one year were such deposed females, which remained only in good territories, whereas the females from poor territories abandoned them and became floaters. The demoted females were on average related to the offspring as to a grandchild (coefficient of relatedness 0.24) and so by their 'grandmotherly' care were able to gain indirect fitness benefits. Indirect benefits seem to drive the evolution of this 'grandparent helping'; such behaviour was unknown in birds until recently and had only previously been described in a few mammal species (Richardson *et al.* 2007). Despite strong competition for dominant positions, males were deposed much less often than females.

Other acrocephalids

There are indications that Rodrigues Warblers, as well as Rimatara and Pitcairn Island Warblers also occasionally breed in groups (del Hoyo *et al.* 2006, Thibault & Cibois 2006). In one study, more than one-third of Henderson Island Warblers bred cooperatively in trios (Brooke & Hartley 1995). These groups could be either polyandrous (i.e. one female breeds together with a dominant male and a younger male helper), or, more rarely, polygynous, with one male and two females laying in the same nest. However, contrary to the situation in Seychelles Warblers, the members of a group were never related to each other. All members of a trio helped to raise the young, even when they were not their own, and mostly after participation in the incubation as well. In this system there are no indirect reproductive fitness advantages. All young in a group were offspring of members of the trio and therefore there was no extra-group paternity; in polyandrous trios, they were offspring of only one of the males, though it is interesting to note that they had the subordinate male as their father more often than the dominant male. By contrast, in birds breeding in pairs, both extra-pair fertilization (EPF) and egg-dumping were recorded.

Although the breeding success of trios was slightly higher than that of pairs (1.43 fledged young *vs* 1.25), the success achieved per adult bird was on average slightly lower than for individuals in pairs.

One advantage of breeding in trios is that the burden of egg and brood care is reduced for any one individual, from which young inexperienced birds in particular profit. Females who are part of a pair must make a greater effort in incubation and provisioning than those who are in a trio. Unlike Seychelles Warblers, the option of remaining in a very good natal territory instead of dispersing appears not to exist for young Henderson Island Warblers, since Henderson Island has a stable climax vegetation and habitat quality hardly varies, at least not as much as on Cousin. It is likely that trios, in the form of coalitions of younger birds, could be helpful in securing a territory. They also represent a 'transition stage' in which the first breeding experience can be gained (Brooke & Hartley 1995).

The helpers are also unrelated to the breeding pair in two other partially cooperative breeders, Moustached Warblers and African Reed Warblers. The latter breed in large isolated areas of reed in Namibia and are almost wholly socially monogamous, but in about 12% of breeding attempts examined, Eising *et al.* (2001) found trios, the breeding pair and one helper that was almost always male (i.e. polyandrous trios). The 'island' habitat was almost saturated so helpers appeared only in higher-quality territories. Helping only increased hatching success, not fledging success, and not through better incubation but through reduced clutch losses. The benefit therefore seems to have been gained from improved detection of predators and from mobbing. The low incidence of helping in this species suggests that birds have some opportunity to disperse within a metapopulation in search of vacant territories.

In a small study, Blomqvist *et al.* (2005) discovered an unusual mating system in Moustached Warblers. In the very dense vegetation inhabited by the species, they recorded secondary males that behaved very secretively and provided care at 17 out of 25 nests. The additional care provided by these extra males is probably responsible for the positive results in nest success, while at the same time preparing their way for a later mating with the assisted female in a further breeding attempt. An analysis of paternity showed that 27% of the young were 'extra-pair' (fig. 10.1), though sired by males that fell into two discrete categories. Two-thirds of the EPY had been sired by the silent and helping males, and the remaining third by close territory neighbours. In the former case we can perhaps presume that females solicit EPCs for direct benefits and in the latter for indirect benefits.

Polygyny

The polygyny threshold model

The territories settled by males of the migratory acro-cephalids on their arrival in spring differ substantially in qualities such as food resources, nest sites, or security against predators, especially in the marsh-nesting species (chapter 6). In general, females arrive on the breeding site after males, so the earliest females can pair with the males that have occupied the best territories (chapter 6). However, a later-arriving female faces a dilemma: she can either mate with one of the remaining unpaired males and raise young with him in a poor-quality unproductive territory, or she can forgo the help of a male during brood care altogether and settle – more or less as a 'single mother' – in an optimal habitat, where she can become the secondary female of a polygynous male. If the sex ratio is equal, then some males will remain unpaired (Emlen & Oring 1977). The polygyny threshold model (Orians 1969) describes such a situation (fig. 10.6), in which it is more advantageous for a female to accept the costs arising from the absence of male support as a secondary female, since these are balanced by the benefits of the higher quality of

territories belonging to already paired males. The polygyny threshold is the environmental situation in which both strategies result in equal fitness (distance 2, horizontal axis in fig. 10.6). Of course, it will be the primary females of polygynous males in top-quality territories who will come off best.

The conditions under which polygynous mating arises, including when the sex ratio is balanced, have been most thoroughly studied in Great and Oriental Reed Warblers. Females appear actively to choose their pair-mates, and those females that join up with already paired males have, presumably, previously visited the territories of unmated males. In a field experiment with females that had been tagged with radio transmitters, Bensch & Hasselquist (1992) found that they had inspected the territories of six males on average (from 3 to 12) before deciding on a breeding partner, and this supports an assumption of the polygyny threshold model.

What's in it for females?

Females that enter a polygynous relationship with already paired males receive little or no assistance from that male in provisioning the young, and this can often result in nestling starvation (Bensch 1996). Polygyny in Great and Oriental Reed Warblers can therefore be costly for secondary females (see e.g. Ezaki & Urano 1995). The same applies to Sedge Warblers (Bell *et al.* 1997) and Black-browed Reed Warblers (Hamao 2003). However, the average fledging and recruitment success of Great and Oriental Reed Warbler secondary females is almost equal to that of simultaneously breeding females mated with

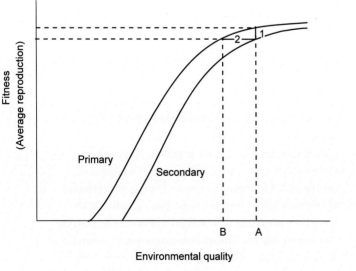

Fig. 10.6 Graphic model of the polygyny threshold, showing the relationship between female fitness and environmental quality and the conditions necessary for the evolution of territorial polygyny. The two curves are for monogamous/primary females and secondary females. Territory A is of higher quality than territory B. Distance 1 is the difference in fitness between females mated monogamously and females mated bigamously in the same environment. Distance 2 is the polygyny threshold (from Bensch 1997). A female that breeds as a secondary female in a good territory with an already paired male has the same or greater breeding success than if she were the only mate of a male with a poorer territory (after Bensch 1997).

monogamous males (Ezaki & Urano 1995, Leisler *et al.* 1995, Bensch 1996). The rate of chick starvation in secondary nests varies in different areas, and is higher in European than in Japanese populations, for example (table 10.2). In Japan, the costs are reduced in two completely different ways. First, in the warmer climate of Japan, it is possible for secondary females to limit the time spent brooding the nestlings and concentrate instead on foraging. Secondly, a longer period of time between broods of primary and secondary females in the same territory means that the polygynous male is able to divide his provisioning effort more effectively between his multiple broods (table 10.2; Urano 1990, Ezaki & Urano 1995). Another mechanism that provides secondary females with a direct reduction of the costs of sharing a mate is random nest predation, which can result in a polygynous male reallocating his assistance to the nest of the secondary female if the nest of the primary female fails (Bensch 1997, Hansson *et al.* 2000, Hamao 2003). It is therefore consistent with these mechanisms that Great Reed Warbler secondary females apparently commit infanticide at the nests of primary females, in order to take over the role of the assisted female. In Sweden and the Czech Republic, experiments with artificial nests containing soft clay eggs provided support for the idea of egg destruction or infanticide as a strategy that was frequently adopted among female Great Reed Warblers (Hansson *et al.* 1997, Trnka *et al.* 2010).

However, costs for secondary females can differ. In Dusky Warblers, a species showing territorial polygyny akin to Great Reed Warblers, Forstmeier *et al.* (2001b) found that secondary females were older than simultaneously breeding monogamous females, which is also the case in Great Reed Warblers. Clearly, prior breeding experience might

help older females to profit more from the benefits and to suffer less from the costs of polygyny than younger females, which ultimately indicates that the polygyny threshold may not be the same for all individuals. A more detailed look at individual variation in Dusky Warbler females also revealed that birds with stronger bills were more likely to breed in territories with high food abundance, where they received less male help – and were therefore mostly secondary females – presumably because they were better adapted to exploit the abundance of large food items in a rich site (Forstmeier *et al.* 2001a). Interestingly, Great Reed Warbler secondary females, who also receive little or no male support, also catch larger prey items than females assisted by males, though possible bill size differences have not been investigated (Sejberg *et al.* 2000). Presumably these females compensate for the lack of male help by using more energy-consuming foraging techniques, such as moving faster, that will yield larger prey. Thus both the interspecific comparison of the six wetland *Acrocephalus* species (section 10.1) and the intraspecific comparison above suggest that female emancipation from male assistance in raising the brood is correlated with female bill morphology and territory quality, both of which allow females to catch larger prey.

What's in it for males?

If we compare both the traits of polygynous males and the characteristics of their territories, with those of males that find themselves only one female, or even remain unpaired, we can make a guess at the cues that determine female choice. Normally the males of the two large reed warbler species that become polygynous are those that arrive in the breeding quarters earliest, occupying relatively large territories with higher-quality vegetation

Table 10.2 Some differences in breeding biology in Great and Oriental Reed Warblers (after Dyrcz 1995)

	Great Reed Warbler	Oriental Reed Warbler
Density of breeding population	lower	higher
Starvation of whole broods	regular	unknown
Overlapping between broods of primary and secondary females	small	large
Desertion of late broods by males	unknown	not rare

(Ezaki & Urano 1995, Hasselquist 1998). The same more or less applies to polygynous Sedge Warbler males (Bell *et al.* 1997). Early-arriving Great Reed Warbler males are usually older, have longer wings, and larger song repertoires. What is decisive is that they occupy the most attractive territories, with extensive and dense areas of better quality reeds (tall with thick stems) and a long water-reed edge, or with good foraging areas nearby (Dyrcz 1986, Ezaki & Urano 1995, Bensch *et al.* 2001). The security of a territory against predators can also contribute to its quality (Ezaki & Urano 1995, Albat 1996, Hansson *et al.* 2000).

As an example, fig. 10.7 shows the positive correlation between male and territory quality in one area. The males with the highest values are most likely to be chosen by more than one female (Catchpole *et al.* 1985). However, many of the characters measured in Great Reed Warbler males become more pronounced with age in some populations (e.g. wing length or song repertoire size; section 7.4). Hasselquist (1998) took this into account and, by standardizing traits according to age, found that age alone determines territorial occupancy order in males (fig. 10.8). He also found that territory quality played the key role in determining the annual and lifetime breeding success of an individual male Great Reed Warbler. The relationship arose as follows: male arrival order was closely correlated with territory attractiveness rank and was the most important factor predicting pairing success, that is, the number of females. Harem size in turn determines fledging success and the number of offspring recruits (standardized postfledging offspring survival; fig. 10.9). Thus, females seem to prefer the most attractive territories, which are occupied by early-arriving males, and these females gain direct benefits through increased production of fledglings and offspring recruits. In addition, they can expect offspring with improved viability from such males, hence there are also indirect benefits (section 7.4).

As a further explanation for the consensus in the choice of particular territories, Bensch *et al.* (2001) discussed the possibility that territory attractiveness results from tradition, whereby inexperienced birds copy the territory preference of experienced ones.

The selective advantage to males in arriving early is enormous. In Oriental Reed Warblers, which moult their

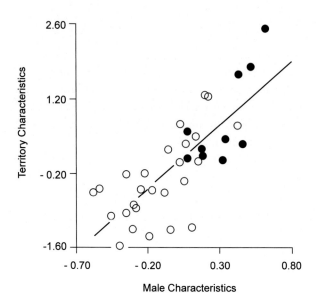

Fig. 10.7 Result of a canonical correlation analysis between 4 characteristics of males and 11 characteristics of their territories in 36 Great Reed Warblers during one study year: 10 polygynous (filled circles), 20 monogamous and 6 unmated males (open circles); from Catchpole *et al.* 1985. Male and territory quality are positively correlated with each other. The males with the highest values are more likely to be polygynous than those with lower values (from Catchpole *et al.* 1985).

entire plumage before the autumn migration (section 11.5), this has led to some males abandoning their late broods in order to start their moult (table 10.2). Another factor which might explain this could be that late-hatched offspring are less likely to survive, and thus males would be better off in reducing the costs of rearing unviable offspring in order to invest in self-maintenance. Deserted nests do not necessarily suffer higher nestling mortality and the males tend to settle earlier and become polygynous in the following spring, thus increasing their fitness (Urano 1992).

Extra-pair paternity

In Great Reed Warblers the EPP rate varies considerably between different areas; for example it is three times higher in Germany than in Sweden (8.9 and 3% respectively; fig. 10.1). The level is similar in Black-browed Reed Warblers, up to a quarter of whose males can be polygynous (Hamao 2003, Hamao & Saito 2005). Given the different interests of at least three individuals in a polygynous relationship, and frequent conflict between males to attract new females or guard their first mate, quite different predictions have been made of the proportion of EPFs in this mating system (Birkhead & Møller 1996, Petrie & Kempenaers 1998, Hasselquist & Sherman 2001). In both the Swedish and German Great Reed Warbler study populations (which differ from each other in several respects, such as breeding density, synchronization, etc.) monogamous and polygynous males were about equally at risk of becoming cuckolded (Leisler & Wink 2000). Likewise, in Sedge Warblers the frequency of cuckoldry did not differ significantly between monogamous and polygynous males (Hasselquist & Langefors 1998, Buchanan & Catchpole 2000b). By comparison, Black-browed Reed Warbler polygynous males lost paternity much more frequently than did their monogamous counterparts, especially when the fertile period of their two females overlapped. Males, which hardly participate in incubation in this species, were commonly able to achieve EPFs when their females were on the eggs (Hamao & Saito 2005).

In contrast to Sedge Warblers, Great Reed Warbler females always chose males with a larger song repertoire than their social mate as an EPC partner (section 7.4) and these were always neighbouring males. Genetic analyses showed that the offspring of these males had an increased survival rate, fuelling the supposition that, by engaging in EPFs, Great Reed Warbler females seek genetic benefits for their offspring (good or compatible gene effect; Hasselquist *et al.* 1996). The Swedish Great Reed Warbler population, which descends from only a few founder birds and thus shows inbreeding depression (section 8.5), was tested to ascertain whether extra-pair males and social males differed in their relatedness to females pursuing promiscuous matings. Contrary to recent findings in some insects, reptiles, and also other birds, no evidence could be found for inbreeding avoidance among females that actively sought EPCs (Hansson *et al.* (2004a). Neither did the Swedish females select their mates on the basis of MHC genes (Westerdahl 2004), a fact that distinguishes them from Seychelles Warblers, an island species.

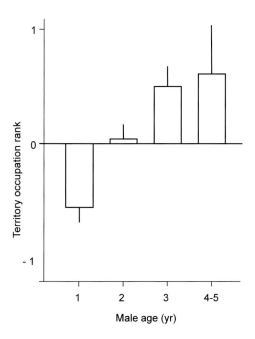

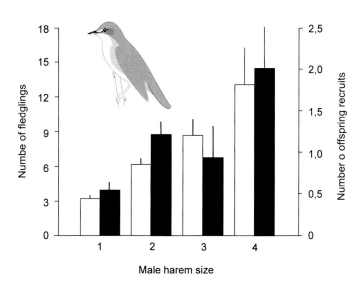

Fig. 10.8 Mean territory occupation rank in relation to age in Great Reed Warbler males. Positive rank values indicate earlier territory occupation dates, negative ones later dates (from Hasselquist 1998).

Fig. 10.9 Average number of fledglings (open columns) and recruits (offspring recorded as adults on breeding grounds in later years, filled columns) of Great Reed Warbler males with different harem sizes (number of females; from Hasselquist 1998). In general, the breeding success of a male increases with the number of females.

Photo 10.8 Aquatic Warblers are the only acrocephalid with a promiscuous mating system. In a single brood nestlings can be the offspring of up to five different males. Species with a high uncertainty of paternity and high sperm competition have large testes, a high sperm production, and large sperm storage organs, as in this male Aquatic Warbler (photo Klaus Nigge).

Offspring sex ratio

In polygynous systems, there are both variable costs for male or female offspring (sons are larger and more expensive to raise) and their expected reproductive success (sons have a higher variability) and also variable resources that are available to mothers. Primary females have better access to resources than secondary females, via male support, even though they are both paired with the same individual male. According to the variable sex ratio theory of Trivers & Willard (1973), when resources are abundant there should be a favourable bias in the brood sex ratio towards the sex that is more costly but promises more future reproduction, that is, sons. It is indeed the case that primary females have a higher proportion of sons in their broods than secondary females in three polygynous songbirds, Oriental and Great Reed Warblers and another marsh-nesting species, Yellow-headed Blackbirds (Patterson *et al.*

1980, Nishiumi *et al.* 1996, Nishiumi 1998, Westerdahl *et al.* 2000). In the first two species, males increased their share of parental care when the proportion of sons in the brood increased, though sons were not preferentially fed (Westerdahl *et al.* 2000).

Polyterritorial behaviour

For males, a further pathway to polygyny is polyterritorial behaviour, which has been thoroughly studied in Sedge Warblers (Zajac *et al.* 2008). After pair formation, a third of breeding males resume their singing activity, almost all of which occupy a second territory, spatially separate from the first. Particularly in years of lower population density, some of these males manage to mate with an additional female in this second territory. Although it is the case that the second territory is poorer than the first, it does not differ significantly in quality from the territories of unpaired or non-breeding males. Within the flexible mating system of the acrocephalids, polyterritorialism is a tactic whereby availability of space enables males to mate sequentially on separate territories.

6 Promiscuity – no pair bonds for Aquatic Warblers

Aquatic Warblers are the only acrocephalid with no pair bond and a promiscuous mating system. In this species, females invariably care for their offspring unassisted and obtain no direct benefits from the males. Their nests are scattered or aggregated in loose clusters amongst the grassy vegetation, in places which have a richer supply of cover and food. Within such nesting clusters the females are aggressive towards each other and use exclusive foraging grounds during the brood-raising phase (Dyrcz 1993, Dyrcz & Zdunek 1993). Though abundant, suitable nest sites and adjacent good feeding areas are unstable because they may change quickly within weeks or days due to water level changes (floods or drying out) or spring burning of the vegetation. This forces the birds to be mobile or to search for new areas to lay replacement and second clutches (Wawrzyniak & Sohns 1977, Leisler 1985, Kozulin *et al.* 1999). Males have defined singing posts and advertise themselves in core areas, especially at dusk (Schulze-Hagen *et al.* 1999). It became clear only by radio-tracking that males frequently change their 'activity ranges' within their large and substantially overlapping home ranges of up to 8 ha. If few fertile females are available, and are distributed over a large area, the males will cover long distances to find them (Schaefer *et al.* 2000). A very high percentage of broods have mixed paternity and within a brood the proportion of young of multiple paternity can be high. In three-quarters of the broods studied, more than one male was involved; up to five different males can fertilize individual eggs in a clutch (Schulze-Hagen *et al.* 1993, Leisler & Wink 2000). The percentage of young sired by at least one 'additional' father (the term 'extra-pair' is inappropriate here, since no pairs exist) was 39% (fig. 10.1).

In a comparative analysis of more than 100 species, Garamszegi *et al.* (2005) were able to show that relatively large testes and high sperm production will evolve when uncertainty of paternity and sperm competition are high in a species. Fig. 10.10 makes very clear the special position occupied by Aquatic Warblers regarding relative testis length in comparison with other wetland *Acrocephalus* species with various mating systems. The differences are even greater when the mass of the seminal glomera is compared. Prolonged copulation is also quite unique in Aquatic Warblers. Instead of the normal mating duration of 1-2 seconds, males and females remain together on the ground for an average of 24 minutes, during which the male sits on top of the female, holding onto her head feathers with his bill (fig. 10.11). In this time he has up to six cloacal contacts (Schulze-Hagen *et al.* 1995). The male inseminates the female repeatedly just before and just after egg laying, and in this way he can win a share of the intense competition for paternity.

Little is known about the criteria for female choice and male mating tactics and their outcomes in Aquatic Warblers. Broods with single paternity were begun slightly earlier and were spatially more isolated than multipaternal broods (Dyrcz *et al.* 2002, Gießing 2002). This indicates that high densities, as in the North American sparrows below, promote the development of clutches with mixed parentage (Hill *et al.* 2010). The number of sired nestlings per individual male ranged from 0 to 8 in Aquatic Warblers, thus reproductive success among males was extremely skewed, with less than 20% of them fathering almost half of the nestlings on a study plot. The highest reproductive success was achieved by males that had longer wings and a better body condition, and that were free of blood parasites (Dyrcz *et al.* 2005).

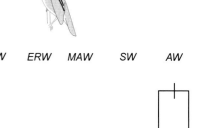

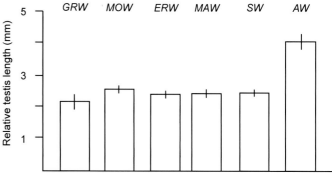

Fig. 10.10 Mean relative testis length (corrected for body mass) in six European reed warblers (from Schulze-Hagen *et al.* 1995).

The unique breeding system of Aquatic Warblers shows many striking parallels to the situation of the North American sparrows, which inhabit a similar habitat (section 14.2).

7 The evolution of mating systems

From an ecological and behavioural perspective we have found a strong correlation between ecological conditions (i.e. habitat quality) and mating systems in various acrocephalid species (section 10.1). Leisler and his team (2002) also examined which type of mating system – monogamy with male parental care or polygamy with reduced paternal care – was ancestral in the group, and which was derived. To this end they used a phylogenetic analysis based on complex statistical procedures. An important additional question is whether mating system evolution in the group has been associated with changes in habitat quality. The hypothetical ancestor in the group was most likely monogamous, living in a poor habitat, with males participating fully in nestling provisioning but much less so in incubation.

Over evolutionary time, a change by some species to better habitats was accompanied by a reduction in paternal care, which facilitated the occurrence of polygynous systems. Polygyny is the ancestral mating system in other groups of birds, such as the cotingas, Neotropical frugivores which exploit the rich food supply of lowland forests (Prum 1998). In these groups, monogamy and biparental care evolved several times where the taxa switched to insectivory and colonized high altitudes.

Cooperative breeding – the ancestral breeding system in various groups (Arnold & Owens 1999) – developed several times independently in the acrocephalids. Phylogenetic analysis strongly suggests that it evolved in poor habitats, in birds with a monogamous mating system, and a sedentary lifestyle. Although the various systems differ greatly from each other (section 10.4), helping always arises intraspecifically only in high-quality territories.

Since the choice of a breeding partner is an individual decision, and natural selection acts on its result, then the same relationship patterns between ecological conditions and mating systems that we have found in *inter*specific comparisons should also be found in *intra*specific studies. A neat confirmation of this comes from the study by Zajac & Solarz (2004) of the Sedge Warbler mating system in a food-poor habitat with a homogeneous distribution of critical resources in Poland. In contrast to populations with a high-quality food supply and very variable territory quality, in which some of the males are always polygynous, males in poor and homogeneous conditions do not succeed in mating with more than one female (except when the sex ratio is very skewed), since the preconditions for polygyny are not fulfilled.

For a better understanding of the relationship between social mating systems and frequency of genetic mating systems (EPFs) in the group, we need more paternity studies, both of individual species and of populations of a single species. With more information, comparative studies could then test hypotheses which might explain variation. For instance, if females seek genetic benefits from extra-pair matings, then genetic variability among males in a population should affect female benefits. Only three studies on this topic are currently available for acrocephalids: one on Sedge Warblers in a stable population in equilibrium, a second on Seychelles Warblers as an insular species whose population is very small and has passed through an extreme genetic bottleneck, and a third on Great Reed Warblers in a semi-isolated Swedish population. Results obtained so far regarding extra-pair mating in reed warblers suggest a general pattern in which males involved in extra-pair copulations are on average larger and older than within-pair ones, although they do not consistently differ in terms of secondary sex traits, condition, or relatedness to the female (*references to be added*).

The crucial questions that remain to be answered are whether, how, and why females obtain benefits from choosing particular males as sires for their offspring, and which benefits these are. It will be some time before we find answers, despite the fact that our present knowledge of sexual strategies is more detailed and precise than could have been imagined in the 1970s. With their great variety of mating systems, the reed warblers have already made an enormous contribution to our understanding of these fascinating topics.

Summary

Reed warblers are an excellent example of how the various forms of cooperation between sexual partners actually reflect conflicting interests. Their breeding systems are greatly varied and extend from social monogamy (with both parents contributing equally to the raising of the brood in food-poor habitats where prey size is often small), through polygynous forms with declining paternal efforts, to promiscuity with uniparental care in highly productive habitats where prey size is often large, in the case of Aquatic Warblers.

It is important to distinguish between 'social' and 'genetic' mating systems, because young of even the socially monogamous reed warblers are often not the offspring of the social male. Females can benefit from mating with multiple males, not only directly, perhaps by obtaining additional resources, but also indirectly, by obtaining genes that increase the fitness of her offspring. Such indirect benefits are often postulated, but the evidence is mixed and controversial. Nevertheless, results from Great Reed Warblers and Seychelles Warblers have demonstrated that females can indeed improve the genetic quality of their offspring by engaging in extra-pair copulations.

A significant correlation between mating system and food supply or habitat type has so far been demonstrated in 17 different species. Those that are the least monogamous (Great Reed Warblers, Aquatic Warblers) have the most powerful bills in terms of greater relative depth. Where food-poor habitats and a more sedentary lifestyle come together cooperative breeding has arisen.

Fig. 10.11 A copulation bout in Aquatic Warblers (drawing by David Quinn).

The cooperative system of the Seychelles Warblers has developed through both habitat saturation and territory quality. It has become one of the most intensely studied helper systems anywhere in the world. On the tiny islands inhabited by these warblers, pairs occupy a long-term territory for up to nine years, have a very low reproduction rate, and a high life expectancy. In high-quality territories, mostly female young remain as helpers instead of occupying new territories in areas of lower quality. This is also the case for birds that were translocated to unoccupied islands. At first there was no cooperative breeding, but after all the high-quality areas had been settled, young birds born on high-quality territories began to stay as helpers rather than occupying breeding vacancies on low-quality territories. The ways in which such behaviour directly and indirectly increases their lifetime inclusive fitness have been precisely analysed, as has the manner in which dominant females succeed in determining the sex of their offspring according to the prevalent ecological conditions. Such exact and detailed information can only be gained in a long-term study in which every single bird is known, by using sophisticated experiments and state-of-the-art molecular-genetic technology.

Female Great Reed Warblers and Oriental Reed Warblers entering a polygynous relationship with already paired males receive reduced assistance in caring for their brood. However, recruitment success of secondary females is no worse than that of other females that are mated with monogamous males. Any disadvantage is compensated for by the better territorial quality of the polygynous males (polygyny threshold). In addition, secondary females are commonly older and hence more experienced than monogamous females. They can catch larger prey, for example, making them more independent of male help. In the promiscuous Aquatic Warblers, males have the lowest certainty of paternity and the highest sperm competition of any acrocephalids, and this is reflected in the largest testes and sperm storage organs of any in the family.

A reconstruction based on phylogenetic, ecological, and behavioural data reveals that the likely hypothetical ancestor of the acrocephalid group was monogamous and living in a poor habitat;, but that the change to better habitats by some species was accompanied by a reduction in paternal care duties, which thus facilitated the development of polygynous systems.

11

Warblers on the move and in moult

Basra Reed Warbler trapped during migration in Ngulia.

Every autumn countless thousands of Palaearctic migrant birds are concentrated in the east African flyway on their way to their winter quarters. Those that come from Ethiopia intending to fly further south follow a narrow route which passes east of the central Kenyan highlands through the Tsavo National Park. Every year since 1969, their arrival has been eagerly awaited by a Kenyan-European ringing team at the Ngulia game-viewing lodge, complete with all the necessary trapping and ringing equipment. On a moonless night, early in December, a thick mist lies over the whole area and the clouds sink ever lower. Now the bright floodlights, which have been erected to illuminate the watering hole for the nocturnal game animals, attract the birds like moths to a streetlight. These are ideal trapping conditions and the nocturnal migrants flutter and swirl around, forced to the ground by the light. Soon the mist nets are sagging with their burden of trapped birds. On a good trapping night up to 3000 birds can be lifted from the mesh. This makes plenty of work and little sleep for the ringers, since even an experienced well-coordinated team can only manage to measure and ring at most 300 birds per hour. By far the commonest species are Marsh Warblers, of which 1100 have been caught in a single night and around 10 000 in a season. To date, more than 180 000 Marsh Warblers have been ringed at this trapping station alone. Together with more than 120 spectacular long-distance ring recoveries, both in and from the whole of Europe, these birds have provided important information on the phenology and physiology of the seasonal migration (D.J. Pearson, pers. comm.).

Such scenes vividly bring home to us the enormous numbers of acrocephalids that move back and forth every year between breeding site and winter quarters and the huge distances they have to cover. Once again, reed warblers have contributed the most to answering the important questions in avian science that are raised by migration. They are among the most frequently ringed passerines in the Palaearctic. There are even ringing programmes designed solely for reed warblers, such as the EURING *Acro*-project and the 'Acrocephalus' project of Operation Baltic. Members of the Acrocephalidae family have become model species for ecological and physiological migration studies, in the same way that those in the genus *Sylvia*, which are easier to keep in aviary and laboratory, have played an important role in investigations into endogenous mechanisms and their genetic foundation (Berthold 2001).

The acrocephalids consist of tropical sedentary species as well as partial, short-distance, and long-distance migrants. Those that inhabit the temperate zones are all migratory species and must make incredible efforts during their long, often transcontinental, journeys. Many of them, commonly weighing just 10 to 15 g, must cross seas, deserts, and mountain ranges twice in the course of a year. Some, such as Marsh Warblers, cover a total of

18 000 km per annum and are on the move for most of the year. Each species has developed its own strategy for the strenuous undertaking that is migration, for the organization of the journey or the actual route to be followed, and for its coordination with other processes in the annual cycle such as breeding (chapter 8) or moult (section 11.5). In a flexible manner, the differing lengths of the migration routes determine the time when moult is best positioned in this annual cycle. Decisive preconditions for successful migration include both the necessity for intact flight feathers and also for a complete or partial moult to take place, either before or after migration or else on staging sites during migration, depending on the route taken and other possible factors.

Many questions have to be answered. Exactly how do Sedge Warblers and Eurasian Reed Warblers reach their winter quarters? What happens to African Reed Warblers, the sister taxon of the latter? Why do Marsh Warblers and Icterine Warblers take such different migratory routes, despite the fact that they often share both breeding grounds and winter quarters? Why do acrocephalids have such different migration strategies at all and what exactly differentiates them? In what ways are moult and migration co-ordinated with each other and what strategies are adopted?

Photos 11.1 and 11.2 The east African flyway passes over Tsavo National Park in central Kenya, where masses of Palaearctic migrants are attracted by floodlights every year in November and December and then mist-netted. Since 1969 nearly 200 000 Marsh Warblers have been trapped here, up to 1100 in a single night. The data collected here, and the many long-distance ring recoveries, provide important information on the phenology and physiology of migration (photos Michel Laplace-Toulouse).

1

Getting organized for migration

Like many songbirds, acrocephalids are nocturnal, solitary migrants, equipped with endogenous time and orientation programmes (fig. 11.1). By migrating at night they are able to forage for food during daylight hours, while they also profit from a reduced threat of predators and above all from better atmospheric flight conditions (Berthold 2001, Alerstam 2009, Mukhin *et al.* 2009). Migration over long distances and over large, hostile ecological barriers is costly. In addition to the energetic costs of transport, the birds must adjust to unfamiliar surroundings, satisfy nutritional demands under time constraints, compete with other migrants and residents for limited resources, avoid predation, cope with unfavourable weather, and determine an appropriate direction for the next stage of their journey (Moore *et al.* 2005b, Bayly 2006). As a result, migratory behaviour is subject to considerable selective pressures. Evolutionary optimization strategies operate at many points, among them orientational abilities, choice of timing, route, length of flight stages, and choice of habitat and diet at stopover sites (Alerstam 2009).

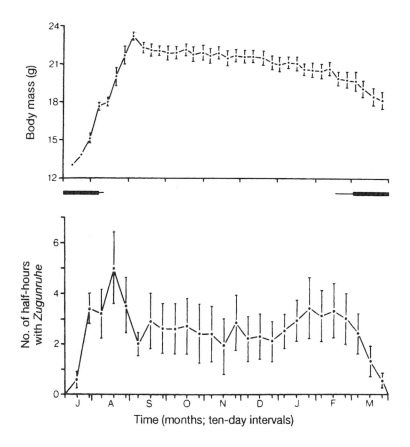

Fig. 11.1 Development of body mass (above) and migratory restlessness or *Zugunruhe* (below) in hand-reared Marsh Warblers under constant conditions. Black horizontal bars: final part of juvenile and initial part of winter moult (from Berthold & Leisler 1980). This endogenous migration pattern correlates well with the field data.

Accumulating fat

Metabolic rate during flight may be five to ten times basal rate (Wikelski *et al.* 2003). Enormous fat reserves must therefore be stored as a fuel supply for long-distance migration. The start and end of fat deposition is determined by the endogenous circannual clock (fig. 11.1). Other energy-consuming processes, such as moult, must either be completed before migration or postponed to a time following arrival in the wintering area (section 11.5). Fat accumulation is accompanied by an increase in both daily food intake (hyperphagia) and utilization efficiency. The photoperiod, here the shortening of day length, is the synchronizing cue (*Zeitgeber*); diminishing food supply is an apparent modifier for fat accumulation when the time for migration arrives, a mechanism that becomes more efficient as the season progresses (Bairlein 2000, Schaub & Jenni 2001a).

Long-distance migrants divide their migration into phases of flight and stopover in order to top up their fat reserves constantly. Most migrating warblers spend much more time refuelling than flying. It is therefore of crucial importance that the route is efficiently organized and this has been studied in great detail in acrocephalids (e.g. Schaub & Jenni 2001b, Balanca & Schaub 2005). The particular suite of features with which each species is equipped (e.g. food spectrum, foraging techniques, or habitat utilization) plays an important role here. Just how different the resulting strategies can be is illustrated by comparing Eurasian Reed Warblers and Sedge Warblers, two long-distance migrants that are morphologically similar, show no relevant differences in wing span, details of primaries, or lean weight, and have a similar migrational route from northwest Europe to sub-Saharan Africa.

Colin Bibby and Rhys Green (1981) were the first to recognize the substantial dissimilarities in the strategies of these two warblers and to work out the causes. Already during their pre- or early migrational fattenings both these long-distance migrants show characteristic differences. Sedge Warblers restrict themselves to less mobile arthropods such as aphids, chironomids, spiders, mayflies and so on, whereas Eurasian Reed Warblers take more mobile prey (chapter 4). Hence Sedge Warblers

Photos 11.3 and 11.4 The great differences in fat accumulation and migrational strategies are illustrated by Sedge Warblers and Eurasian Reed Warblers, which are morphologically similar and have similar migrational routes from NW Europe to sub-Saharan Africa. While many Sedge Warblers feed on reed aphids – a superabundant but unpredictable late-summer food source – Eurasian Reed Warblers exploit instead more predictable though less abundant widely distributed dipterans. Sedge Warblers can therefore fuel up very quickly and are able to fly directly from central Europe to sub-Saharan Africa. Eurasian Reed Warblers by contrast cannot build up large fat reserves, and hence must frequently interrupt their southward journey to refuel (Sedge Warbler feeding on aphids; Eurasian Reed Warbler catching a soldier-fly [Stratiomyidae]; photos Mark Nollet, David Cookson).

make much more use than Eurasian Reed Warblers of reed aphids, which can attain high densities in northern and central European reedbeds – a superabundant but unpredictable late-summer food source. Sedge Warblers aggregate in reedbeds with a high abundance of less mobile prey, whereas Eurasian Reed Warblers do not. They exploit instead more predictable though less abundant arthropods – mainly dipterans – that are widely distributed in marshes and other wet sites.

Such divergences exert considerable influence on migratory strategies: superabundance of reed aphids enables Sedge Warblers to fuel up very quickly, but due to the spatiotemporal unpredictability of this resource they start migration early (in August). Moult of body feathers is

therefore shifted to a period after arrival in the winter quarters (Redfern & Alker 1996). Sedge Warblers fuel up as soon as they encounter a site with superabundant food anywhere in northern and central Europe. In southern Europe, where *Phragmites* matures and wilts earlier because of higher temperatures and earlier start of growth, Sedge Warblers cannot expect any opportunity to refuel close to the Sahara. Having gained enough fat they are able to fly directly from central Europe to sites south of the desert. Their fuel loads often reach 60% and may exceed 100% of lean body mass (Bibby & Green 1981, Kullberg *et al.* 2000, Schaub & Jenni 2000a, 2001a, b). If rich food supplies are unavailable, Sedge Warblers generally switch to a strategy with shorter flight intervals, more stopovers, and longer stopover times, similar to the pattern in Eurasian Reed Warblers (see also Bayly 2007). Alternative strategies involving shorter flight intervals are more usual, at least for the Sedge Warblers which originate in northeast Europe, as they have to traverse dry regions in the Balkans, Italy, or northern Africa (Basciutti *et al*. 1997, Schaub & Jenni 2000a).

For Eurasian Reed Warblers, whose food is more evenly distributed and predictable, though occurring at low density and declining with season, the best strategy is also an early departure from breeding grounds in northern and central Europe, even when they are still moulting body feathers (up to 86% of early migrating Eurasian Reed Warblers are in body moult; Herremans 1990). With this strategy they are not in a position to build up large fat reserves, hence must frequently interrupt their southward journey to refuel. Stopover duration at this stage is long so overall migration speed is low. With the moult completed, fat can be deposited faster, stopover duration decreases and migration speed increases, but body masses of Eurasian Reed Warblers remain relatively constant along their migration route. Their fuel loads rarely exceed 20-25% of lean mass. Only when approaching the Sahara as an extreme ecological barrier do they increase their fat reserves (see below). For this purpose, Eurasian Reed Warblers in the Mediterranean region are able to exploit a much broader and drier habitat spectrum than Sedge

Warblers (Schaub & Jenni 2000a,b). The migratory strategies of both species are therefore completely differently organized (fig. 11.2).

Another opportunity for rapid premigrational fattening is offered by frugivory, to which *Sylvia* species, for instance, can switch during autumn migration. Thanks to their mixed diet, which includes many fruits, Garden Warblers are able to use a widely distributed, abundant, and easily accessible food supply, which facilitates rapid fat deposition at many sites. As a result they can easily moult all their body feathers prior to departure and master very long stretches during their journey (Schaub & Jenni 2001a,b). However, frugivory is rare among the acocephalids (with the exception of *Hippolais* species; Bairlein 2006; see also section 4.1).

Different again are the strategies employed by short-distance migrants such as Moustached Warblers. In a Mediterranean population of this species in southern France only a fraction of the birds was migratory (partial migrants). From late July until early December the birds did not significantly accumulate fat, independent of age, sex, and moult status. Hence they were only able to undertake short migrational flights that mostly started in November or December. Since the probable overwintering area of this population was in northern Spain, the

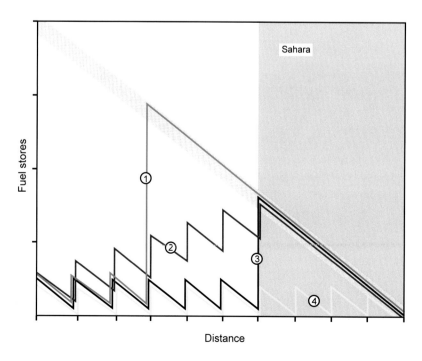

Fig. 11.2 Strategies for autumn migration from Europe to sub-Saharan Africa in terms of fuel stores. The diagonal grey bar indicates the fuel load needed for flying to the southern border of the Sahara without refuelling. The different strategies are 1: start of non-refuelling migration well before the Sahara (western populations of Sedge Warblers); 2: steadily increasing fuel load along the route (Garden Warblers); 3: large fuel deposition just before the Sahara (Eurasian Reed Warblers); 4: refuelling in the Sahara (Spotted Flycatchers; from Schaub & Jenni 2000b).

entire migratory distance would be between 200 and 400 km and could thus be covered without significant energy stores (Balanca & Schaub 2005). The same is true for the Balkan population, which winters in Italy and Dalmatia (Vadasz *et al.* 2007).

Time is of the essence

A further factor shaping overall migration strategy is time minimization. Crucial roles are played by staging stopover frequency and duration, as well as efficiency of fuel deposition. Exact stopover durations and weight increases can be worked out only by using complex capture-recapture analyses and precise calculations of fat deposition rate (Schaub *et al.* 2001). These parameters have been calculated for Eurasian Reed Warblers, since this migrant has been trapped in large numbers (e.g. Schaub *et al.* 2001, Bolshakov *et al.* 2003, Rguibi-Idrissi *et al.* 2003, Yosef & Chernetsov 2005, Bayly 2006). At 17 European and north African ringing sites, average autumn stopover duration of non-moulting first-year Eurasian

Reed Warblers was 9.5 days. As they moved through Europe, the fat deposition rate remained relatively constant, and rose significantly only towards the end of the season and in the south, reaching its highest level in northern Morocco (Schaub & Jenni 2000a,b, 2001b, Jenni & Schaub 2003, Balanca & Schaub 2005). The long stopover durations can be explained by the low fuel accumulation rate typical of this species, which accounts for the low overall autumn migration speed. The threefold reasons why Eurasian Reed Warbler fat deposits reach their highest value shortly before the crossing of the most formidable hindrance, the Sahara Desert, are an expansion of the habitat spectrum, increased feeding frequency and a rise in fat deposition efficiency (Schaub & Jenni 2000a,b). Long-range migrants are clearly able to modulate their energy budget both seasonally and according to their current location. In this they utilize information from the geomagnetic field, which acts as an important external cue, as has been experimentally demonstrated for Thrush Nightingales (Henshaw *et al.* 2008).

Photos 11.5 and 11.6 Migration over long distances and over extensive, hostile ecological barriers is costly. In the hot and arid regions every source of shadow and water is welcome. In the extreme late summer heat of Kuwait these migrating Great Reed Warblers come into close contact with a Little Crake at a defective irrigation pipe, or can be threatened by a Rufous Bush Robin (photos Mike Pope).

Photos 11.7 and 11.8 Even today migrant birds are hunted, captured, and killed by people in many regions. In Oman, until the 1990s, there was a traditional 'Marsh Warbler Festival', when in April/May returning Marsh Warblers resting in the few bushes of the arid parts of the country were netted and eaten in large numbers. Following intervention and explanations by ornithologists, this practice is now prohibited (photos Gerhard Nikolaus).

The decision when to leave a staging site is also of great importance for a successful migration. The use of capture-recapture models for three passerine migrants, among them Eurasian Reed Warblers has once again demonstrated that this decision depends on fuel deposition rates and fuel stores. The birds follow simple rules that are in accordance with the time-minimizing hypothesis: when migrating over a continental area with many potential stopover sites, warblers depart to another site if their fat deposition rate is not sufficient. If this rate is positive, they shorten their stay by increasing the rate. When faced with an ecological barrier such as the Sahara, the rule is to look for a good stopover site and attain the threshold amount of fuel stores necessary for crossing it (Schaub *et al.* 2008; see also Bayly 2006, 2007). Stopover durations and weight gains also depend on habitat quality, particularly if stopover sites are only of limited availability. Stopover durations of first-year Eurasian Reed Warblers are longer than those of adults, not only in autumn but also in spring.

It has been calculated that acrocephalid migrants make flights of 450 km over continental Europe (10 h flight/night; ground speed 45 km/h). Yet because of their long stopover durations, the effective progress of migration would only be 39-56 km/day for Eurasian Reed Warblers and 45-89 km/day for Sedge Warblers (calculated from ring recoveries; Schaub & Jenni 2001b). However, the further south the birds, the more marked the increase in overall migration speed. It could be calculated that 'fat' migrants, such as Great Reed Warblers, are capable of

non-stop flights of up to 70 hours, and that to cross the Sahara at a width of 2000 km a flight of 40 hours would be necessary at an average songbird ground speed of 50 km/h. However, the reality is rather different (Biebach 1990, Schmaljohann *et al.* 2007). In one study using radar equipment in Mauritania, it was clearly demonstrated that songbirds do not cross the Sahara non-stop, but fly only through the night, sometimes into the early hours of the morning. They then land anywhere in the desert in order to rest and wait, ideally in the shade of rocks or stones out of the heat of the day, before resuming their journey after sunset. This intermittent migration with its flight-and-rest strategy thus prevents the excessive energy and water loss that would otherwise be caused by daylight flight through hot, dry, and turbulent air. The average flight altitude during autumn is less than 1 km above the ground and in spring between 2 and 4 km (perhaps because of the different wind conditions; Biebach 1990, Schmaljohann *et al.* 2007).

Migration in autumn often lasts substantially longer than in spring because of longer staging stops; southbound migration is apparently under less time pressure than migration towards the breeding grounds. In Marsh Warblers, which make a long autumn stopover in eastern Africa (section 11.2), migration southward lasted four months, whereas the return migration took only six weeks (Pearson 1990). Yohannes *et al.* (2009a) found no difference in the northern section of the journeys but the southern section was covered at different speeds in both migrations.

In spring returning warblers are obviously in a hurry to reach their breeding grounds as soon as possible (Schaub & Jenni 2000a, Rguibi-Idrissi *et al.* 2003). Stopover durations are therefore shorter, long-distance bouts of flying more usual, routes more direct, all of which result in an overall higher migration speed. Together with trapping projects, radar studies in the Sahara also point to diurnal migratory activity being greater in spring than in autumn (Rguibi-Idrissi *et al.* 2003, Schmaljohann *et al.* 2007, Yohannes *et al.* 2009a). The closer reed warblers get to their breeding grounds, the shorter become their flight bouts (4-6 hours) and stopover intervals (1 day). A large proportion of Baltic Eurasian Reed Warblers reach their breeding sites with still considerable fat levels, which will be advatageous to them in the early phases of reproduction (Bolshakov *et al.* 2003, Moore *et al.* 2005a). Young Eurasian Reed Warblers arrive later on the breeding grounds than adults (Rguibi-Idrissi *et al.* 2003), and males arrive earlier than females (protandry).

This male-first arrival pattern is the norm among passerine birds. One hypothesis postulates that protandry arises because sexual selection acts on males to increase their chances of mating opportunities by early arrival. In this regard, an important factor must be the sex ratio (and also mating system) in a species, since it determines whether or not all males can get a social mate (Kokko *et al.* 2006). Furthermore, the carry-over effects brought by males from their winter territories can also play a role, as well as indications that the later-arriving females are similarly equipped with reserves and hence better prepared for breeding (B. Helm, pers. comm.; see also Raess 2008 for Eastern Palaearctic migrants). The timing of spring arrival in some species can therefore be subject to both natural selection (environmental conditions for the earliest possible arrival) and sexual selection. Because the milder spring weather resulting from global warming could weaken the effects of natural selection acting against males that arrive too early, the spring arrival date

Map 11.1 Topographical barriers that must be overcome or avoided during autumn and spring migrations of north-temperate breeding Old World passerines (yellow = bare ground, blue = water bodies, grey = altitudes above 2000 m (compiled based on GlobCover http://ionia1.esrin.esa.int/ and WorldClim http://www.worldclim.org/).

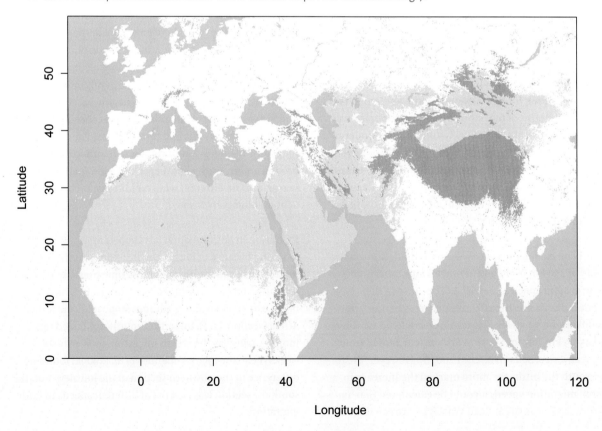

of species in which sexual selection is strong should become even earlier than in species with weak sexual selection. This prediction was confirmed by Spottiswoode *et al.* (2006) using a comparison of the long-term passage dates of various species, including several reed warblers. For instance, the arrival date of Sedge Warblers advanced to a greater degree than that of Marsh Warblers, a species with lower rates of polygyny and extra-pair fertilizations (chapter 10).

2 ▪▪▪▪▪▪▪▪
Migration routes and wintering grounds
When we look on a map of the world at the breeding and wintering quarters of the acrocephalid long-distance migrants, it is striking just how far apart they are and how they are separated by a daunting barrier of seas, deserts, and mountain ranges. Such hindrances extend without interruption from the Atlantic coast of Europe almost to the Pacific coast of Asia (map 11.1). They include the Mediterranean Sea, the Sahara Desert, the Arabian Peninsula, Iran, and the deserts, uplands, and mountain chains of central Asia stretching to Tibet. To overcome these obstacles many reed warblers often have to cross more than 2000 km of hostile desert and mountainous regions or undertake lengthy detours.

Three large regions provide the wintering grounds for these long-distance migrants: sub-Saharan Africa, the Indian Subcontinent, and southeast Asia (map 11.1). Acrocephalids typically reach their winter quarters by the most direct route; in other words, species with a more westerly breeding distribution migrate to Africa, those breeding in east Asia move to southeast Asia, while those in between migrate to the Indian Subcontinent. However, if we look in greater detail, we find some deviations from this rule. For example, Asian populations of Great Reed Warblers and Sedge Warblers migrate – just like

Map 11.2 Differing sizes of breeding and winter quarters. Blyth's Reed Warblers have a very extensive breeding range, but all populations are concentrated in very restricted winter quarters (Kennerley & Pearson 2010).

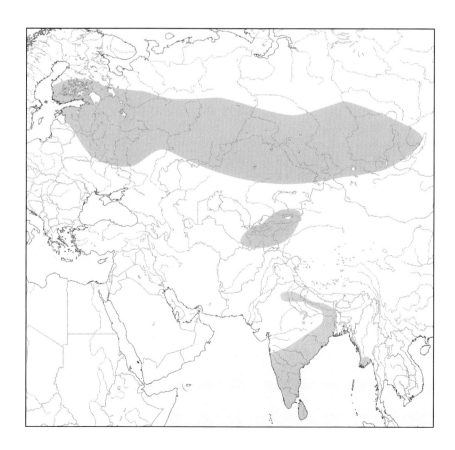

Map 11.3 Mirror-imaged autumn migration routes to Africa (and migratory divide) of western and eastern populations of Eurasian Reed Warblers (Zwarts *et al*. 2009).

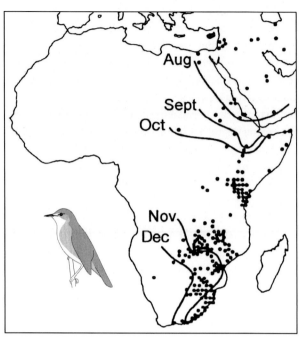

Map 11.4 Main monthly progression of Marsh Warblers on their southward passage through the African continent. This long outward migration includes stopovers in Ethiopia and follows the tropical rainbelt (from Dowsett-Lemaire & Dowsett 1987b) see Map 11.7.

their European conspecifics – to southern Africa, whereas Blyth's Reed Warblers and Booted Warblers, whose breeding distributions extend westwards to Scandinavia, move to southeast India without exception (Kennerley & Pearson 2010). As it happens, these two species have unusually broad ranges averaging almost 90 degrees of longitude. Their entire populations migrate from this enormous expanse to the Indian Subcontinent, where they are concentrated in relatively small-sized winter quarters (map 11.2). Paddyfield Warblers represent a further Asian species that overwinters in India, but which has recently expanded its range into Europe (section 13.2). Their Bulgarian population is a good illustration of the way in which the most recent changes in breeding distribution are reflected in migration routes. Instead of taking the shorter route directly southwards this population recapitulates in reverse the order of its geographic expansion, as birds head northeast round the Black Sea in autumn before migrating in a southeasterly direction (Zehtindjiev *et al*. 2010). By contrast, most European acrocephalids, or European populations of particular

species, migrate either eastwards or westwards, thus avoiding the central Mediterranean and Sahara where the barriers are widest.

Migrating towards Africa

Eurasian Reed Warblers in Europe are examples of populations which mostly retain their west-east breeding distribution in their winter quarters north of the equator in Africa: British birds overwinter on the west African coast (centred on Guinea), those from eastern France go further east (mostly into Côte d'Ivoire), and those from the Czech Republic move even further east (centred on Ghana). From eastern Nigeria eastwards a second main wintering region begins, that of Eurasian Reed Warblers from southeast central Europe. However, both wintering grounds are reached via two completely different principal migratory routes. The west European, Fennoscandian, and parts of the central European populations (including the Czech birds) move in a southwesterly direction, while the Slovakian birds migrate with populations from Hungary, as in a mirror image, to the southeast (migratory divide). The latter cross the eastern Mediterranean, changing direction in eastern north Africa to head south and later southwest. Only a few Eurasian Reed Warblers of unknown provenance cross the Mediterranean on a central route (map 11.3 ; Dowsett-Lemaire & Dowsett 1987, Schlenker 1988, Procházka *et al.* 2008, 2011; for migratory connectivity, see below).

The route to the winter quarters is thus not always direct, reflecting both historical developments and ecological constraints. One of the most intriguing species examples are the Marsh Warblers, whose autumn migration includes some striking changes of direction and is divided into two stages (map 11.4). Western and central Euro-

pean Marsh Warblers leave in July on a southeast heading and travel round the eastern Mediterranean, crossing Turkey and the Near East, followed by crossings of the northern Arabian Peninsula and the Red Sea. Having covered this stage rather quickly by the end of August or September, they take a pause of up to three months in southwest Ethiopia, during which time they undergo a partial moult. Marsh Warblers do not complete their migration until November-January, a further 1 000 - 4 000 km to the south. During this stage practically the entire world population follows a narrow corridor 100 - 200 km wide through southeast Kenya to southern Africa, where their wintering and moulting quarters extend deep into the Republic of South Africa and occasionally as far as

Map 11.5 Crossroads of migration routes in western central Asia. Here the NW to SE routes (Blyth's Reed Warblers, Booted Warblers) and the NE to SW routes (Sedge Warblers, Great Reed Warblers) of migrating acrocephalids cross each other (after Chernetsov *et al.* 2007).

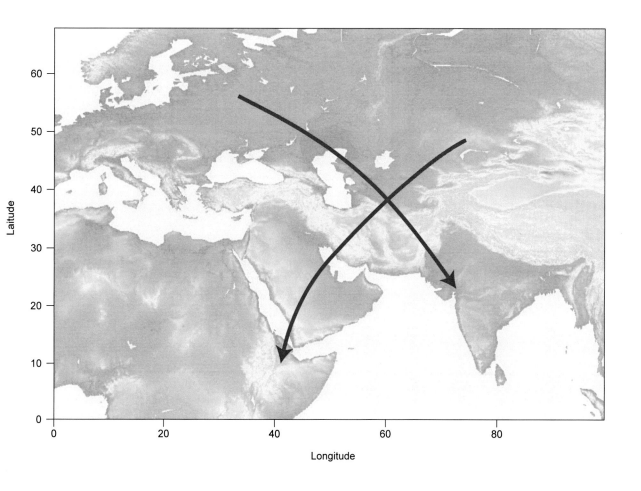

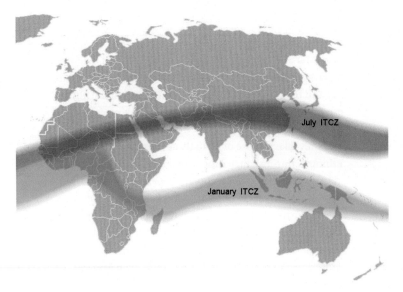

Map 11.7 The map shows the differing conditions encountered by overwintering Palaearctic birds in Africa and in the east of the tropical Oriental Region. The beneficial effects enjoyed by overwinterers in the dry eastern and southern half of the African continent, thanks to the advance of the ITCZ (a rain-bringing low pressure trough), are markedly reduced for the wintering birds in the tropical Oriental Region because extensive landmasses are lacking, and because a maritime-type climate dominates its peninsular and insular topography throughout the year.

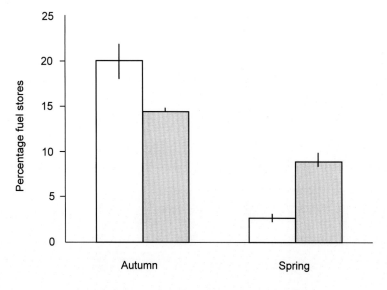

Fig. 11.3 Fuel stores of Blyth's Reed Warblers (as an example of a Palaearctic-Indian migrant; open columns) and Garden Warblers (a Palaearctic-African migrant; filled columns) trapped during autumn and spring migration in western Kazakhstan. The differing fuel stores are explained by the different migration routes (see Map 11.5; from Chernetsov et al. 2007).

Namibia. The autumn stopover area in Ethiopia is just as important to Marsh Warblers as their southern winter quarters (Pearson 1990, Yohannes *et al.* 2007, Kennerley & Pearson 2010).

The reasons for such a two-stage migration (and also for moult itinerancy) are to be found in ecological adaptations to seasonal resources. In their first staging area the birds benefit from humid conditions and fresh vegetation following the rains of July-August. It is only when conditions there begin to dry out during late October and November that the warblers resume their journey. Their route, through Kenya and Tanzania to the south of the continent, now follows the Intertropical Convergence Zone (ITCZ), which from November brings the dry season of several months to a close (map 11.7). Similar patterns in southward migration with a long pause (which is also used for moult) are shown by Basra Reed Warblers and Olive-tree Warblers, less so by Upcher's Warblers or Eastern Olivaceous Warbler populations, but also by some non-acrocephalids, such as River Warblers or Thrush Nightingales (Yohannes *et al.* 2007, Kennerley & Pearson 2010). On the other hand, Icterine Warblers, an arboreal species that favours green trees, take an entirely different route, even though both breeding and wintering grounds partly overlap with Marsh Warblers. These birds reach their overwintering areas in Zambia, Botswana, and South Africa on a western route via forested west and central Africa (Pearson 1990).

Crossing central Asia

Huge numbers of Eastern Palaearctic acrocephalids have to traverse the so far little-studied region of western central Asia, which represents a particularly unfavourable zone. This vast arid area consists of steppes, semi-deserts, extensive uplands, and mountain deserts. Moreover, the migrants face the necessity of crossing the world's largest montane region, including the Tien Shan, Pamir-Altai, and the Tibetan Plateau with the Himalayas. The overall length of their migratory route over such inhospitable areas may reach 3500 km. The migrants heading for India must cross up to 2000 km of this terrain at an altitude of 5-6000 m. Because some of these Palaearctic birds migrate to the Indian Subcontinent while others head for Africa, a busy crossroads of avian migration exists in western Kazakhstan. Of the nine acrocephalids that move through this region, five migrate in a southwesterly direction to Africa (the *fuscus*-group of Eurasian Reed Warblers, Great Reed Warblers, Marsh, Sedge, and Eastern Olivaceous Warblers) and four on a southeast heading to India (Blyth's and Clamorous Reed Warblers, Paddyfield and Booted Warblers; map 11.5). Given that they must overcome quite different handicaps, we are not surprised to see that they employ quite different strategics to deal with them (Bolshakov 2003, Chernetsov *et al.* 2007).

This is dramatically illustrated by a comparison of the fuel stores found on captured birds during autumn and spring migration at various trapping stations in the central Asian region. In autumn, the fuel loads of Blyth's Reed Warblers before their desert and mountain crossing are much greater than in Palaearctic-African migrants, which face a less broad barrier. In spring, Blyth's Reed Warblers carry less fat than in autumn, and even less than the African migrants, because at this time the arid regions of western central Asia have an ephemeral vegetation after the winter rain and hence a high abundance of arthropods, thus offering significantly better stopover and refuelling possibilities for migrants (fig. 11.3). To Palaeartic-Indian migrants, this dry desert belt represents a greater ecological barrier in autumn than in spring. On the other hand, Palaearctic-African migrants make a detour to the northwest around this hindrance and migrate north of the Caspian Sea. That way they do not need such large fat stores as those necessary for birds which traverse the Sahara. While the dry desert belt is more inhospitable for Palaearctic-Indian migrants in autumn than in spring, in the mountains the reverse is the case. Here, not only are most regions still snow-covered in spring but also the weather is dominated by an extremely strong headwind from the west. The pressure to avoid the arid belt as far as possible in autumn, and, conversely, to avoid the uplands in spring, leads to the loop migration routes of some Palaearctic-Indian migrants. Adverse climatic conditions and low abundance of invertebrate prey at stopover sites in central Asia substantially reduce the average speed of return migration for many migrants (Dolnik 1990, Bolshakov 2003, Chernetsov *et al.* 2007, Raess 2008).

Other geographical factors

Populations of the same species can consist of either non-migratory birds or short-distance and long-distance migrants. An example would be Eastern Olivaceous Warblers with their extensive breeding range. Of the subspecies, the birds from Europe and Asia (*elaeica*) migrate to central Africa and Arabia, while of the Egyptian nominate subspecies (*pallida*) only the northern populations migrate short distances, whereas the southern birds are believed to be sedentary, as are the members of the subspecies *laeneni* which breed in the south Saharan oases (Svensson 2001, Kennerley & Pearson 2010, map 11.6).

In the southern hemisphere migration has, so to speak, a different 'mathematical sign', in that birds migrate from south to north (austral migration). In the austral winter, populations of African and Australian Reed Warblers move in a northerly direction, but both species are only short-distance migrants.

Until recently, migratory connectivity, the link between breeding and non-breeding populations, could only be studied by using ringing recoveries. Recent advances in geochemical and genetic analyses, together with ongoing work on geolocators and satellite telemetry, now allow us to determine the geographical origin of individual breeding or wintering birds (Webster *et al.* 2002). Understanding these links has important ecological and conservation implications, for a globally threatened species such as Aquatic Warblers, for instance. Their winter quarters were only found as recently as 2007 and the exact location subsequently refined through the application of geolocators in 2011. This was greatly helped by the analysis of stable isotopes from winter-grown feathers, which was used to narrow down the possible areas to an approximate region (Pain *et al.* 2004). A search of the area soon confirmed the laboratory results: a large part of the total world population of Aquatic Warblers winters in a narrowly delimited flood region on the lower River Sénégal on the Senegal-Mauritania border (Flade *et al.* 2011).

By using stable isotope profiles in feathers as part of a Europe-wide study of 17 Eurasian Reed Warbler populations, it was shown that the wintering areas of southwest-migrating and southeast-migrating populations are not only almost completely separate (see above) but also that the wintering grounds of the latter are much drier (xeric); at the same time, three populations with unknown migra-

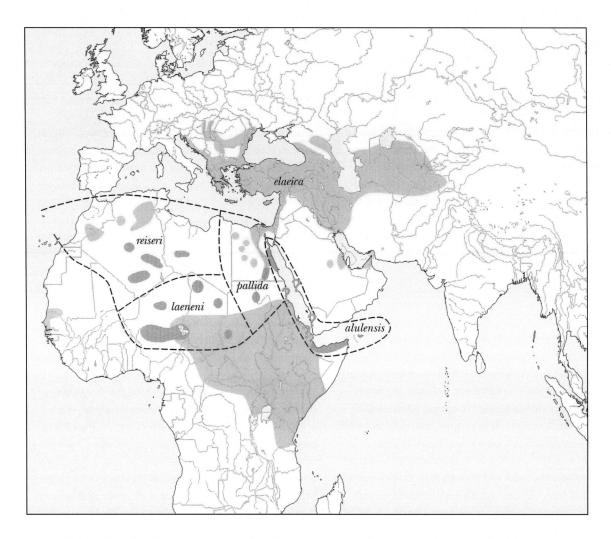

Map 11.6 Distribution map of Eastern Olivaceous Warblers as an example of a species with resident, short-range migrant, and long-range migrant populations. Yellow: migrant breeding range; blue: winter or 'non-breeding' range; green: resident breeding range (from Kennerley & Pearson 2010).

tory direction were assigned to the southeast-migrating populations. In general there is an astonishing connectivity between breeding and non-breeding grounds (Procházka *et al.* 2008, 2011). With the aid of multiple recaptures of Great Reed Warblers at Lake Kvismaren in southern Sweden over a period of six years, it was discovered that the isotope profiles of feathers of individual birds appeared to be consistent between years. This suggests that individuals tend to return and replace their feathers in the same Afrotropical wintering habitats, indicating a strong year-to-year fidelity to their African moulting sites. As a result, warblers can benefit from acquired experience in finding better feeding sites and avoiding predators in patchily distributed habitats (Yohannes *et al.* 2008a).

3 Post-fledging movements, dispersal, and site fidelity

While migration is a process characterized by seasonal outward and return movements between more or less fixed destinations, dispersal is an exploratory one-way movement leading from a natal or breeding site to future breeding grounds. This movement might take place via various intermediate staging sites but, in contrast to migration, no innate orientation or seasonal periodicity has been shown to be involved. 'Natal dispersal' is used for young birds that settle away from their birthplace, 'breeding dispersal' for the movement of adults from one breeding locality to another (Berndt & Sternberg 1968). Dispersal is influenced by both endogenous and exogenous factors such as resource distribution, population density, or competition. A major function of dispersal is the redistribution of individuals in space; that of natal dispersal – as an 'intergenerational movement' – is the avoidance of inbreeding and intraspecific competition (section 8.5). Dispersal is a life-history trait, influencing evolutionary processes such as population dynamics, population genetics, and range expansion.

The study of dispersal is still in its infancy, simply because we still do not have the technical means that allow us to follow those members of an avian population which undertake the greatest movements. At the same time, it is also very difficult to establish the extent to which the long-distance trips made by immature birds after they become independent translate later into natal dispersal distances, or if these exploratory movements are in fact a preparation for migration itself.

Nocturnal exploratory flights

The well-designed experiments using Eurasian Reed Warblers which have been carried out by Andrey Mukhin and co-workers from the Russian Rybachy Biological Station (on the Courish Spit in the Kaliningrad Region on the Baltic) have greatly contributed to the study of dispersal. Since reedbeds are mostly small in area and patchily distributed, reed warblers are forced to leave them during their post-fledging movements. The researchers managed to retrap many young warblers, ringed as nestlings, in their post-fledging period outside reedbeds. Some of the birds were equipped with transmitters and followed telemetrically for up to three weeks. A control group was kept in special orientation-registering cages to record their activity patterns (Mukhin 2004, Mukhin et al. 2005, 2008).

As a result, Mukhin and his team were able to show that all the Eurasian Reed Warblers make exploratory movements of considerable length during their post-fledging period on leaving their natal reedbeds and that these always occur in the middle of the night. Eurasian Reed Warblers that were clearly on migration had covered stretches of over 200 km along specific routes, and were older than eight weeks when they were retrapped, whereas the post-fledging movements were of a maximum of 44 km and had no discernible preferred heading. Shuttle movements, with a later return to the natal site, were common. Post-fledging excursions have been demonstrated in birds only four weeks old and can last for up to

three weeks. In this period, flight lengths and activity increase continuously. While clearly migrating Eurasian Reed Warblers in the Baltic region were, without exception, fat and had more or less completed their body-feather moult, the post-fledging dispersers had stored no fat and were in heavy body moult. From this we can conclude that nocturnal activity is not an exclusive feature of migration but has already begun long before the onset of autumn departures (Bulyuk et al. 1999, Mukhin 2004, Mukhin et al. 2005; see below).

These studies illustrate that autumn migration and post-fledging movements are two quite different phenomena. The latter serve several purposes: they contribute to natal dispersal (i.e. selection of a future breeding site), to familiarizing the birds with their immediate surroundings for forthcoming departure, and to creating the knowledge of the natal area which will be necessary for successful navigation during the return migration of the following spring (Baker 1993). However, we might expect that evaluation of habitat quality and its suitability for breeding would be undertaken during daytime foraging movements, which normally do not exceed several hundred metres in range. Explorations made at greater distances are not necessarily translated into following season settlement, as demonstrated in a mathematical model by Vitaly Grinkevich et al. (2007). According to this, 90% of all Eurasian Reed Warblers breed within 4 km of their natal locality, although their post-fledging movements extended up to ten times as far, or, as the authors put it 'juvenile reed warblers see the world but settle close to home'.

It is possible that nocturnal premigratory activities by juvenile Eurasian Reed Warblers also have the function of training flight experiences and of developing orientation abilities (stellar compass) for the first stages of migration. In captive reed warblers of the same age, such premigratory night flights are expressed as weak nocturnal restlessness, which indicates an endogenous basis. Recent work has shown that nocturnal flights well before migration are not unique to Eurasian Reed Warblers but are a common phenomenon among many long-distance migratory songbirds (Mukhin 1999, Mukhin et al. 2005). Dispersing reed warblers preferred to make nocturnal movements in high pressure systems, in still air, or with light winds (Bulyuk 2003). These short-duration exploratory flights have recently been noted among adult birds on stopover sites during migration, and in fact appear to be a common form of behaviour in nocturnal migrants, serving to evaluate meteorological conditions aloft before the birds take off on the next stage of their journey, for example (Schmaljohann *et al.* 2011).

Dispersal and homing of adult birds within the breeding season also take place nocturnally (Mukhin *et al.* 2009). Several pairs of adult Eurasian Reed Warblers were marked with transmitters and forced to leave their breeding sites by simulating nest predation. Soon after nest loss they left their reedbeds, exclusively at night. Those birds which had been caught and translocated experimentally by up to 20 km also returned to their nests at night, crossing inhospitable habitats in one or even two nocturnal flights, and only returning after a delay of several days. Although at this time their usual lifestyle is predominantly diurnal, it would appear that the birds adopt nocturnal activity during the breeding season in order to be able to cover longer stretches.

During nocturnal migration and dispersal warblers have to fly over unsuitable habitats and therefore need to assess the suitability of stopover habitats during landfall, especially when visual cues are reduced or absent, as in darkness or fog. Distant cues may then be of help in habitat selection decisions. Species song might serve as an acoustic cue and Mukhin *et al.* (2008) were in fact able to demonstrate that species-specific song propagated by loudspeakers did attract moving Eurasian Reed Warblers and Sedge Warblers. The reactions of acrocephalids as wetland habitat specialists were clearly stronger than those of Pied Flycatchers and Redwings, both of which are classed as habitat generalists. Adult Eurasian Reed Warblers tended to be more attracted by acoustic cues than juveniles, which suggests that previous experience may play a role in habitat recognition using acoustic stimuli. Further studies are required if we are to gain a better understanding of the proximate (mechanistic) and ultimate (evolutionary) processes behind this interesting behaviour.

Shall we go or shall we stay?

Acrocephalids fit well into the general patterns of songbird site fidelity and dispersal. First, based on ring recovery data, it is usually the case that natal dispersal distances are greater than breeding dispersal distances, as shown for Eurasian Reed Warblers and Sedge Warblers

by Paradis *et al.* (1998). Put another way, breeding site fidelity is always stronger than natal site fidelity and hence can be highly pronounced: 92% and 96% respectively of resited or recaptured Great Reed Warblers and Sedge Warblers came from the same locality. In other words, individuals commonly return year after year to the same spot (Hansson *et al.* 2002a,b, Prochazka & Reif 2002). One Eurasian Reed Warbler was still monitored near its breeding site 14 consecutive seasons after it was ringed (BTO website, longevity records).

When compared with larger wetland species such as gulls, ducks, or rails, which disperse further than species of dry habitats (Paradis *et al.* 1998), dispersal has conversely been found to be reduced in an isolated Swedish population of Great Reed Warblers studied by Hansson *et al.* (2002a,b), even though the species actually has high dispersal potential (Hansson *et al.* 2003b). What is in fact typical for wetland acrocephalids has yet to be decided – on the one hand it is known that isolated populations often show higher philopatry than non-isolated ones (Weatherhead & Forbes 1994), but on the other, we currently have no dispersal distances for species from other habitats at our disposal for comparison.

Secondly, we know that female birds in general disperse further than males. In the Great Reed Warbler population on Lake Kvismaren in southern Sweden far more female than male immigrants to the study area are recorded. Using genetic methods, it was possible to estimate the relatedness between any given individuals in the population (Hansson *et al.* 2003a). This study showed that female immigrants had a lower genetic relatedness but, on the other hand, carried the majority of novel alleles into the population. The females that were genetically most different originated from populations furthest from the study area, in fact from populations outside Scandinavia. Since these immigrants do not have a reduced lifetime reproductive success, the small number of female long-distance dispersers make a substantial contribution to the gene flow. In addition, a cohort analysis from capture-recapture data revealed a remarkable offspring-parent resemblance in dispersal behaviour, indicating that variation in dispersal of Great Reed Warblers has a partly genetic basis (Hansson *et al.* 2003b). It was also shown that the avoidance of inbreeding in the species is achieved not via kin discrimination but exclusively via female dispersal (Hansson *et al.* 2007).

In studies of long-distance dispersers emigration data may be distorted by mortality and unmonitored individuals. However, the case of the Seychelles Warblers on the tiny island of Cousin (chapter 8) allows us to exclude confounding variables and to obtain unbiased results on various aspects of natal dispersal. As with the Swedish Great Reed Warblers discussed above, every individual of this island is known to the researchers, and, what is more, the population is closed with no immigration or emigration. Even in the small and saturated population of this territorial species, females dispersed and forayed further from their natal territories than did males (Eikenaar *et al.* 2008a). However, this did not reduce the chances of females mating with relatives (Eikenaar *et al.* 2008b). This demonstrates yet again how important it is that each individual case is carefully examined. We only have access to such information today thanks to the application of the very latest molecular genetic techniques in precise long-term population studies of reed warblers. In this way genuinely new horizons are being opened up.

Thirdly, natal dispersal distances in migrants are greater than in resident species (Paradis *et al.* 1998, Thorup 2006). This characteristic might have played a role in the settlement of Pacific islands by reed warblers, which are most likely descended from migratory Asiatic species (chapter 12).

Site fidelity can be observed not only on the breeding grounds, but also in moulting areas and in stopover localities during migration, as well as in overwintering areas (Cantos & Tellería 1994, Yohannes *et al.* 2007, 2008b). High winter site fidelity has been recorded regularly in Marsh Warblers (Kelsey 1989), but also in Eurasian and Basra Reed Warblers and Melodious Warblers (Kennerley & Pearson 2010).

4

Mortality, predation, and parasites

Migration is costly; it is possible that only raising young and moult are more energy-consuming (Bennett & Harvey 1987, Gill 2007). During migration, individual acrocephalids can die in large numbers – exhausted and starving in the desert, blown off course by storms, weakened by parasites, taken by predators, and killed by humans. Some examples stand for many: in the most hostile zones of the Nubian Desert in Sudan what can only be described as

bird cemeteries have been found, with acrocephalids common among the avian mummies (Nikolaus 1983). In the list of prey species taken by populations of Eleonora's Falcons that breed on Mediterranean islands, and also by Sooty Falcons along the Red Sea coasts, acrocephalids are fairly prominent, corresponding to the large proportion of passage birds through these areas for which they are accountable (D. Ristow, G. Nikolaus, pers. comms). At one stopover locality in the Nubian Desert, four shrike (*Lanius*) species specialize on passerine prey, and their most frequent prey are Marsh Warblers, which represent the commonest passage species in August-September. When there were great numbers of migrants the shrikes, also on migration, nearly always fed on the brain of their dead victims. It was quite noticeable that a wave of shrikes arrived one or two days after a 'warbler wave'. If the weather was unfavourable and fattening conditions for migrants limited, then it was easier for the shrikes to gain weight because they could so easily take the numerous birds that succombed. Conversely, the shrikes had a harder time when fattening conditions were better for their potential prey (Nikolaus 1990).

Photo 11.9 A variety of predators find easy pickings among the masses of migrating birds. Sedge Warblers, with their heavy fat deposition, fall victim more frequently than Eurasian Reed Warblers, which carry less fat and so possibly remain more manoeuvrable In the Nubian Desert four migrant shrike species specialize on co-migrant passerine prey, and the commonest on their menu are Marsh Warblers. Shown here is a Steppe Grey Shrike feeding on a Spotted Flycatcher in Kuwait (photo AbdulRahman Alsirhan).

Many passerine migrants, such as Eurasian Reed War-blers, carry only light fuel loads. This is not only a result of low prey availability, but also expresses the trade-off in keeping the risk of predation low (Bayly 2006). In other species with heavy fat deposition, such as Sedge War-blers, manoeuvrability and take-off ability are negatively affected and result in a higher predation risk from birds of prey. On the island of Lesvos in the Aegean, Sedge War-blers are often very fat prior to crossing the Mediter-ranean and Sahara (Kullberg et al. 2000). Fat Sedge Warblers behave in a more secretive way and are more reluctant to fly. The heavier the warblers, the greater the reduction in their flight velocity, acceleration, and angle of ascent, thus putting them at greater risk of being pre-dated (fig. 11.4). In fact Sedge Warblers are the common-est acrocephalid among the prey remains of Eleonora's Falcons in a colony off the island of Crete, despite the fact that far more Eurasian Reed Warblers are on passage here.

Parasites can impair a bird's viability, especially at times of additional stress such as migration, moult, or breed-ing. Common blood parasites are *Haemoproteus* spp. or *Plasmodium* spp., of which eight and ten respectively have been recorded in Great Reed Warblers (Bensch et al. 2007). If transmitted during migration, these haemospo-rides can induce a 'migration cost' with negative impact on host population size and viability (Bensch et al. 2007, Hasselquist et al. 2007). However, the full extent of these effects remains poorly understood, and in different spe-cies fitness effects range from seemingly negligible, as in the case of Swedish Great Reed Warblers (Bensch et al. 2007), to severe, as in Aquatic Warblers, where infected males sired fewer offspring, weighed less, and arrived later on breeding grounds than uninfected males (Dyrcz et al. 2005). The susceptibility to and outcome of avian malaria is influenced by genetic factors (e.g. MHC charac-teristics; section 10.2). The prevalence of the malaria pathogen is higher, for example, in Great than in Eur-asian Reed Warblers (Shurulinkov & Chakarov 2006), and in the case of *Haemoproteus payevskyi* was at a level of *ca* 20% in Great Reed Warblers over a period of six years. No difference in prevalence was found between age or sex classes in Great Reed Warblers, but there were clear dif-ferences in parasite intensity, which was higher in one-year-olds and older females than in adult males. The common pattern is that females have a higher prevalence of malaria than males, and that prevalence decreases with age. It has yet to be explained why parasite intensity in infected adult birds decreases during the course of the breeding season. As none of the juvenile birds shows infection by *H. payevskyi* before autumn migration, the transmission clearly occurs on wintering sites or at stop-over sites during migration. In addition, *H. payevskyi* has been found to be common in resident west African bird species. The fact that the infection occurred in Africa would also explain why one-year-olds and females show a higher parasite burden, since they remain for a longer period of time in their winter quarters than the males (Hasselquist et al. 2007). The risk of blood parasite trans-mission is not the same everywhere. Using isotope signa-tures in winter-moulted feathers, it was shown that moulting sites with a higher incidence of malaria are gen-erally drier and further to the north in west Africa than sites with a lower incidence (Yohannes et al. 2008b).

5 Moult

Together with reproduction and migration, moult is the third crucial process that has to be 'fitted into' a bird's annual cycle. Because each of these three phenomena makes its own particular demand on a bird's time and energy, they are more or less mutually exclusive. The effects of moult have far-reaching impacts on a bird's behaviour, as illustrated by our captive Aquatic Warblers, which, although living for over nine years in a large aviary, moulted very quickly and so lost their primary feathers every year from mid-December. After 15 to 20 days, both the large gaps between primaries and secondaries and the almost synchronous loss of all tail feathers led to flightlessness. This is a strategy that is known from other skulking species in thick tussocky vegetation (Round & Rumsey 2003), and meant that, for the Aquatic Warblers, most activity in the following two weeks was restricted to the ground. During this time they crept through the undergrowth like mice, were more timid than usual, and spent much of their time hidden in dense vegetation.

As a result of wear and bleaching, feathers have to be renewed at regular intervals if their quality is to be main-tained. The constant contact with sharp, silicic-acid-rich reed and grass leaves contributes substantially to plumage wear, particularly in reed warblers. Similarly, Willoughby (1991) found that in another group of closely related but ecologically diverse species, the New World emberizid sparrows of the genus *Spizella*, of all the environmental factors such as solar radiation, dust, and so on that can abrade plumage, exposure to abrasive vegetation had the greatest direct effect. The plumage of adult Eurasian Reed Warblers appears to be among the most worn of any passerine within Europe – more than two millimetres of abrasion in wing and tail feathers (Thorne 1975, Brensing 1985; fig. 11.5). In the Tuamotu Warblers, which live on unprotected South Sea atolls, feather wear is so excessive (presumably through the intense sunlight and abrasion from particles in the strong wind) that their flying ability is reduced (Bruner 1974).

Like most other passerines, reed warblers undergo a complete annual moult of all their plumage, as well as a partial moult which is commonly restricted to areas of the body plumage. Their moult follows the pattern typical of the majority of passerines, beginning with the head feathers, followed quickly by the first primaries; with several wing feathers always replaced simultaneously. While the moult pattern is the same in all reed warblers, the seasonal

timing, place, extent, and rate of moult varies considerably, thus exhibiting great flexibility and plasticity depending on environmental circumstances (Weber *et al.* 2005, Barta *et al.* 2008, de la Hera *et al.* 2010).

Strategies for when and where to moult

Complete moult is seasonally determined, and in most songbirds it begins directly after the end of the reproductive period (Svensson & Hedenström 1999). In the resident reed warblers of the tropics and subtropics the timing of their complete moult is variable. Some Pacific species interrupt their virtually year-round slow moult during breeding, while in others the two processes overlap, at least partly, while a third group moults immediately after breeding (postnuptial; Bruner 1974, Holyoak & Thibault 1984, Thibault & Cibois 2006). In the Seychelles Warblers complete moult ends immediately prior to the start of breeding (prenuptial), and birds that have been translocated to the island of Aride with its 'paradisical' abundance of food can even breed and moult simultaneously (Komdeur 1996a).

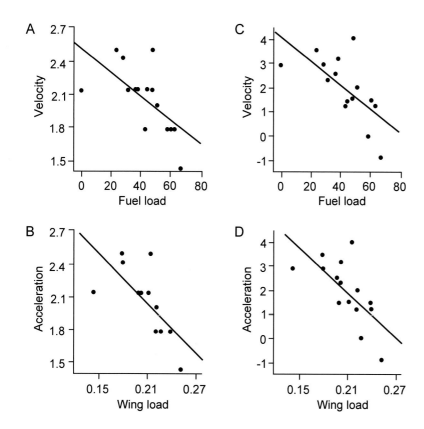

Fig. 11.4 Reduced manoeuvrability relative to fat load in Sedge Warblers. In this experiment at a trapping station, a large transportable aviary was set up to film and analyse the take-off behaviour of captured Sedge Warblers when confronted with the silhouette model of a hunting falcon. Velocity (m/s) measured at 60 cm from the start and acceleration (expressed as m/s²) from the first measurable velocity to 60 cm from the start. Fuel load is given in percentage of body mass; wing load = g/cm² (from Kullberg *et al.* 2000).

Among the acrocephalids that breed in the temperate zones, all of which are migrants, there are by contrast only five species known to undergo an obligate moult on the breeding grounds (summer moult, postnuptial moult), namely Oriental and Black-browed Reed Warblers, Moustached Warblers, at least the nominate subspecies of Blunt-winged Warblers and the majority of Thick-billed Warblers. None of them are extreme long-distance migrants; they are short- and middle-distance migrants and all of them winter in the Oriental Region (although this applies only to the race *mimicus* in Moustached Warblers, as the other races winter in the Palaearctic; Leisler 1972a, Ezaki 1984, Round & Rumsey 2003, Kennerley & Pearson 2010). In these five species the young

also moult their flight feathers while still at the natal site, in other words while they are just two or three months old (complete postjuvenile moult; see below).

All other acrocephalids that migrate long distances (*Hippolais, Iduna, Acrocephalus*) delay their moult until reaching either staging sites much further south on their route or their final winter quarters (winter moult). In this they differ from many other long-distance migrants in related sylviid groups (e.g. the leaf warblers *Phylloscopus*) or unrelated ecological equivalents such as the North American *Dendroica* wood warblers, in which not a single species moults in its tropical winter quarters, but all replace their feathers on their breeding grounds after the end of

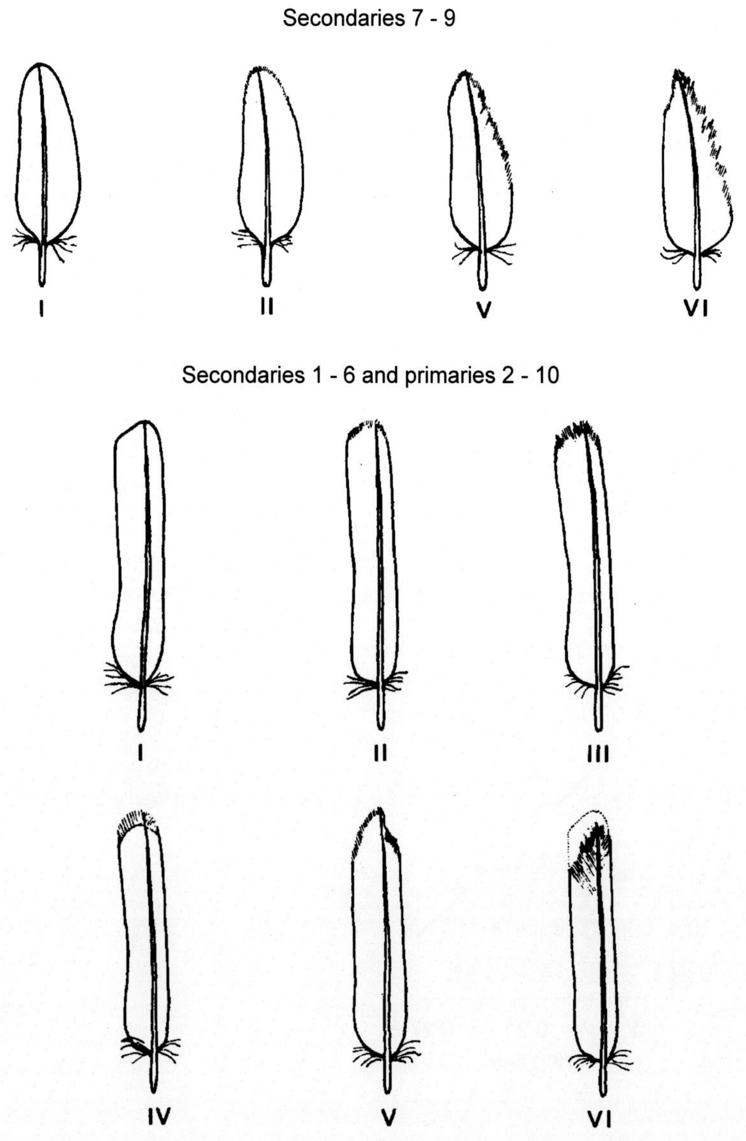

Secondaries 7 - 9

I   II   V   VI

Secondaries 1 - 6 and primaries 2 - 10

I   II   III

IV   V   VI

Fig. 11.5 Different stages of the abrasion of primaries and secondaries in Eurasian Reed Warblers (I to VI: abrasion stages; from Brensing 1985).

Photo 11.10 To maintain their high quality, feathers have to be renewed at regular intervals. Bleaching by sunlight and wear and tear resulting from frequent contact with sharp-edged blades of grass lead to feather abrasion, often by more than 2 mm. The plumage of adult Eurasian Reed Warblers appears to be among the most worn of any passerine in Europe. Such abrasion can be clearly seen on this Clamorous Reed Warbler in Kuwait (photo AbdulRhaman Alsirhan).

breeding. Many migrants in these latter two groups typically winter in tropical forests (a much buffered, non-seasonal, and comparatively less productive environment), whereas the acrocephalids winter in highly seasonal productive habitats (section 3.5).

Phylogenetic comparisons in over 140 taxa of sylviid warblers (*sensu lato*) show that summer moult is the ancestral state and winter moult the derived (Svensson & Hedenström 1999, Hall & Tullberg 2004), and also that the main moulting strategy is correlated with migration distance and with migratory habits as such. As taxa increased their migratory distances, summer moult disappeared and winter moult evolved. However, it was only when sophisticated modelling calculations on the evolutionary 'decisions' of the birds were made by Barta and co-workers (2008) that the causes of the variation in moult-migration patterns and the possibility of evolutionary transitions were revealed: in both cases the decisive role is played by the temporal and spatial distribution of food. Summer moult occurs in migrants, (e.g. in the east Asian acrocephalids mentioned above and in *Phyllosco-*

pus or *Dendroica),* when food has a long-lasting peak on the breeding site in the summer, but is less seasonal elsewhere during the non-breeding season. Winter moult occurs if there is a short period of high food availability in summer and a strong peak at staging or wintering locations. Winter moult occurs only in long-distance migrants (Jenni & Winkler 1994), and can only evolve if food resources on the breeding and wintering grounds are both strongly seasonal and out of phase with each other.

Most migrating acrocephalids, both wetland- and scrub-dwellers, have developed exactly this mode of life: they briefly exploit productive, seasonal habitats for breeding, and during the non-breeding season habitats of a similar nature for moult (section 2.3 and chapter 3). Most prominent among them are those that winter in Africa; they can utilize the seasonal, productive conditions there for their moult, when the ITZC moves southward across the continent in the second half of the year. Reed warblers are able to start their moult at very different times (see below). The moulting opportunities for those species that winter in India are more limited. All of them – Paddyfield, Large-billed, Sykes's and Booted Warblers, Blyth's Reed Warblers and Clamorous Reed Warblers of the race *brunnescens* – start their moult soon after arrival in the northern tropics. Conditions for the development of a winter moult in tropical southeast Asia are unfavourable because landmasses that extend far to the south (as does Africa) and that are therefore ideal for overwintering birds, simply do not exist, and also because a maritime climate dominates throughout the year. Only two species – Manchurian and Streaked Reed Warblers – renew their flight feathers here, using reedbeds or tall rank grassland for the purpose. The other reed warblers that overwinter in the same region – Oriental and Black-browed Reed Warblers, Thick-billed Warblers – have already undergone a summer moult after breeding (see above). These warblers differentiate themselves from typical winter-moult species by their considerably longer sojourns in their breeding areas and by shorter, more direct migration routes to their winter quarters.

Interestingly, it has recently been demonstrated that up to 20% of individuals in the Mediterranean populations of Great Reed Warblers, which are close relatives of the Oriental Reed Warblers, also have a complete summer moult while still on their breeding grounds, while others,

Photos 11.11 to 11.13 Complete moult is seasonally determined, beginning in most passerines directly after the breeding season. Only five migrating reed warbler species, most of them from eastern Asia, undergo their moult on the breeding grounds. By contrast, all long-distance migrant acrocephalids of the Palaearctic-African migration system moult after autumn migration, mostly in places where there is a temporary abundance of food on account of the rainy season. This full moult must proceed quickly, being at the mercy of unpredictable environmental

including first-year birds, renew at least some remiges (e.g. Maragna & Pesente 1997, Copete *et al.* 1998). Mediterranean subpopulations can apparently afford a summer moult because the climate allows them to remain longer in their breeding areas; what is more, their migrations are shorter than the Great Reed Warbler populations which breed further north, and which overwinter much further south in Africa (fig. 11.6). Additional examples are provided by (semi-)resident birds of the *A. scirpaceus/baeticatus* complex in Morocco, around half of whose adult birds and some of the first-calendar-year birds undergo a complete moult north of the Sahara (Amezian *et al.* 2010), and also by the northern and southern populations of the *brunnescens* subspecies of Clamorous Reed Warblers, where the former undergo a winter, and the latter a summer moult (Hall & Tullberg 2004, Kennerley & Pearson 2010). These examples are certainly not unique. The transition from summer to winter moult and vice versa need not therefore be the result of a long evolutionary process but can rather take place flexibly and at 'short notice'.

Scheduling the time and duration of moult

There is substantial variation in the timing of moult in the winter quarters. The following rules can be established for those reed warblers that winter in Africa: the further north the winter quarters, the earlier the moult, which can even take place in moulting or stopover sites before the final overwintering grounds are reached. The further south the final wintering latitude, the later the moult, which at times can even take place shortly before the onset of return migration. For instance, Marsh Warblers, the species with the longest migration and the most southerly winter quarters, arrive in worn flight feathers in their southern African overwintering areas, to moult there from the end of January until departure around the end of March (Dowsett-Lemaire & Dowsett 1987b; fig. 11.6).

conditions, since a bird's flight abilities are drastically affected by the simultaneous loss of several wing and tail feathers. This Basra Reed Warbler in Kuwait is in heavy (partial) moult of its head feathers; the Moustached Warbler in Hungary undergoes a rapid moult on its breeding grounds, and on this Clamorous Reed Warbler in India numerous fresh tail and primary feathers are growing in at the same time (photos Mike Pope, Anon., Sunil Singhal).

The further south or the later a reed warbler population moults, then the longer the process lasts. It is not only the moult of Moustached Warblers or Oriental Reed Warblers on the breeding grounds that is rapidly completed but also that of species which moult in the northern tropics. The post-juvenile primary moult of Oriental Reed Warblers (at the very early age of around two months) lasts not much more than 30 days (Ezaki 1984), thus approaching the shortest known time for primary moult

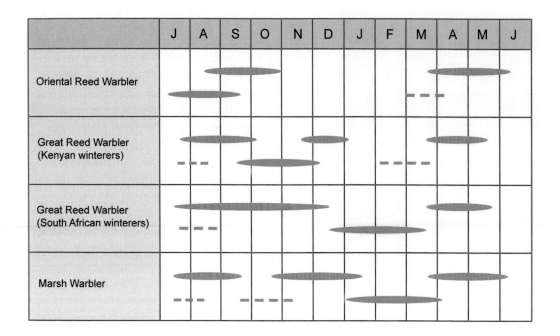

Fig. 11.6 Comparison of the timing of complete (solid brown) and partial (dotted brown symbols) moult relative to periods of migration (solid grey) in three *Acrocephalus* species (from Kennerley & Pearson 2010).

in any passerine bird, that of Snow Buntings in Greenland, at 28 days (Ginn & Melville 1983). The moult of Moustached Warblers on the breeding grounds, and of Aquatic, Sedge and Great Reed Warblers in the Sahel, takes around 40-60 days (Bensch *et al.* 1991, Leisler 1991a,b, Hedenström *et al.* 1993, Schulze-Hagen unpubl.). On the other hand, the moult of those reed warblers that winter much further south lasts considerably longer – up to 92 days. Those reed warblers that migrate the furthest have nowhere near so much time at their disposal as in their final winter quarters. For a species such as Marsh Warblers it appears to be important to leave the breeding grounds early in order to secure a good winter territory, thus allowing plenty of time for moult and formation of fat reserves before return migration and breeding (Pearson 1973). Both sexes of this species hold winter moulting territories in southern Africa, which they defend by singing and which have the function of safeguarding food resources during that critical period (Kelsey 1989; sections 3.5 and 6.1). Such moulting territories are as yet unknown among the reed warblers that moult in the northern tropics. Under the time con-

straints of unfavourable environmental conditions, the duration of moult can be remarkably short (Bensch *et al.* 1991), though this might be at the cost of feather quality, since higher feather quality is an important precondition for a successful and efficient return migration (Pearson 1973, Holmgren & Hedenström 1995, Barta *et al.* 2008, de la Hera *et al.* 2010).

We have seen just how variable the scheduling of moult in the acrocephalids can be, and that moult in the non-breeding season has obviously evolved as an adaptation under pressure of time constraints for breeding, migration, and survival (Svensson & Hedenström 1999, Barta *et al.* 2008). The timing of the rainy season that varies with latitude and longitude, and the resulting rainfall pattern, may have an overriding effect and be capable of explaining both migration patterns and the timing and duration of moult in long-distance migrating acrocephalids (Jones 1995). It is highly significant that a genetically based, flexible moult-timing mechanism that can be fine-tuned to local conditions has been found in forms of Eurasian Stonechats (Helm & Gwinner 2006). Such a potential for flexibility resulting from both genetic variation and individual plasticity in moult performance is widespread in the sylviid warblers and may be a basal trait (Hall & Tullberg 2004). The diversity and flexible timing of moult can be seen as a buffer in the annual avian cycle in which breeding has priority.

Summary

Every year enormous numbers of reed warblers move back and forth between their breeding and wintering sites, some of them covering a total of 18 000 km in the process. They are among the most frequently ringed migrants and serve as model species for ecological and physiological migration studies. From the results of these, one fascinating finding has been that Sedge Warblers and Eurasian Reed Warblers have completely different optimal migration strategies, despite being the same size and living together in the same reedswamps. When accumulating fat for migration, Sedge Warblers tend to feed on less mobile prey such as aphids – a superabundant but unpredictable late-summer food source. In this way they are able to fuel up quickly and fly directly to sites south of the Sahara. By contrast, Eurasian Reed Warblers feed on prey (mostly dipteran flies) that occurs at low densities but which is more predictable and more evenly distributed. They cannot therefore build up substantial fat reserves and so must frequently interrupt their southward journey in order to refuel.

The three main wintering regions of long-distance migrating reed warblers are sub-Saharan Africa, the Indian Subcontinent, and southeast Asia, but the final destination of any one species will depend upon the location of its breeding ground and the history of its settlement.

There is an astonishing connectivity between breeding and non-breeding grounds, with individuals commonly returning year after year to the same spot for overwintering and moulting, thus enabling them to benefit from acquired experience.

While migration is a seasonal process characterized by outward and return movements, dispersal is an exploratory one-way movement leading to a redistribution of individuals in space. During their post-fledging period, Eurasian Reed Warblers make nocturnal exploratory movements over considerable distances, in the course of which they familiarize themselves with their immediate surroundings and develop the internal map necessary for successful navigation during return migration the following spring. In Great Reed Warblers, a remarkable offspring-parent resemblance in dispersal has been shown, indicating a genetic basis for the behaviour.

Moult is another crucial process that has to be fitted into a bird's annual cycle. Studies of reed warbler moult have made important contributions to our understanding of the evolutionary pressures determining the timing of such a crucial event in the cycle as well as the greater flexibility of the process than was previously thought possible; it can now be seen as a buffer in a cycle in which breeding has priority.

While most passerine migrants undergo a complete moult on the breeding grounds, this is true of only a few acrocephalid species, most of them living in the east Asian temperate zone. Others, the longer-distance migrants, delay their moult until reaching either staging sites on their route or their final winter quarters.

12

Our island home

Tahiti Warbler.

In the course of our studies in the American Museum of Natural History (AMNH) in New York we were most impressed by the enormous number of reed warbler skins originating from the Polynesian islands. They fill tray after tray in the huge cupboards, lying in rows on their backs with their strikingly yellow-coloured bellies upwards. They were all collected for the AMNH in the course of the famous Whitney South Sea Expedition in the early 1920s. Series totalling 1500 reed warbler skins were accumulated from southeast Polynesia alone (Murphy & Matthews 1928, 1929).

Reed Warblers occur on islands in two different ways. When we think of reedbeds populated by large numbers of acrocephalids, it is clear that they are simply insular habitats isolated in the surrounding landscape (chapter 3). Yet a whole group of acrocephalids has managed to colonize oceanic islands, and not just any islands, but those lying far from continental land masses (see fig. 1.1 map). Around one-third of reed warbler species are found on islands in the Atlantic, Indian, and Pacific Oceans and they are all oceanic island endemics. What is more, this group does not even include those species on large continental islands such as Madagascar and New Guinea, or those on shelf islands which, though geographically separate from a mainland, are relatively easily reached via island stepping-stones, such as the Bismarck Archipelago and the Solomon Islands.

According to David Steadman (2006), a leading authority on tropical Pacific birds, the ability of reed warblers to disperse over the vast oceans is 'unmatched among passerines'. Only monarch-flycatchers of the genus *Pomarea* have reached a similar number of very remote islands in Oceania. On several of these islands (Tuamotu, Rimatara, Kiritimati, Nauru) an acrocephalid is often the only songbird, and in some cases the only terrestrial bird at all. The extraordinary success of reed warblers in colonizing such islands is illustrated not only by their wide distribution but also by the population numbers present on many islands, at least in earlier times.

In this chapter we shall discuss why acrocephalids have been so successful in colonizing these remote places. In particular we shall ask a series of important questions. Which islands were settled by the birds? Which habitats do they occupy? How do they differ from their continental congeners? How are they related to each other? How have they managed the colonization, and why are they good survivors on islands? What adaptations have they evolved for island life? Have they all followed a common path or has each species its own evolutionary history?

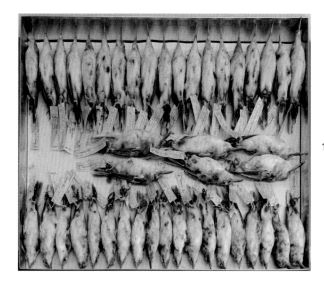

Photo 12.1 A drawer full of skins of Southern Marquesan Warblers in the American Museum of Natural History in New York. In the course of the famous Whitney South Sea Expedition in the early 1920s, around 1500 island-dwelling reed warblers were collected in SE Polynesia (photo Hans Winkler).

1 Island species – relatives, distribution, and colonization history

If we look at island reed warbler distribution (table 12.1; map 12.1) the first thing that strikes us is that in all oceans only a single representative of the Acrocephalidae family occurs on any one of the colonized islands. The pronounced territorial behaviour of any already established acrocephalid species has probably prevented new arrivals from settling. All island acrocephalids are descended from the clade of the large reed warblers, from the *Acrocephalus* as well as the *Calamocichla* subgroups. Within the latter, the very close relationship between the continental Greater Swamp Warblers and the insular Cape Verde Warblers allows us to conclude that the volcanic

Cape Verde Islands were colonized by reed warblers only in recent times. They lie just 460 km off the African coast but were never joined to the continent by a land bridge. By contrast, the divergence between Seychelles Warblers and Madagascar Swamp Warblers in the western Indian Ocean lies in the much more distant past (chapter 2).

Although the idea that the insular acrocephalids in the Pacific derive from an ancient invasion of large reed warblers from Asia was accepted by Tristram and Hartert at the end of the 19th century, and later modified by Mayr (1941), it was only recently that it could be confirmed and refined by the use of molecular techniques (Leisler *et al.* 1997, Cibois *et al.* 2007, 2011a; chapter 2). Based on the genetic research undertaken by Alice Cibois, Robert Fleischer, and their co-workers (2011b), it appears to be beyond dispute that in the course of the radiation of this group into the vastness of the Pacific, multiple colonizations of archipelagos took place, starting both from the mainland and from previously settled island groups.

Map 12.1 Map indicating the two main lineages of reed warblers in the Pacific Ocean. The islands or island groups occupied by each of them are shown in red or blue. The spreading lineages meet in the area of the Marquesas. Guam (green) was independently colonized (the species that occurred here, *Acrocephalus luscinius*, is now extinct. NB: the forms earlier placed under the name *A. luscinius* are not closely related to each other). The question mark indicates the unknown common ancestor of the Pacific radiation (from Cibois *et al.* 2011 a).

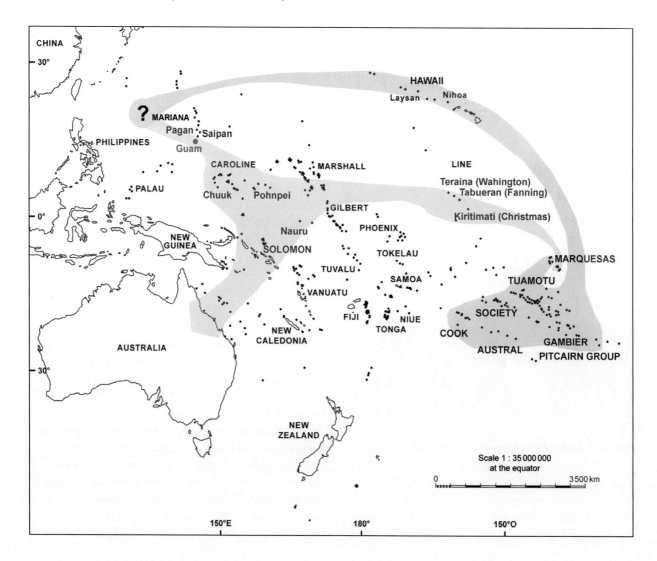

Table 12.1 List of the insular *Acrocephalus* species (M = extant, H = extinct in historical times, P = extinct in prehistoric time).

Archipelago/Is.	Record	Status	Species	Vernacular name
Cape Verde Is.	several	M	A. brevipennis	Cape Verde Warbler
Seychelles	several	M	A. sechellensis	Seychelles Warbler
Rodrigues	Rodrigues	M	A. rodericanus	Rodrigues Warbler
Mariana Is.	Saipan Guam Pagan	M H H	A. hiwae A. luscinius A. yamashinae	Saipan Warbler Guam (Nightingale) Warbler Pagan Warbler
Caroline Is.	several	M, H	A. syrinx	Caroline Islands Warbler
Marshall Is.	Majura Atoll	P	A. spec. (undescribed).	
Nauru	Nauru	M	A. rehsei	Nauru Warbler
Hawaiian Is.	Laysan Nihoa	H M	A. familiaris A. f. kingi	Laysan Millerbird Nihoa Millerbird
Line Is.	Kiritimati Tabueran, Tereina	M	A.aequenoctialis A. a. pistor	Kiritimati Warbler
Cook Is.	Mangaia Mitiaro	M, P M	A. kerearako A. k. kaoko	Cook Islands Warbler
Society Is.	Tahiti Moorea Leeward Is.	M H H H	A. caffer A. longirostris A. musae A. m. garretti	Tahiti Warbler Moorea Warbler Leeward Islands Warbler
Austral Is. (Tubuai)	Rimatara	M	A. rimitarae	Rimatara Warbler
Tuamotu Is.	several atolls	M, H	A. atyphus polytypic	Tuamotu Warbler
Gambier Is.	Mangareva	H	A. astrolabii	Mangareva Warbler
Marquesas Is.	Northern group Southern group	M	A. percernis polytypic A. mendanae polytypic	N Marquesan Warbler S Marquesan Warbler
Pitcairn Is.	Pitcairn Henderson	M M	A. vaughani A. taiti	Pitcairn Island Warbler Henderson Island Warbler

The path taken by reed warblers during their penetration of Oceania is therefore not as simple as was once supposed (Baker 1951). It was thought, for example, that the birds spread ever further in a one-way 'stepping-stone' fashion, from the continent to an island and from there to the next. In fact, most reed warblers reached their new island homes by frequent, independent, long-distance dispersals. For instance, the Mariana Islands have been colonized at least three times independently and not via continuous diversification within the archipelago. Reed Warblers on Saipan and Pagan are each independent descendants of the large monophyletic group of Pacific species, whereas the species on Guam falls outside this group, in turn having its origin in an independent colonization from the mainland (Cibois *et al.* 2011a).

The main Pacific radiation started approximately 2.4 million years ago in two lineages, the first an 'east Polynesian' group, within which the warblers of Hawaii diverged at an early period and which includes all species in Polynesia, except those in the Line and southern Marquesas Islands (Îles Marquises). Representatives of the second younger 'Micronesian' group not only advanced as far as those two archipelagos but also diversified, principally in

Micronesia, reaching Australia in a sort of 'reverse' colonization, from islands to a continent (map 12.1).

Double or repeat colonization of archipelagos led to the present complex situation in several instances, where neighbouring islands, or island groups, are settled by reed warblers that are not immediately related to each other. In the Marquesas Islands the reed warblers of the northern islands (Northern Marquesan Warblers) are most closely related to the Tuamotu Warblers, while the southern populations (Southern Marquesan Warblers) are closer to the Kiritimati Warblers. The two colonization events occurred simultaneously *ca* 0.6 mya (Cibois *et al.* 2007, 2011a). A further example is that of the acrocephalids of the Society Islands that colonized or are colonizing Tahiti, Mo'orea, and the Leeward Islands (Îles Sous le Vent). Contrary to expectation, none of them is closely related to any other (Cibois *et al.* 2008). In all of these cases, parallel or convergent adaptations to the local environment on an archipelago have resulted in forms of differing lineage becoming so similar in appearance that they were previously assumed to be a single taxon, and it is only now – by using molecular methods – that we can recognize them as different and independent species.

Of the various factors that normally influence colonization, the size of islands and their distance from the nearest continent or to the next archipelago apparently played only a minor role, since the birds also inhabit small islands and 'distances between islands are not a limiting factor *per se*' (Cibois *et al.* 2011a).

Also of importance are the age and geological type of the islands concerned. Eroded volcanic and raised limestone islands are more suitable for long-term settlement than active volcanoes or atolls, on which landbird populations are obviously vulnerable to eruptions, changes in sea level, or tsunamis (Steadman 2006). Reed Warblers occupy all four of the generally distinguished island types – volcanic islands, raised limestone islands, atolls, composite islands. Most high islands in the region are around 1-10 million years old, and were available for settlement by reed warbler lineages, apart from the Pitcairn group, where Henderson emerged only 0.4 mya and could only then be colonized from Pitcairn Island (Cibois *et al.* 2011b).

Lowered sea levels in the Pleistocene created larger stepping-stones, which may have aided the warblers' colonization process. During such episodes there was both an

Photos 12.2 and 12.3 More than one-third of reed warbler species are found on oceanic islands lying far from continental land masses. The ability of acrocephalids to disperse over vast oceans and to colonize islands is unmatched among passerines. On several of the Pacific Ocean islands a reed warbler is the only terrestrial bird. One can hardly imagine that the low scrub of the barren-looking volcanic island of Nihoa in the Hawaiian archipelago is home to the Millerbird (photos WikiCommons, Jack Jeffries).

increase in island area and also in the extent of suitable vegetation. In this situation the resulting higher densities of landbirds could have served as source populations (Steadman 2006, Cibois *et al.* 2007).

By contrast, interglacial sea level highs have repeatedly extinguished reed warbler populations on low-lying atolls. Cibois *et al.* (2010b) were able to reconstruct such a case most precisely on the basis of geological data and the genetic structure of the reed warblers of the Tuamotu archipelago. When the sea level rose 125 000 years ago and was around 6 m higher than today, three elevated carbonate islands acted as refugia for the landbirds. When it fell again a recolonization followed, mainly from one of those islands (Makatea).

At present, reed warblers have a disjunct distribution in the Pacific (map 12.1) and are found discontinuously in a long strip between Micronesia and the Pitcairn group, together with an extension from the Line Islands northward to Hawaii. However, they are completely absent from the Solomon Main Chain, which still harbours the *sumbae* subspecies of Australian Reed Warblers, across the 5000 km to the Cook Islands, and are apparently avoiding the large and species-rich islands of Vanuatu, Tuwalu, Fiji, and Samoa. Neither do they occur on the central Hawaiian Islands. Even if acrocephalid species did once inhabit some central Pacific islands but have since became extinct (Graves 1992, Steadman 2006), the best explanation for their disjunct distribution is that, as so-called 'super-tramps' – species specialized for rapid over-water coloni-zation and present only on small or species-poor islands (Mayr & Diamond 2001, McNab 2002) – reed warblers were unable to establish themselves amidst the richer biotas of larger islands (Cibois *et al.* 2007, 2011b).

For many years, our picture of the colonization of the Pacific by Old World bird groups was shaped by the scenario of a northeast-southwest settlement and of a consequently steady decrease in species diversity ('faunal attenuation'; a downstream 'sink' for continental forms). The latest molec-ular studies have gradually traced the immigration routes of various avian groups (white-eyes, monarch-flycatchers) and have thus revealed a variety of differing patterns and ages (Slikas *et al.* 2000, Cibois *et al.* 2004, Filardi & Moyle 2005). On the whole, the far-reaching spread of reed war-blers through small Pacific islands can be regarded as recent diversification into areas where resources are sparse.

Island habitats

The great majority of the islands or island clusters that have been settled by reed warblers lie within the tropics, a few slightly outside; only Pitcairn, Tubuai, and the northwest Hawaiian group are subtropical. The climate is tropical under oceanic influence on the Cape Verde Islands, under monsoon influence in the western Indian Ocean, and under the influence of the trade winds in the Pacific, where the islands have moderately wet non-sea-sonal climates, although those outside the trade-wind belt on the equator can experience extreme droughts.

On these various islands acrocephalids have settled in completely different habitats from those inhabited by their ancestral continental forms, and now live in wood-land, dry bush, and upland areas, though if wetlands are present then they will utilize them. An idea of the variety of environments present on different archipelagos and of the different habitats occupied by reed warblers is given in table 12.2 (based on Pratt *et al.* 1987, del Hoyo *et al.* 2006, BirdLife International 2004-2010). Island habitats of reed warblers are generally drier and have more bushes and trees than those of continental reed warblers. They range from very open habitats (Kiritimati Warblers) to extremely dense thickets (Rodrigues Warblers), and from low scrub (Nihoa Millerbirds) to tall ravine forests (several Pacific species). Wetlands of native reed, mangroves, and cane thickets are settled, as are different types of native forest, such as coastal and inland lowland, limestone, and moun-tainous forests on steep ridges up to more than 1000 m. For all three species on the Society Islands, one habitat of extreme importance is, or was, the groves of Polynesian bamboo up to 18m high (Cibois *et al.* 2008). Other tree species that are commonly utilized, both for nest sites and foraging, are the *ca* 15 to 20-m high sea hibiscus or purau, *Ficus* spp., rosewood, and the widespread nutritious morinda, as well as a series of non-native plant species, such as strand tangantangan, rose-apple, and guava. While it is apparently the case that reed warblers have increasingly moved into woodland in the course of their immigration, they have retained the group-specific prefer-ence for forest edges and clearings (section 34).

While Cape Verde Warblers occur in various anthropogeni-cally altered habitats on the three islands they still inhabit, it is assumed that before people arrived they would have inhabited the tall scrub that originally covered most of the total surface of the Cape Verde Islands (Hering & Fuchs

Table 12.2 Habitats, population sizes, and status (+ = stable, - = endangered) of the extant island reed warblers.

Species	Habitat	Population estimate	Status
Cape Verde Warbler *A. brevipennis*	variable, scrub on slopes, reed bed in valleys, plantations	1 000 - 1 500	+
Seychelles Warbler *A. sechellensis*	tall scrub-like vegetation, mangroves	2 500	+
Rodrigues Warbler *A. rodericanus*	dense thickets, non-native woodland	150	+
Saipan Warbler *A. hiwae*	thicket-meadow mosaic, reeds, forest edge, mangroves	2 000 - 2 500	+
Caroline Islands Warbler *A. syrinx*	tall grass, gardens, secondary growth, montane forests	?	−
Nauru Warbler *A. rehsei*	remnant forests, gardens, coastal vegetation, scrub	2 500 - 10 000	+
Nioha Millerbird *A. familiaris kingi*	bushy hill sides, bunch grass	480 - 1 150	+
Kiritimati Warbler *A. aequinoctialis*	open areas with scattered trees, dense brush	10 - 20 000	+
Cook Islands Warbler *A. kerearako*	variable, herbaceous vegetation, reed beds, gardens, brushy woodland	?	−
Tahiti Warbler *A. caffer*	bamboo groves, riverine woodland at moderate altitude	300	−
Rimatara Warbler *A. rimitarae*	natural forests, undergrowth, mixed horticulture, swamps	680 - 2 600	+
Tuamotu Warbler *A. atyphus*	brushy woodland, gardens, plantations, xerophytic thickets	?	−
N. Marquesan Warbler *A. percernis*	wet upland forests, brushy hill sides, plantations	?	−
S. Marquesan Warbler *A. mendanae*	wet upland forests, brushy hill sides, plantations	?	−
Pitcairn Island Warbler *A. vaughani*	patches of tall forests, scrubland	250 - 1 000	+
Henderson Island Warbler *A. taiti*	native forest, lush undergrowth	11 000	+

2009). Today most of the habitats utilized by these island endemics are vegetation forms that have been greatly influenced by humans (section 12.6 and table 12.2).

3 Island conditions and phenomena

Islands can be regarded as natural laboratories and so have always held a special fascination for biologists, as they can provide important insights into ecological processes such as divergence and convergence. Particularly well known is the phenomenon of 'island marvels', such as the flightless birds and beetles that result from eons of insular life. Various organisms that have colonized very different islands respond to the altered pressures of these isolated environments in a surprisingly similar fashion; in other words, there are generalized ecological conditions or problems on islands that different immigrants solve in the same way, or convergently.

In table 12.3 we list the ecological conditions for a whole series of specific and well-known island phenomena and evolutionary trends among island endemics, that, taken together, are known as the 'insular syndrome' (Williamson 1981, Grant 1998, Blondel 2000, Kikkawa & Blondel 2002). The supply of resources on islands, of habitats and food in

Photos 12.4 to 12.6 On various islands, most of them lying within the tropical belt, reed warblers have settled in completely different habitats from those occupied by their continental relatives. They range from very open habitats to extremely dense thickets and tall ravine forests. Open scrub and trees on a Tuamotu atoll, groves of Polynesian bamboo and bush in the valleys of Tahiti, and mountainous forest on steep ridges on Grande Comore in Comoros (photos Jean Claude Thibault, Jens Hering [2]).

particular, are generally depauperate and scarcer than on mainlands. For example, southeast Polynesia is characterized by an almost complete lack of native gymnosperms, while ferns are relatively common, which means that herbaceous plants are underrepresented. The trees and shrubs that dominate there are restricted to only a few families of perennials (Mueller-Dombois & Fosberg 1998). Such 'floristic disharmony' can be repeatedly observed, and results in habitats and food resources for the fauna becoming so limited and specialized that it is difficult for some animal groups to colonize these islands at all. Oceanic islands are typically poor in insect variety and abundance – in particular there are fewer pollinating insects than on the mainland (Olesen & Valido 2003). One way for birds to respond to a scarcity of arthropod food, occasional food shortages, and other resource limitations, is to broaden their niche. Minimizing energy expenditure is another elementary solution to the problem that has helped sustain the presence of vertebrates on oceanic islands (McNab 2002).

An important characteristic of islands is the small number of species which inhabit them, a phenomenon formerly known as 'impoverished species ensembles'. This typically means, for example, that a reduction in predator pressure results from the fact that few predators have reached islands, which are generally free of carnivorous mammals. This in turn has allowed insular birds to dispense with antipredator behaviour, to limit their flying ability, and to strengthen their exploratory behaviour,

so that they become inquisitive and fearless (Mettke-Hofmann 1999, Winkler & Leisler 1999). Furthermore, conspicuous plumage abnormalities such as leucism (partial white colouration) are not selected out (section 12.5). The depauperate species community on islands also means that interspecific competition is relaxed, thus allowing the niche breadth of species to expand (see above). 'Ecological release' from predation and from interspecific competition is thought to be the reason why the inhabitants of species-poor islands commonly reach higher population densities than their mainland equivalents.

Colonizers of oceanic islands also live in a relatively disease-free environment. However, a life in pathogen-poor surroundings does not necessarily lead to an overall attenuation in the immune responses of island endemics when compared with continental species (Matson 2006). This was confirmed in a comparative study using *Acrocephalus* species by Beadell *et al.* (2007). They compared parasite prevalence and various measures of immune response in two Pacific island species, Rimatara and Kiritimati Warblers, with the results from a mainland counterpart, the closely related Australian Reed Warblers. Although, in contrast to the latter, the two insular species had no (malaria) blood parasites, the immunological reactions observed in both species were extremely divergent.

From this we can conclude that we should not assume that immunological changes in parasite-impoverished environments are invariably uniform, even among related

Photos 12.7 to 12.10 Acrocephalid species are often confined in their distribution to a single, sometimes tiny island, which in itself means that the risk of extinction is high. Top from left to right: Grande Comore Brush Warblers live on Grande Comore, and Moheli Brush Warblers on Moheli, another island in the Comoros group. Until 1988 Seychelles Warblers were restricted to the 29-ha small island of Cousin, but since then conservationists have translocated them to three neighbouring islands. Below: Tahiti Warblers have recently been classified as 'Endangered' because of their small, declining population. The main threat to the species is the exploitation of bamboo, their favoured habitat (photos Jens Hering, Jan Kube [Seychelles Warbler]).

taxa with similar island residence times and a similarly reduced genetic diversity. Small isolated populations apparently develop unique immunological profiles that make it difficult to predict the impact of introduced diseases (section 13.6).

A further common and repeated response by organisms to their new insular environments is a decrease in their dispersal abilities. The high population densities on many islands can increase intraspecific competition, promoting either dominance, through an increase in body weight for instance, or greater toleration of conspecifics (McNab 2002). Large population sizes also represent a certain security against the risk of extinction, which is otherwise particularly high for island species because of stochastic processes (section 12.6).

4 Morphological adaptations and the expansion of niches

A well-studied theme in the ecology of island species is that of niche expansion. Continental species can frequently use resources only in a narrowly bounded resource space because other resource spaces have been

Table 12.3 Ecological conditions on islands and the consequences for their inhabitants.

Ecological factor			'Insular syndrome'
reduced (absent)	resources		niche expansion energy saving
	predation	} ecological release	tameness exploratory behaviour
	interspecific competition		niche expansion
	parasitism		reduced disease resistance
enhanced	risk of extinction		reduced dispersal

successfully occupied by competitors. On islands, this space *can* be expanded by ecological release or *must* be expanded by resource limitation, in that individuals behave more as generalists. Increased generalism is made possible by the body sizes of island birds converging on an optimal size, around 100 g, and by their bills converging on a medium length. Consequently, larger species should become smaller and long-billed species should develop shorter bills, and vice versa. This phenomenon is known as the 'island rule' (Clegg & Owens 2002, Scott *et al.* 2003). As small passerines, reed warblers should become larger following colonization of an island, and develop longer bills. This would correspond to an adaptation to foraging generalism, since larger bills allow the taking of large as well as small prey (Herrera 1978, Grant 1998).

Next we must ask if island reed warblers actually are larger, if their bills really are longer, and if they diverge in other ways from their continental relatives? To test this, we measured 19 island and 9 mainland species to see which morphological traits might have drifted apart. Starting from an ecomorphological position (chapter 5), we measured skins for ecologically significant external characters relating to hind limb, flight, and feeding apparatus, in other words, the total morphology (Leisler & Winkler unpubl. ms). Let us first of all compare only body weights and bill lengths of island and continental birds (fig. 12.1).

In both cases, variation is greater among island than among continental species. Since this variation is more (bill length) or less (body mass) biased towards larger dimensions, an average net increase in both traits results. With respect to a specific ancestral continental species, Leisler & Winkler (unpubl. ms) explain this by the fact that island species diverge adaptively to a wide range of unique island ecologies and do not change unidirectionally. Nevertheless, in broad comparisons this may result in the trends described by the island rule, which states that smaller species on islands become larger and vice versa (see above).

In birds, the island rule shows only a very general trend. Both during the period of immigration and later, body size is influenced by a range of different factors, such as island size, the distance between island and mainland, island richness, or population density (Grant 1998, Lomolino 2005). The evolution of acrocephalid warblers does not fit well into these general models of size evolution on islands. This is confirmed by Cibois *et al.* (2007), who found that warblers are bigger on the large rich islands of the Marquesas group, while those on the Society Islands group are smaller on similar large rich islands (Cibois *et al.* 2008). This suggests in turn that evolutionary trends in the genus *Acrocephalus* are influenced by a great variety of topographical, climatic, and floristic conditions, all of them idiosyncratically island-specific.

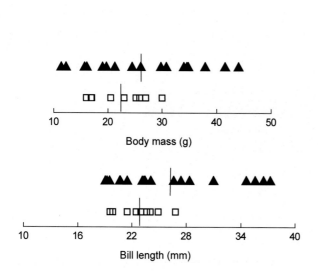

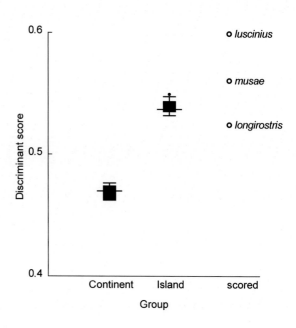

Fig. 12.1 Comparison of body mass (above) and bill length (below) of mainland (n = 9, open squares) and island (n = 16, filled triangles) *Acrocephalus* species. The increase in bill length in island forms is even more pronounced than the increase in body weight.

Fig. 12.2 Result of a discriminant analysis of the total morphology (17 external characters) of 16 island *vs* 9 mainland *Acrocephalus* species, together with the scores of three island forms of which only a few individuals were available. Both groups are clearly separated from each other, with the island forms having thicker legs, shorter rictal bristles, and blunter wings.

Despite this, there are traits that distinguish all insular species from their relatives on the mainland (Leisler & Winkler unpubl. ms; fig. 12.2). Somewhat surprisingly, the most important variables contributing to the clear separation of insular forms from continental forms turn out to be tarsus cross-section, the flight apparatus, and the rictal bristles: birds on islands possess thicker legs, shorter rictal bristles, and rounder, more slotted, and broader wings. In these characters island warblers converged on a common morphology, while body size and bill dimensions were unimportant. Our analysis shows that their anatomy is affected more by altered habitat utilization and foraging conditions on islands than by the novel food supply itself. The morphologies of island warblers point towards terrestrial living and a more acrobatic use of the substrate (tarsus diameter), a reduction in longer flights but greater manoeuvrability (wing traits), as well as a diminished role for aerial feeding and closer contact with various substrates in the course of extractive foraging (rictal bristles; Leisler & Winkler unpubl. ms).

When compared with that of their continental counterparts, field observations confirm that habitat use by island warblers is not only altered but expanded, through the forced change from wetlands to a more variable exploita-

tion of scrub and trees. On the archipelagos the reed warblers were observed in all vegetation strata, from birds feeding on the ground and in low bushes to individuals singing or foraging in the fronds of the tallest coconut trees. Island warblers utilize all substrates: foliage, twigs, branches, trunk, bark crevices, and often the ground and litter layers as well (Graves 1992, del Hoyo *et al.* 2006, Alice Cibois *in litt.* 2011, BirdLife International 2004-2010).

The longer and more massive bills of the island warblers vary in their dimensions according to their more catholic diet, which often consists of large prey and prey types, many of which differ from the usual diet on the continents. Island species feed, for instance, on land snails, nectar, seeds, fruit pulp, and small reptiles (del Hoyo *et al.* 2006, Mosher *in litt.* 2009). Longer and more massive bills are also better suited for obtaining hidden and difficult-to-access prey, which appears to be true for insular birds in general (Winkler & Leisler 2006). Several anecdotal observations report unusual and newly-acquired feeding techniques in Polynesian reed warbler species. For example, birds were observed to act as acrobatically as tits or parrots (Cook Islands Warblers, Tahiti Warblers; Holyoak & Thibault 1984, Jens Hering pers. comm.), as they turned leaves on the ground (Tuamotu Warblers), stretched their neck in an extreme fashion to reach for prey under leaves, or probed into litter and tangles of dead leaves (several species; Alice Cibois, Jens Hering pers. comm.). Tahiti and Tuamotu Warblers fix prey with their feet and treat it like shrikes; Southern Marquesan Warblers are capable of gaping or prying, that is, forcefully opening the bill when inserted into the substrate, in the same way that starlings do. They also probe a lot and skilfully kill large prey items such as small geckos, which has earned them the local name of 'komako', originally from a legend about a yellow bird pecking at a shark (Bruner 1974). Such feeding innovations in island birds seem to be stimulated by both food shortages and a tendency to opportunism. The discussion around the function of a larger bill in island species has focussed on its role in increasing generalism, but has tended to overlook the significant fact that specific feeding techniques, such as probing and gaping, require long bills, independent of prey size.

Greenberg & Mettke-Hoffman (2001) predicted that island colonists are likely to be less neophobic and to increase exploration behaviour in order to cope with a variable and cryptic food supply on islands. It does indeed appear to be the case that all island warblers are described as curious, tame, and explorational (Pratt *et al.* 1987, A. Cibois, J. Hering pers. observations).

Photos 12.11 to 12.13 In an insular situation, species expand their niche because of both resource limitation and ecological release, since they are not confronted by competitors. Furthermore, following island colonization, reed warblers search more for hidden prey. This leads to the development of stronger legs, shorter vibrissae (bristles around the bill and face), rounder wings, and longer bills; a number of species – as small passerines – have also become larger. The longer bills and more powerful legs can be seen in the singing Saipan Warbler (right) and Southern Marquesan Warbler (top left). An increased variability in plumage, with partial leucism or colour morphs in some species, is illustrated by this dark-coloured Tahiti Warbler (photos Jack Jeffries, Alice Cibois [2]).

5

Communication, breeding biology, and further idiosyncracies

The voices and song patterns of insular bird species differ from those of their closest mainland relatives (e.g. Marler 1960, Price 2008). The biggest differences are to be found in the Pacific reed warblers; the songs of the island species of the *Calamocichla* group rather resemble those of their African and Madagascan relatives. The songs of this group have a strong phylogenetic significance (chapter 7). According to Bell (2001), who compared the songs of 11 Pacific insular species with those of Oriental and Australian Reed Warblers, all the island birds have a simpler song than their continental congeners. They vary from the complex, long, melodic song-phrases of the west Pacific species, such as Saipan Warblers, to the extremely reduced call-like songs of the Pitcairn and Henderson Island Reed Warblers (Pratt *et al.* 1987, Graves 1992). Shorter, less powerful and less elaborate songs are also delivered by Rimatara and Cook Island Warblers (Holyoak 1974, Thibault & Cibois 2006) and by the Millerbirds of Hawaii (Morin *et al.* 1997). Several of the island warblers insert calls of varying length into their songs, so that complex combinations of extremely variable warbles and whistles are created. Some authors emphasize the large repertoire of additional call notes with which island warblers communicate (Baker 1951, Bruner 1974, Holyoak & Thibault 1984, Milder & Schreiber 1989, Thibault & Cibois 2006).

It is well established that island songs are simpler and shorter, and that individual birds can vary considerably (Baker 1996, Price 2008). There are two main hypotheses to explain the simplicity of island songs. First, the colonization of an island could have been initiated by young birds whose songs were not fully developed and, given the lack of models with full song, these were passed on to the next generation in an incomplete form ('withdrawal-of-learning hypothesis'; Thielcke 1973). Secondly, the few adult birds which might have colonized an island could, by chance, have been birds possessing only a fragment of the source population's song repertoire. In the course of transmission to later generations the song types that reached an island could additionally undergo random changes because of the limited number of model singers. In such a situation song units can be lost, but new ones can also be created during song development, or calls can be gradually inserted ('drift hypothesis'; Schottler 1995, Price 2008).

Photos 12.14 and 12.15 With their longer and more massive bills, island warblers are better equipped for obtaining hidden and difficult-to-access prey. Generally, the diet is more catholic and the prey larger than the usual diet on the continents. The unusual feeding techniques in Polynesian reed warblers are well illustrated by this young Tahiti Warbler searching for prey on a palm tree (photos Jens Hering).

However, these hypotheses cannot account for the marked trend towards simplicity in the songs of the Pacific reed warbler forms. Altered intraspecific social situations on islands represent a selection pressure that could have stimulated the creation of shorter signals, making it easier to recognize individuals. Like other tropical species they probably live in stable social networks, in which neighbouring pairs get to know each other individually over a long period. In such cases songs are simple and are more strongly coded for estimating distance to the sender and for individual recognition, with a much-reduced function of attracting a partner through greater song complexity (Stutchbury & Morton 2001; chapter 7). Further factors will have contributed to the formation of island species song, such as altered conditions for sound dispersal in the new habitats or changes in body size and bill form, as has been demonstrated for Darwin's finches (Bowman 1983, Podos & Nowicki 2004; section 7.5).

Very contradictory observations have been made regarding how island reed warblers cope with the often very high population densities. For example, whereas Milder & Schreiber (1989) stress the *low* intraspecific interactions in Kiritimati Warblers, by contrast Holyoak & Thibault (1984) draw attention to the *high* intraspecific aggression within Southern Marquesan Warblers.

An immediately striking peculiarity in some Polynesian acrocephalids is the partial leucism (reduction in pigmentation) of their flight and body feathers, the frequency, extent, and location of which vary widely. The leucistic feathers are more or less asymmetrically distributed, affecting principally the wing and tail feathers. In the three species in which leucism is most strongly pronounced (Pitcairn Island, Henderson Island and Rimatara Warblers) it progresses with age in both sexes, so is often absent in juveniles and most pronounced in adult birds several years old. In Henderson Island and Rimatara Warblers it is not only age- but also sex-related, with males exhibiting more white than females (Holyoak 1978, Graves 1992, Brooke & Hartley 1995, Thibault & Cibois 2006). Possible causes and mechanisms for this phenomenon, that might be related to small population sizes on these islands, remain completely unresearched. However, the latest molecular relationship studies show that the island taxa most closely related to each other can be very different when it comes to the prevalence of leucism. Plumage leucism is common and advanced in Rimatara Warblers, but completely lacking in their closest relatives, Cook Island Warblers; it is present in around 14% of Tuamotu Warblers but is almost absent in their closest relatives, the Northern Marquesan Warblers (only 0.6%). Regarding the possible function of such colour variants, Holyoak (1978) speculates that this polymorphism possibly correlates with the reduction in, or loss of, song in these species and 'serves complementary functions to song in allowing recognition of individuals in encounters over territories and other behavioural

contexts'. Graves (1992) found no evidence that leucism is a consequence of songlessness or vice versa, but concedes that it may function as a signal in dominance hierarchies. A definitive answer to this problem will most likely be found by employing the methods of behavioural ecology and genetic studies. (For further aspects of population genetics in island species see section 13.6).

Some males of insular species do help with nest building and a high level of paternal brood care appears to be typical of all island reed warblers. The males of Tuamotu and the Marquesan Warblers have an incubation patch (Bruner 1974); the male Nihoa Millerbirds, Henderson Island Warblers, Cape Verde Warblers, and Rodrigues Warblers all share the incubation of the clutch to varying degrees. Only in Seychelles Warblers and Kiritimati Warblers do females incubate exclusively, though information is lacking for the remaining species. When the female takes an incubation break, Seychelles Warbler males guard the nest with its sole egg against predators, which also represents a considerable paternal investment. That they incur certain costs in doing this was shown when some of them were translocated to a neighbouring island without predators; they stopped guarding and foraged more for themselves, thus improving their physical condition (Komdeur & Kats 1999). The males of all island species studied play an important role in provisioning the offspring, both in the nest and after fledging, which may be related to the scarce food supply (chapter 10).

Most species are socially monogamous with strong pair bonds (e.g. Nihoa Millerbirds; Morin *et al.* 1997) or even pair for life (e.g. Kiritimati Warblers, Seychelles Warblers) and occupy permanent territories. Polygyny has never been reported, even in those island reed warblers that breed in large territories, such as Kiritimati and Saipan Warblers (Milder & Schreiber 1989, Craig 1992). However, some species do breed cooperatively, while there are indications that others breed not only in pairs but occasionally in trios.

6

Population trends of island warblers

More than 90% of the bird species that have become extinct in the last 500 years lived on oceanic islands (Johnson & Stattersfield 1990). In other words the risk of extinction on such islands is disproportionately high; species are made vulnerable by small ranges, irrespective of whether they are on the mainland or on islands. However surprising it may appear, if we correct for range size, more continental species than island species are endangered at the present time (Manne *et al.* 1999). Small populations restricted to a single island face a greatly increased risk of extinction if their environmental conditions undergo a rapid change, while large populations, by contrast, can more easily compensate for such fluctuations in numbers. Natural catastrophes such as weather anomalies, volcanic eruptions, fire, earthquakes, or tsunamis have frequently contributed to the decimation or extermination of species. In 1981 the volcanic eruption on the island of Pagan in the Mariana Islands almost destroyed all of the vegetation, which led to the extinction of the entire population of the endemic subspecies *yamashinae* of the Nightingale Warblers (Reichel *et al.* 1992).

In historical times the eradication of endemic island species has almost always been a direct or indirect result of human colonization. Since the arrival of humans in recent times species richness on oceanic islands has decreased by 30-50% (Olson 1989).Yet even prehistoric

peoples drove insular species extinct, as indicated by discoveries of fossil bones (Olson & James 1982). The appearance of Europeans in the Pacific at the end of the 18th century resulted in a fresh wave of exploitation, which led to further extinctions of some insular acrocephalids (BirdLife International 2004-2010). Increasing settlement and population pressure, above all on the South Sea islands, resulted in extensive changes to the native vegetation and indirectly to the displacement and loss of species. A wide variety of cultivated plants was introduced, and indigenous woody stands were cleared, burned, and degraded in the course of agricultural improvement. The biggest changes were to atoll ecosystems, and they remain the most threatened. Conversely the inaccessibility of very steep mountain slopes and eroded volcanoes delayed the pace of extinction which resulted from deforestation, cultivation, and hunting (Steadman 2006). The interaction of various factors frequently seals the fate of a population. On the Cape Verde island of Brava, a combination of habitat destruction, high predation pressure, and increasing desertification extinguished the local population of Cape Verde Warblers (Hering & Hering 2005, Hering & Fuchs 2009).

Of the 16 extant island reed warbler species, 11 are classed as Vulnerable, Endangered, or Critically Endangered (table 12.2), while at least a further four have disappeared in recent times. For a long time the entire world population of Seychelles Warblers was confined to the 29-ha island of Cousin and in 1968 it was less than 30 individuals (section 13.4). The entire population of the *kingi* subspecies of Millerbirds on the 63-ha island of Nihoa (40 ha with vegetation) has fluctuated between 480 and 1150 birds (BirdLife International 2004-2010). The only species which have relatively large ranges extending over several islands are Tuamotu Warblers, Marquesan Warblers, and Caroline Islands Warblers. No threat currently exists for the populations of these species, nor for that of the Cook Island Warblers (BirdLife International 2004-2010).

Most insular warblers are able to utilize a broad habitat spectrum (table 12.2), can generally cope well with anthropogenic habitats, and tolerate some forest clearance. However, reed warblers and pigeons are the only species to have survived the devastating environmental destruction on those islands that have been subjected to the decades-long practice of industrial phosphate mining

Photo 12.16 Island reed warblers are socially monogamous with strong pair bonds, often for life, and they occupy permanent territories. Their songs are simply structured and short, being more coded for individual recognition than for attracting a partner (Tahiti Warbler; photo Jens Hering).

(Thibault & Guyot 1987, Steadman 2006). Of the *ca* 16 island species, 11 today inhabit secondary vegetation. Those that prefer wetlands appear to be less flexible (though the forest specialists can also be very sensitive); populations or subpopulations of Nightingale Warblers have thus died out on those islands where they were restricted to wetlands (Guam and Pagan), while those that also bred in forests have done rather better (Reichel *et al.* 1992). By contrast, there has been a dramatic level of local extinctions in the woodland-dwelling Tahiti Warblers (Monnet *et al.* 1993).

One catastrophic development has been the introduction in many places of non-native plants and animals that can wreak havoc with an island's fragile ecosystem in a frighteningly short time. An example is the plant *Miconia calvescens* (Meyer & Florence 1996). Introduced from South America, it currently poses the greatest threat to the local reed warbler habitat in the forests of the Society Islands because it forms a dense canopy that shades out all but its own seedlings. Even more disastrous has been the deliberate introduction of animals such as feral goats, pigs, cattle, and cats, as well as the accidental arrival of tree-climbing, nest-robbing black rats or brown tree snakes. Common Mynahs are also nest predators which have contributed to the decline and regional extinction of several acrocephalid populations (Thibault & Cibois 2006). Feral goats have destroyed most of the forest understorey on the Mariana Islands, thus preventing the regeneration of many tree species, and are probably also responsible for the disappearance of the Nightingale Warblers on Aguijan. The enormous increase in brown tree snake numbers on Guam has been held responsible for the extinction of the local Nightingale Warbler population, as well as those of other endemic species. It is alarming that this snake has also spread to the nearby island of Saipan and is, increasingly endangering the reed warblers there, whose population declined by more than 50% between 1982 and 2007 (Reichel *et al.* 1992, Camp *et al.* 2009).

The case of the Laysan and Nihoa Millerbirds

Millerbirds are the only Old World warblers known to have colonized the Hawaiian Archipelago, the most remote group of islands in the world. Both forms (*familiaris* and *kingi*) which are regarded as distinct subspecies, were confined to the tiny islands of Laysan and Nihoa respectively (Morin *et al.* 1997, Fleischer *et al.* 2007).

When the subspecies *kingi* was first described by Alexander Wetmore in 1924 after he had examined study skins, the Laysan Millerbirds were already extinct. The 400-ha island of Laysan possessed a dense vegetation cover of 187 ha and was home to large seabird colonies. Among the few endemic landbirds its reed warblers were by far the commonest species; they nested in tall *Eragrostis* grass and were relatively tame. Their preferred prey were the ubiquitous – but subsequently also extinct – miller moths, from which the warblers took their name. The fate of Laysan was sealed with the start of guano extraction in 1890, although the extinction of its Millerbirds, together with that of many other endemic animals and plants, was ultimately caused by the introduction of rabbits in 1903. With no predators, the rabbit population exploded and stripped the island of vegetation. In 1915 the population of the reed warblers was estimated at 1500 individuals, but in the following year only a handful of birds could be found. In that time the miller moths also disappeared, along with the *Eragrostis* vegetation nesting habitat. In the few small stands of tobacco that had to serve the Millerbirds as cover, their nests were easily found and almost completely plundered by Ruddy Turnstones and Laysan Finches. Following the extinction of the miller moths the only remaining food was a saltwater fly (*Neoscatella sexnotata*), for which the reed warblers had to compete with the Laysan Ducks.

In contrast to Laysan, Nihoa has no rabbits, but its small reed warbler population of the *kingi* subspecies of Millerbirds is teetering on the brink of extinction. Although the current population of *ca* 800 individuals has remained relatively stable over the last 30 years, it is highly vulnerable to chance events such as severe storms and droughts, as well as accidental introductions of alien species or diseases (Foote 2009). In the 1980s a non-native grasshopper (*Schistocerca nitens*) arrived on Nihoa; in 2002–2004 a plague of these insects aggravated the drought and destroyed most of the island's vegetation (Latchininsky 2008), which almost led to the demise of these warblers. In order to restore a Millerbird population to Laysan and reduce the chance of extinction of the Nihoa population, 24 Nihoa Millerbirds were transferred 650 miles to the Island of Laysan in September 2011 as reported on the website of the American Bird Conservancy (accessed 23.9.2011 http://www.abcbirds.org/ newsandreports/releases/ 110919.html).

7

What it takes to be a successful reed warbler colonizer

There is no question of calling island-dwelling reed warblers 'strays', that is, birds that occur as accidental vagrants, as they must be regarded as 'colonizers' of oceanic spaces (section 12.1). Thus we think it appropriate to close this chapter by speculating about those requirements which allow reed warblers to colonize islands so successfully, and even to deal with man-made habitat alterations. In general, birds of scattered and unstable habitat patches, such as reed warblers, appear to disperse well (section 11.3). Compared with other passerines, the perfectly appropriate body mass of the large reed warblers perhaps gave them an advantage, unlike the lineages of smaller *Acrocephalus* warblers, which, as we would expect, have never colonized islands (Lomolino 2005). A further advantage might be their generalized diet and foraging techniques, principally gleaning, while their excellent flying and climbing skills enable them to exploit many different habitat structures.

Moreover, their chances of success as island settlers were undoubtedly increased by their ability to build well-hidden nests in the emergent structures of a variety of coastal woody plants, whether they were screw-pines, beach heliotrope, *Messerschmidia*, *Barringtonia*, or sheoaks (casuarinas). This supposition is supported by the results of a study by Thibault *et al.* (2002) concerning the relationship between nest placement and predation risk in songbirds on Polynesian islands. Without exception, the researchers found that reed warblers build their nests in emerging vertical twigs and significantly closer to the top of trees than the endemic monarch-flycatchers (*Pomarea*) which occur on the same islands. By contrast, the monarch-flycatchers place their nests mostly on horizontal branches and inside the canopy, or at its base, a practice which has fatal consequences. *Pomarea* nests are extremely endangered by introduced black rats, which are accomplished climbers, while those of the reed warblers remain untouched.

This chapter has provided some insights into the unusual success story of an unassuming group of birds and their conquest of the planet's most remote islands, a group that in general has been stereotyped as a specialized inhabitant of wetlands. Studies on island reed warblers have considerably deepened our understanding of adaptation in insular bird species. They have shown that the selective demands common to most islands are mainly related to changes in habitat utilization by new arrivals and affect all morphological complexes. However, there are many issues which still remain unstudied, such as the environmental conditions common to different island groups that have contributed to the emergence of parallel adaptations in plumage coloration.

Above all else, what we hope this chapter in particular has emphasized is the overwhelming importance of the need to analyse these phenomena very carefully, case by case, before making any generalizations. Islands have their individual histories, which means they may have very little in common.

Summary

More than 40% of the species in the genus *Acrocephalus* are endemics on tropical islands in the Atlantic, Indian, and Pacific Oceans. The ability of reed warblers to disperse over vast distances at sea is unmatched among passerines; some of them have settled the planet's most remote islands, on several of which they are the only terrestrial landbirds, often at high population densities. All island reed warblers descend from the clade of large reed warblers.

The story of their complex colonizations of the Pacific was only very recently deciphered using a molecular-genetic analysis of relationships between the species concerned. The main radiation – not involving the species on Guam – began approximately 2.4 million years ago. In the course of their colonization of the Pacific, the reed warblers did not continuously spread further and further in a one-way 'stepping-stone' fashion, but rather reached their new homes by frequent and independent long-distance dispersals. Double and repeat colonizations led to the present complex situation of neighbouring islands in three archipelagos (Marquesas, Society, and Mariana Islands), which have been settled by species that resemble each other through evolutionary convergence, even though they are not directly related. Although several waves of colonization have reached these archipelagos, only a single species exists on each island. Surprisingly, Australia has been settled from islands to the continent – a form of 'reverse' colonization.

These island endemics can be regarded as 'supertramps': species with the ability to colonize small and isolated islands with limited food resources. The insular situation is characterized by a variety of habitats, limited or specialized food resources, and depauperate species communities. Given such conditions, interspecific competition is relaxed, which allows the possibility of the niche expansion that is necessary for new arrivals in such an environment.

Some general morphological adaptations have also occurred: island species possess thicker legs, shorter rictal bristles, rounder wings, longer bills, and are generally larger than their relatives on the continent, reflecting their altered habitat utilization. All island forms have simpler songs than their mainland congeners and many live on permanent territories, exhibiting strong or even life-long pair bonds. A high level of paternal brood care is typical and some species breed in trios.

Although reed warblers can cope well with the extreme conditions found on small oceanic islands, a number of them have become extinct, mainly because of anthropogenic disturbance, and most of the species are classified as endangered.

Photo 12.17 Independent young often remain for a prolonged time in the parental territory, like this young Tahiti Warbler (photo Jens Hering).

13

A life of change – population and conservation issues

Eastern Olivaceous Warbler, a species that has expanded its breeding range in the 20th century.

When Phil Round removed a strikingly long-billed *Acrocephalus* species from his mist net in a grass filter bed near Bangkok in March 2006, he soon realized that it must be a Large-billed Reed Warbler, despite the fact that the species had not been seen for 140 years and was thought by many to be extinct. The very existence of the enigmatic Large-billed Reed Warblers was considered doubtful. Indeed, the only known individual was the type specimen collected in northern India in 1867 by A. O. Hume, now in the Natural History Museum in Tring, UK, that some had thought was a hybrid or some other aberrant form until its status as a genuine species was established beyond doubt by Staffan Bensch and David Pearson using DNA analyses and precise morphometric measurements in 2001. Not surprisingly, Large-billed Reed Warblers remain one of the world's least known birds (Bensch & Pearson 2002, Round *et al.* 2007).

In 2008 and 2009, the discovery in various museum collections of over 20 new specimens of Large-billed Reed Warblers, which had formerly been misidentified as Blyth's Reed Warblers, supplied the clues to their distributional range, by indicating that the species breeds in central Asia, possibly migrating along the Himalayas to winter in northern India and southeast Asia (Svensson *et al.* 2008, Koblik *et al.* 2011). In 2009, a sensational discovery was made in the Pamir Mountains – in the easternmost panhandle of Afghanistan and in southeast Tajikistan – when 29 Large-billed Reed Warblers were trapped in mist nets, among them newly fledged young (Timmins *et al.* 2009, Ayé *et al.* 2010, Timmins *et al.* 2010). In the riparian scrubby bushland along the Amu Darya and other mountain rivers at an altitude of 2 600 - 2 800 m, these closest relatives of Blyth's Reed Warblers appear to be the second most common species after Mountain Chiffchaffs. Since these montane riverine woodlands are under intense pressure from humans who collect firewood and fodder there, conservation action is most urgently needed to ensure that the least-known reed warbler of all does not once again disappear into the darkness of history.

Almost as mysterious, and perhaps even rarer today, are the Streaked Reed Warblers, which were still very common on autumn passage in small millet fields in Hebei Province, China, around 1925 (Kennerley & Pearson 2010). We also know very little about the scarce Manchurian Reed Warblers, with their tiny population numbers, in great contrast to our knowledge of the widely distributed species such as Sedge Warblers, Blyth's Reed Warblers, and Paddyfield Warblers.

The environments inhabited by animals and plants undergo continuous long-term and short-term changes.

Photo 13.1 6 of the 33 continental acrocephalids are classified as 'Vulnerable' in the IUCN Red List, and the number of threatened island species is even greater. Many of the remaining species show at least regional long-term population declines. Against this background, it was a great surprise when Large-billed Reed Warblers, thought by some to be extinct, were 'rediscovered'. The size of the breeding population in southern central Asia remains unknown and is probably small. Their montane riverine woodland habitat is under pressure from firewood collection and grazing. Conservation action is therefore urgently needed (photo Phil Round).

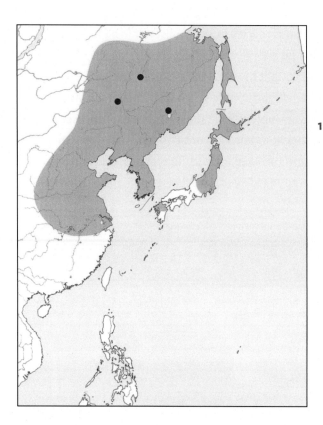

Map 13.1 Distribution ranges of Black-browed (yellow) and Manchurian Reed Warblers (dots), which occur sympatrically, but whose ranges and population sizes differ strikingly (after Kennerley & Pearson 2010).

Unfortunately for acrocephalids, in recent times the wetland and reedswamp habitats of many members of the family have been particularly vulnerable to intense man-made pressures. For example, the impact of now ubiquitous eutrophication has been both positive and negative. In the early days of artificial fertilization the growth of *Phragmites* beds was accelerated, but now the pendulum has swung to the negative side, with habitat degradation and reed die-back among the consequences. A further factor is climate change – another Janus-faced issue with conflicting results. While some reed warblers have been able to expand their range in Scandinavia, others have been decimated in their sub-Saharan winter quarters by long periods of drought. Increasing pressures of use and constant disturbance of water regimes have meant an enormous loss of wetlands and hence of habitats for their inhabitants. The process is likely to continue unrelentingly.

All these factors have drastic effects on a whole range of aspects of acrocephalid population biology, including abundance, distribution, and protection issues.

1 Range, abundance, and species numbers

Why are some acrocephalid species so rare while others are so strikingly common? In recent decades many of the Acrocephalidae family have undergone significant population changes, rarely positive and mostly negative. The family has a very extensive geographic distribution. In general, migratory species have considerably greater ranges than sedentary species. The three most northerly species (Sedge, Booted, and Blyth's Reed Warblers) have the largest continuous ranges.

One explanation of why the distribution area of a species is almost always greater the further north it occurs, is based on the supposition that such species were able to colonize extensive homogeneous regions of the northern landmasses following the end of the last glaciation (Dynesius & Jansson 2000). Ultimately it is periodical changes in the Earth's orbit that drive the dynamics of climate and species distribution, and this dynamic is more pronounced towards the poles than near the equator. The comparatively large ranges of northern bird species are subject to rapid contractions and expansions, which also promote a tendency to mobility and generalism (Dynesius & Jansson 2000). By contrast, species ranges in lower latitudes are clearly smaller (Rapoport's Rule), while species diversity is higher because of the heterogeneity of landscapes and habitats. The frequent altitudinal differences in these geographic regions also promote heterogeneity and diversity. Among the continental acrocephalids with the smallest ranges are Papyrus Yellow Warblers and Basra Reed Warblers, but species which occur further north can also have limited ranges, such as Aquatic Warblers or Manchurian and Streaked Reed Warblers, the latter with a probable distribution range of just 2400 km². We can probably assume that these areas of occurrence represent fragments of a formerly more extensive range (map 13.1; del Hoyo *et al.* 2006, BirdLife International 2004-2010).

As long as their habitats remain intact, the breeding densities of reed-dwelling acrocephalids can reach levels that are among the highest of all bird species, colonial breeders excepted (chapter 6). Just one impressive example will suffice to illustrate the point: for a 20-ha *Phragmites* 'island' in the otherwise inhospitable Namib Desert, the estimated breeding population of African Reed Warblers was calculated at 1270 (Eising *et al.* 2001). Such a value illustrates the great potential of reed warblers to reach high densities, even though large-scale populations in reedswamps range only between 3 and 30 breeding pairs per ha (Schulze-Hagen 1991c). In their winter quarters reed warblers can also be found in high densities in optimal habitats, as seen in counts of 103 Sedge Warblers per ha in the floating *Polygonum senegalense* vegetation of the Inner Niger Delta (Zwarts *et al.* 2009).

Photo 13.2 The habitats of most acrocephalids consist of various stages of ecological succession, and are therefore by definition subject to alteration. For acrocephalids, this commonly results in short-term and long-term population changes. In the 19th century, several reed warbler species in central Europe showed marked increases in numbers after the introduction of fertilizers and the resulting eutrophication of water bodies; since the 20th century, high-intensity farming has had an increasingly negative effect. Nevertheless, many species in this situation have displayed a certain flexibility and tolerance in their habitat preferences. This Great Reed Warbler is foraging in a sunflower field in Tuscany (photo Daniele Occhiato).

With a good knowledge of regional conditions, the warbler populations of entire regions or countries can be extrapolated. Using capture-recapture techniques during autumn migration, the total Eurasian Reed Warbler population of one country, Sweden, has been calculated at 500 000 to 600 000 pairs for the period 1988-1993. Taking the entire reedbed area in Sweden into account, this figure gave the equivalent of a country-wide mean population density of 3.7 breeding pairs per ha of reed (Stolt 1999). In another study based on the mapping of territories in different parts of the reedbed and in different subtypes of the reed fringe, with corresponding extrapolations, the population of this species on the Neusiedler See alone was estimated to be 130 000 pairs (Grüll & Zwicker 1993).

In Europe, with its high density of ornithologists, there are estimates of bird breeding populations for the entire continent, made possible by the activities of the European Bird Census Council (Hagemeijer & Blair 1997). Nowhere else in the world does such detailed information exist for such large areas, although we can assume that the population figures for several countries are considerable underestimates. The two commonest acrocephalids are Sedge Warblers with 5.5 million breeding pairs and Icterine Warblers with 4.7 million (geometric means). If we take the upper estimates of their numbers, then each of these species would have 10 million pairs in Russia (west of the Urals) alone (table 13.1). Given its size, and the fact that the landscape is only extensively utilized over enormous areas, Russia possesses great potential for high population densities. Those who only know the situ-

Table 13.1 Estimated mean population size of acrocephalid species in Europe (Russia to the Urals, and Turkey; after Hagemeijer & Blair 1997 and BirdLife International 2004).

Species	Europe	Russia	Turkey
Aquatic Warbler	4 700	3 100	
Sedge Warbler	2.3 mil.	3.2 mil.	
Moustached Warbler	33 000	160 000	9 000
Paddyfield Warbler	146 000	32 000	
Marsh Warbler	1.7 mil.	320 000	
Eurasian Reed Warbler	3.1 mil.	32 000	15 000
Great Reed Warbler	950 000	1.6 mil.	32 000
Blyth's Reed Warbler	12 000	320 000	
Olivaceous Warbler (*I. opaca* & *I. pallida* in 1997 NOT separately listed)	151 000		1.6 mil.
Olive-tree Warbler	10 000		3 000
Icterine Warbler	1.5 mil.	3.2 mil.	
Melodious Warbler	1.7 mil.		

ation in western Europe, with its decade-long continuous decline in the population of these two warblers in particular, can hardly comprehend such numbers (section 13.4). At the other end of the scale, the smallest total populations, each less than 25 000 singing males, are those of Aquatic Warblers, whose range is restricted to eastern Europe, and Olive-tree Warblers, whose distribution is limited to some areas of southeast Europe, the Turkish Mediterranean coasts, and the Near East (BirdLife International 2004-2010, Aquatic Warbler Conservation Team pers. comm. 2011).

Reasons to be positive

The habitats of most acrocephalids consist of various stages of ecological succession and are therefore especially liable to change (map 13.2). Accordingly, in the last 200 years alone, many acrocephalids have undergone substantial fluctuations in both population and range. These changes are all closely tied to intervention by humans in the water regime and also to agricultural practices; until around 50 years ago such changes tended to

be positive. Following the agricultural reforms of the 19th and early 20th centuries, tremendous alterations in farming practice occured, first in England and western and central Europe, later in northern and eastern areas of the continent. The enclosure of common land and the widespread reclamation of 'wasteland' led to almost a doubling of the area used for agricultural production. Together with the planting of clover, the adoption of methods of rearing livestock indoors all year round resulted in a great increase in agricultural yields. The introduction of artificial fertilizers made levels of production possible that were many times higher than before, but the runoff from the amount of fertilizer necessary for such increases naturally had huge effects on all kinds of wetlands. The narrow belts of *Phragmites* that used to fringe generally oligotrophic water bodies now expanded enormously with the increasing levels of nutrients; their biomass rose substantially, including both vegetation mass and invertebrate density. Large-scale land reclamation on the one hand, and the general eutrophication of rivers and wetlands on the other, had a stronger impact on the avifauna of the 19th century than we can ever imagine today (Schulze-Hagen 2004). Given the growing supply of habitat at that time, the subsequent reed warbler population increases and range expansions are easily understood.

Marsh Warblers can serve as a good example of how a species expanded to fill its mosaic-like disjunct distribution range. Until the beginning of the 19th century the species was more or less confined to valleys and lowlands in central Europe. With the spread of agriculture and the use of artificial fertilizers in previously undeveloped regions, new habitats were opened up for the species by the creation of tracts of herbs, and by extensively husbanded cereal fields where the crop was intermingled with various tall herb species. From the middle of the 19th century the birds' range extended over a large continuous area, within which its population density greatly increased. During this expansion phase the species began to colonize higher-altitude regions of Germany, where it had not previously occurred, gradually moving ever higher. Together with Sky Larks, Marsh Warblers were the commonest birds of agricultural habitats in some regions in the first half of the 20th century. This strong population base enabled the species to spread eventually from central Europe into Scandinavia (Schulze-Hagen 1991b).

Perhaps the most impressive example of a population increase is that of the Eurasian Reed Warblers, a species that has benefited from the expansion of reedswamps. From the end of the 1800s the species spread into formerly unoccupied northern and northeast Europe; in the 1930s the region around Stockholm was settled, in the 1950s southern Sweden. The growth in numbers in Scandinavia was rapid; for example, the Swedish province of Dalarna, where the first nest was found in 1951, was completely settled by 1965. On Lake Vättern there was an eight-fold population increase between 1965 and 1982 (Svensson 1978, Persson 1983). Eurasian Reed Warblers first bred in Norway, near Oslo, in 1947; today they occupy most of south Norway. An expansion on a broad front into Finland and the Baltic states began in the 1930s, and today the latter are home to a uniform and dense population (Schulze-Hagen 1991c). In 1976, an estimate of 100 000 pairs was made for the whole of Sweden; by 1988-1993, this had increased to 500 000 pairs (Stolt 1999; see above). This means that Sweden now holds the second largest Eurasian Reed Warbler population in Europe after Romania. The numbers of Great Reed Warblers have also increased greatly in central Europe, and they have also similarly expanded into southern Scandinavia (Holmbring 1973, 1979, Leisler 1991b, Berndt & Struwe-Juhl 2004, Hansson *et al.* 2007).

In eastern Siberia, Thick-billed Warblers were able to expand and saturate their distributional range following an increase in deforestation, road-building, and agriculture in the region, as a result of which thick shrubbery was able to develop in many places (Ivanitskii *et al.* 2005). Since the middle of the 20th century the eastern race of Olivaceous Warblers (*elaeica*) – a taxon that was formerly subtropical to Mediterranean – has undergone a marked range expansion from the Balkans and western Ukraine to the north, where in many places it breeds exclusively in human settlements, ignoring areas of seemingly suitable habitat outside them. One explanation for this might be that Olivaceous Warblers develop local traditions via birds that have become imprinted on the particular vegetation structure of gardens and parks. Another possibility is that Olivaceous Warblers breed in villages and cities mostly because of the presence of human activity *per se*, thus mitigating the impacts of nest predation and brood parasitism (Antonov *et al.* 2007b). Other success stories have been those of the Paddyfield Warblers in southeast Europe, which have colonized the northern and western shores of the Black Sea following the increasing spread of

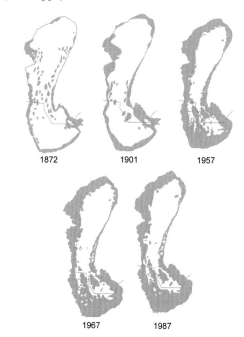

Map 13.2 Increase in the size of the *Phragmites* belt in the Neusiedler See between 1872 and 1987. This reed zone now covers 56% of the lake surface and is, at 180 km², the largest unbroken area of reed in central Europe. The reasons for the *Phragmites* expansion are the lowering of the water level and the abandonment of grazing by cattle (Anon. 1996).

1872   1901   1957

1967   1987

Photo 13.3 In west and central European countries, habitat loss is caused by the large-scale decline of reedbeds as a result of several factors, of which the central problem is human intervention in water regimes. The fragmentation of originally unbroken expanses of *Phragmites* can be clearly seen along the Havel near Berlin (photo Manfred Krauß).

251

rice cultivation there since the 1960s, and the recent range expansion of Sykes's Warblers following the introduction of irrigation systems in the semi-arid regions of central Asia (Kennerley & Pearson 2010). In Egypt, Clamorous Reed Warblers occasionally breed in crops of maize (Castell 2001), while various island reed warblers nest in secondary vegetation (chapters 3 and 12). All these examples demonstrate that acrocephalids can co-exist with humans, given the right circumstances.

As a consequence of the demise of agriculture over large areas of the former Soviet Union since the 1970s, huge regions have reverted to the dense scrubland that is ideally suited to the needs of Booted Warblers, which have now extended their range west as far as Finland, where breeding was first confirmed in 2000 (Kennerley & Pearson 2010). In southwest Europe, the range expansion towards the northeast by Melodious Warblers has created a contact zone with the previously allopatrically separated Icterine Warblers (sections 2.4 and 6.2). In just a few years in areas of northern France, Melodious Warblers have become the second-commonest bird after Garden Warblers in forestry plantations and clearings following tree-felling. The most likely cause for this expansion is increasing scrub invasion and afforestation on calcareous grassland, heathlands, and other rough open spaces as a result of changes in agricultural practice. Bushy successional stages, often of anthropogenic origin, are the ideal habitat for this species (Ferry & Faivre 1991b).

3 Negative population trends and their causes

Every conservationist knows that with increasing human population density the pressure on wetlands is also increasing. The number of people on the planet has recently passed the seven billion mark. Development and drainage on a huge scale are everywhere the common causes of wetland loss. Fifty per cent of all wetlands that once existed have been destroyed, and the scale is global. The pressure on wetlands in tropical regions such as south Asia and in Africa has for some time been much stronger than in Europe. Mangrove and papyrus ecosystems are shrinking rapidly, water abstraction from rivers and lakes is continually increasing, as is the pollution of water bodies by industrial effluents and runoff from the ever greater levels of agricultural fertilizers and pesticides. A further problem

is the widespread salinization of soils in drained regions (Zwarts *et al.* 2009). Entire landscapes have become desiccated, for example the Aral Sea or Lake Chad. In the 1980s and 1990s, by the construction of up-stream dams and when a punitive dictator took political action against the Marsh Arabs of his country, the draining of the Mesopotamian marshes also brought Basra Reed Warblers to the brink of extinction. However, they are not the only ones, and it is true to say that most acrocephalid species in the 21st century are being affected more than ever by habitat degradation and climate change.

The example of European reed-dwelling acrocephalids

Habitat loss

Although the populations of Eurasian Reed Warblers and Moustached Warblers have been more or less stable over the last 50 years, numbers of Marsh, Sedge, and Great Reed Warblers have suffered medium to drastic declines and the case of Aquatic Warblers has caused particular concern (see below). The principle cause of the negative population trend of Marsh Warblers, whose decline so far has been rather slow, has been the destruction of tall herb habitat in agricultural areas. Following the ubiqui-

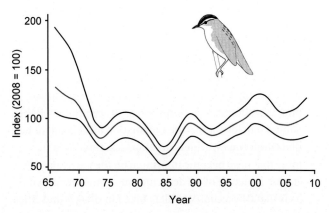

Fig. 13.1 Long-term trend of British Sedge Warblers since 1965. Much of the year-to-year variation in population size is driven by changes in adult survival rates, which in turn are related to changes in rainfall on their wintering grounds (BritishTrust for Ornithology 2010).

tous intensification of agriculture, with its increase in field area as well as loss of all marginal structures, numbers in some regions fell by up to 90%, while Marsh Warbler losses in wetlands unaffected by agriculture have been far less (Schulze-Hagen 1991b).

Compared with the population declines of Marsh Warblers, those of Sedge Warblers in many parts of central Europe have been much greater, though mainly as a result of problems in their winter quarters (section 13.4). Nevertheless, the drop in their numbers has also been exacerbated by habitat loss in their breeding areas; large-scale manipulation of the water table has converted many marshlands into scrubland and marshy woodland, a process accelerated by eutrophication. Agrarian intensification has considerably amplified habitat degradation, so that 60-90% of habitats have disappeared in many regions; for instance in England over 90% of East Anglian fenland was lost between 1934 and 1984 (Peach *et al.* 1991). Consequently Sedge Warbler populations have been completely extinguished in many areas, while in others their present density is a mere shadow of past numbers. It is to be feared that similar losses resulting

from the drainage of wetlands will continue to occur in future when the European Union Common Agricultural Policy takes effect in the largely intact habitats of eastern Europe (Koskimies 1991a, Zwarts *et al.* 2009; fig. 13.1).

Since the 1970s the western populations of Great Reed Warblers – the smallest part of the total European population – have collapsed. While breeding areas in central Europe have been abandoned, the east European population has been less affected. The decline in numbers in

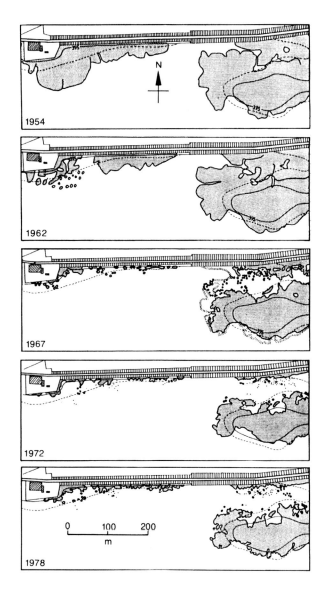

Fig. 13.2 Clear decline in the extent of reed on the Reichenaudamm (Bodensee/Lake Constance) between 1954 and 1978 (by kind permission of W. Ostendorp).

Table 13.2 Changes in land use and water table management in western and central Europe as causes of negative population trends in European acrocephalids. The impacts of these factors are often combined.

Agricultural intensification
Cessation of extensive agriculture
Loss of agrarian diversity
Loss of small-scale and edge structures (herb vegetation)
Extensive monocultures
Fertilization
Pesticides

Artificial water level regulations
Lowering of the water table
Drainage and water abstraction
Reduction of water table fluctuations
Eutrophication
Acceleration of marshland succession
Die-back of water reed

Photos 13.4 to 13.6 The cessation of traditional farming methods, in combination with manipulation of water regimes and increasing eutrophication, soon results in a successional process in the form of scrub invasion of meadows and reedbeds. This excess of biomass makes habitats unsuitable for reed warblers. Increased density and height of vegetation negatively affects arthropod density. While the vegetation in habitats occupied by Aquatic Warblers is low and open to the light, in unmanaged meadows it soon becomes a tangled mass (Pomerania). Self-seeding birch illustrate the rapidly proceeding sucession (Biebrza/Poland; photos Franziska Tanneberger, Lars Lachmann).

western Europe, often up to 95%, has its origins in many factors, but in general can be said to result from habitat loss and the 'Atlanticization' of the summer climate (Leisler 1991b, Graveland 1998, Berndt & Struwe-Juhl 2004). As inhabitants of the open-water edge of reedswamps, Great Reed Warblers are more sensitive to habitat degradation (i.e. the disappearance of *Phragmites*) than, for example, the more flexible Eurasian Reed Warblers, which have a broader habitat spectrum, utilizing both the open-water and landward edges of reedbeds.

Reedswamp decline

In the last 50 years *Phragmites* stands on most water bodies in central and western Europe have shrunk or even disappeared altogether. For instance, the Zürichsee has lost 94% of its reed area, the Havel near Berlin 70%, and the Norfolk Broads in East Anglia, UK, 80%. In a period of just five years (1962-1967) the width of the *Phragmites* fringe along the Reichenaudamm in the Bodensee (Lake Constance) contracted by 50 m (Ostendorp 1993; fig. 13.2). The causes of this process are many, but six factors

play a decisive role: 1| Manipulation of the water table by artificial regulation of the water balance and alteration of the annual water level fluctuation. 2| Direct destruction of the banks through building activities. 3| Mechanical disturbance by wave pressure and flotsam, made worse on the open-water side of the reedbed by erosion of the banks and boat traffic. 4| An excess of nutrients due to eutrophication and introduction of waste water. 5| Increase in climate changes leading to heavier precipitation, high water, and storms, all of which damage the growing *Phragmites* in particular. 6| Grazing by Greylag Geese, muskrats, and coypu, all three of which are neozoans in the area (Ostendorp 1993).

Large flocks of Greylag Geese, whose numbers have grown rapidly in recent decades, visit the water edge of reedbed to feed on young *Phragmites* shoots and to roost. The damage caused by their grazing and trampling is substantial. In the places where they feed on the water-facing side, the reed fringe becomes narrower every year. Muskrat and coypu, both alien species introduced in the 20th century, consume large amounts of *Phragmites*. The great increase in these three herbivores is an important factor in the reduction of reedswamp in many places. Their grazing activities have reached landscape-shaping proportions, since the reed fringe, when completely eaten away on shorelines, is replaced by willow scrub and other secondary vegetation.

The process of reed die-back, which is triggered by disturbance of the water regime, initially affects the 'water reed', the reed vegetation which is permanently inundated (van der Putten 1997, Graveland 1998, Belgers & Arts 2003). The loss of natural water level fluctuations caused by water management measures leads to an interruption in nutrient exchange and the washing-out of inorganic material. By eutrophication of surface water, additional organic matter, such as thick mats of filamentous green algae, enters the system. This process increases the accumulation of litter and the decay of organic matter in the reedbed, which triggers anaerobic processes leading to poisoning of the root systems. The final result is degeneration of the reedbed, which changes from continuous vegetation to clumps with fewer flowering stems, lower seed production, and the increased susceptibility of weakened stems to mechanical disturbance. Lack of water table seasonality in dry littoral zones also prevents germination of seeds, leading to a decrease in genetic diversity, and even monoclonal reed stands that are much more prone to stress. Lower water levels may also allow waterfowl to graze on roots in winter. Although *Phragmites* on the landward side of the swamp is less affected by die-back, it gradually looses its value as a habitat for reed-dwelling birds because of increased matting of the stems (section 13.4). This matting is a result of both over-fertilization and a lack of management by cutting or grazing by livestock. Just how little die-back processes are currently understood is illustrated by the phenomenon of reedswamps that are disappearing being situated next to others that are expanding.

Abandoning agricultural management

While the pressures of population and over-exploitation on wetlands and their ecosystems are still rising in many regions, depopulation in remote and underdeveloped parts of Europe and Russia has meant that an extensive land use is no longer practised. This chiefly involves grazing and mowing of fens, wet meadows, reedswamp, and arable field edges. The abandonment of such areas means the loss of large expanses of the habitat that several acrocephalids depend on, posing a serious problem for the protection and conservation of these species. The cessation of traditional farming methods quickly leads to large-scale successional processes in the form of scrub invasion of sedge meadows or reedbeds (section 13.5).

Photos 13.7 and 13.8 Habitat restoration and conservation management are necessary in order to halt large-scale succession in wetlands. After the breeding season on the Biebrza in Poland, several thousand hectares of fen mire were cut and invading scrub removed using rejigged snowfield vehicles. The restored habitats are not only attractive to Aquatic Warblers but mean a general increase in biodiversity (photos Lars Lachmann).

Thus an excess of biomass is created that is unsuitable as acrocephalid habitat for two main reasons: the structure of the vegetation – increased density and height, as well as a shift in plant species composition – and the substantial decline in arthropod density and hence potential prey supply (Poulin & Lefebvre 2002, Tanneberger *et al.* 2008). Fens in eastern Europe that have lain unmown for years are abandoned by Aquatic Warblers, just as matted and tangled *Phragmites* strips along drainage dykes in Hungary are now only suboptimal for Great Reed Warblers.

The case of Aquatic Warblers, part one

Aquatic Warblers are specialists of large open sedge fen mires and wet grasslands, in which they were very common at the beginning of the 20th century. In Brandenburg, for instance, their nickname was 'the sparrow of the fens'. At that time their breeding distribution

extended along a broad belt of fen mires between 47° and 59°N (map 13.3), from the Atlantic coast to western Siberia. Following the large-scale and systematic embankment and drainage of mires for peat extraction or conversion of the 'improved' areas into meadows and arable fields throughout the 20th century, Aquatic Warblers have disappeared from all but a few remaining tiny habitat fragments. Meanwhile they have been classified as the rarest migratory songbirds in Europe and categorized in the official Red List of the IUCN as Vulnerable, which means they are considered to be facing a high risk of extinction in the wild (BirdLife International 2004-2010). The species is now extinct in at least 11 countries, namely France, Belgium, the Netherlands, western Germany, Austria, the Czech Republic, Slovakia, Italy, the Balkans (the former Yugoslavia), Bulgaria, and Romania.

Map 13.3 Historical distribution range and present distribution (red dots) of Aquatic Warblers. The detail map shows the situation in 2010 (by kind permission of M. Flade).

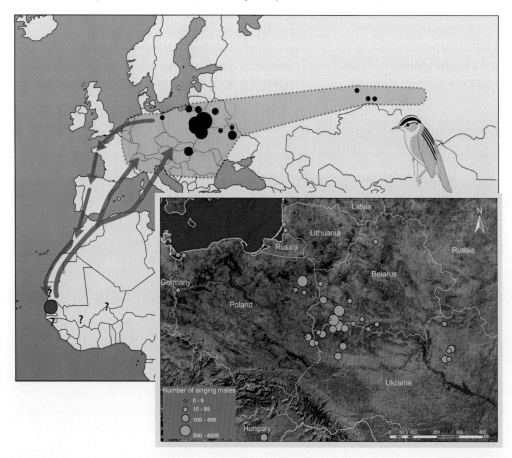

It was long thought that Aquatic Warblers had also disappeared in Ukraine and Belarus, but expeditions in 1995 unexpectedly found substantial populations in these two countries, which had been inaccessible during the Soviet era (Aquatic Warbler Conservation Team 1999, Flade 2008). On the one hand this was of course very encouraging, but on the other, the scale of loss of open fen mires, the habitat of Aquatic Warblers, was found to be even greater than had been feared. Between 1960 and 1995 more than 90% of all fens and other peatlands (15 000 km²) had been drained. Open fen mires had been reduced from 3800 to 440 km². Following the reactor accident at Chernobyl in 1986, 50 km² of intact mire were drained and converted to farmland as compensation for the radioactively contaminated agricultural land that had to be abandoned. Yet not long afterwards it became clear that the enterprise was ecologically and economically senseless, for within just a few years following reclamation the valuable peat layer disintegrated and became unproductive, leading to a drastic reduction in agricultural yield. Only devastated soils worthless for farming remain. Because of the drainage after 1960 alone the habitat of at least 100 000 Aquatic Warblers was lost. Nowadays, the world population of only 11 500 – 16 400 vocalizing Aquatic Warbler males is confined to fewer than 40 regularly occupied breeding sites in only six countries (Belarus, Poland, Ukraine, Hungary, Lithuania, and Germany), together covering less than 500 km² or – perhaps easier to envisage – roughly 22 x 22 km. Four sites alone support about 75% of the global population.

Aquatic Warblers prefer large open wet grasslands, not only on their breeding grounds but also on migration routes and wintering sites. Along their migration routes through western Europe and northwest Africa, and also on their wintering sites south of the Sahara (presently proven in the Sénégal Delta and the Inner Delta of the Niger) only fragments of the original wet grasslands remain, making their position potentially even more precarious (Zwarts et al. 2009, Flade et al. 2011).

4 Some consequences of climate change

Weather events can suddenly and directly influence the population dynamics of birds. They can abruptly affect the availability of food or habitat quality in breeding, staging, or wintering areas. By contrast, only over long time periods can changes in climate have an impact on populations and on developments in their summer, passage, and wintering ranges. However, the current period of climate change, influenced as it is by so many anthropogenic factors, is now occurring at such a speed that the population numbers of many bird species have been affected within the last three decades. Since 1900 the mean global temperature has risen by almost 1°C, but since 1975 the rate of increase has accelerated. Global warming has changed the climate on every continent, leading to a redistribution of precipitation and the spread of desertification over large geographic areas, mainly in the southern hemisphere, that is, the stopover and wintering areas of transcontinental migrants. Rising temperatures in the temperate zones have resulted in earlier plant growth and earlier animal reproduction (life history events), and the responses of different organisms have been varied and dissimilar (Both et al. 2006).

Changes on the breeding grounds

The consequences of climate change on central European reed warblers have been the subject of several long-term studies (e.g. Peach et al. 1991, Foppen et al. 1999, Schaefer et al. 2006, Peron et al. 2007, Halupka et al. 2008). It is extremely difficult to isolate the effects of these changes as there are always a whole range of confounding seasonal or regional variables present. Just how differently each individual species responds can be seen from their arrival dates on the breeding sites. Eurasian Reed Warblers arrive 14-21 days earlier than they did 40 years ago; in Britain this has amounted to a rate of 0.25 days per year. In contrast, Great Reed Warblers and Marsh Warblers have brought their arrival date forward by only a few days. In Mecklenburg-Vorpommern (northeast Germany), the late-arriving Icterine Warblers actually now arrive a few days later than 50 years ago (Peintinger & Schuster 2005, Schmidt & Hüppop 2007, Sparks et al. 2007, Halupka et al. 2008). Eurasian Reed Warblers profit from the temperature rise in March and April in their passage areas, while Icterine Warblers, as extreme long-distance migrants, are more subject to their rigid internal clock than species that migrate shorter distances (Both & te Marvelde 2007).

Reproductive responses to climate change by Eurasian Reed Warblers have been striking (Schaefer et al. 2006, Halupka et al. 2008). Between 1973 and 2002, in Franken, southern Germany, the start of laying has advanced

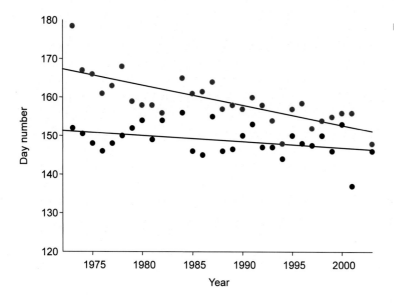

Fig. 13.3 Earlier occurrence of the median laying dates in Great Reed Warblers (black dots) and Eurasian Reed Warblers (red dots) from 1973 to 2003 in northern Bavaria under the influence of climate change. The two species show differing reactions. While Eurasian Reed Warblers have brought their start of laying date forward by 15 days over 30 years, this figure is lower in Great Reed Warblers (from Schaefer *et al.* 2006).

by 15 days, and by 18 days between 1970 and 2006 in Silesia (southwest Poland; fig. 13.3). This has lead to a rise in mean clutch size, since early clutches are usually larger than later ones (laying date regression) and the proportion of early clutches is now higher (fig. 13.3), and also because early broods suffer lower predation losses, probably helped by the extra safety provided by faster reed growth. Furthermore, the end of the laying period in Silesia remained unchanged, thus resulting in a lengthening of the total laying period. A longer breeding period, enlarged clutch size, lower brood losses, the increase in second broods from *ca* 10% to 35% (in Silesia), and finally more replacement clutches following loss of broods, all indicate an increase in reproductive outcome (Schaefer *et al.* 2006, Halupka *et al.* 2008). Accordingly, in several countries the population trend of Eurasian Reed Warblers is positive, or at least constant, despite the loss of habitat. This could be a possible explanation for the steady expansion of the species to the northeast since the middle of the 20th century (section 13.2).

The effects of climate change can differ between closely related and coexisting warblers. In Franken and Silesia, Great Reed Warblers inhabiting the same reedswamps as Eurasian Reed Warblers, but having longer migrational routes, showed only modest changes in breeding phenology. There was also an advance in first-egg-laying dates, but not as pronounced as in Eurasian Reed Warblers (fig. 13.3). The end of the laying period remained unchanged, while clutch size increased in Franken but not in Silesia. A slight rise in breeding success due to a fall in losses caused by poor weather could be accounted for mainly by higher May temperatures and the higher frequency of dry summers (Schaefer *et al.* 2006, Dyrcz & Halupka 2009).

The marked change in breeding phenology has its origin in the increase of local surface temperature in spring, which is correlated with earlier phyto-phenological data and with earlier growth and development of insects. *Phragmites* has also been shown to have developed a similar dependence on growing season and ambient temperature (Dykyjovà *et al.* 1970). Today Eurasian Reed Warblers, which are dependent on a particular developmental stage of *Phragmites* that allows them to conceal their nests adequately, are able to find suitable nest sites earlier. Warmer temperatures and the earlier development of the reed vegetation in turn creates the conditions required by arthropods, which therefore also emerge earlier from the water and develop on the reeds. Because of their high renewal rate of food resources, wet marshlands are less seasonal habitats, offering continuous food abundance throughout the entire breeding period of reed warblers from early May until late August (chapter 3).

In contrast to marshes, deciduous forests exhibit a much stronger seasonality in insect availability. Deciduous woodland is characterized by a short burst of mainly herbivorous insects that forage on the young leaves (Both *et al.* 2010c). As these food resources appear earlier with local

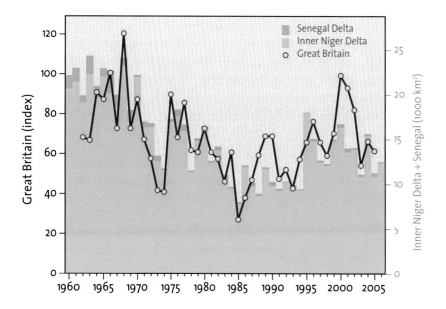

Fig. 13.4 Population trends of British Sedge Warblers in relation to the extent of flooding in their wintering regions, the Inner Niger Delta and the Sénégal Delta, in the preceding winter. Adult survival rates dropped sharply during the great droughts in the early 1970s and in 1983-1984 (Zwarts et al. 2009).

warming, long-distance migrating and late-arriving species, such as Pied Flycatchers, Spotted Flycatchers, Wood Warblers, and Icterine Warblers, find themselves in trouble. Their arrival date on the breeding grounds has not advanced that much, which means that insect abundance is already high in their nest-building and egg-laying period; but the downside is that insect numbers are already falling when nestlings need to be fed. This mismatch has led to severe population declines that are strongest in forests with early and narrow food peaks (Both et al. 2006, Both et al. 2009, Both 2010a,b, Carey 2009).

Although Icterine Warblers are a woodland edge species, they are just as seriously affected as the other forest-dwelling, long-distance migrants; for example, there was a fall in numbers of 85% in the Netherlands between 1984 and 2004 (Both 2010 a,b). The decline of this group of migrants was stronger in western Europe, where spring warming is pronounced, than in northern and eastern Europe, where spring temperatures are increasing only slightly (Both et al. 2006). Thus Icterine Warblers are among the climate change losers, whereas Eurasian Reed Warblers are favoured by the advanced start of their breeding season, which allows them higher fledgling production. As extreme long-distance migrants, species such as Icterine Warblers suffer from a serious 'handicap' in that they have a fixed annual cycle with a predetermined winter moult in southern Africa that does not allow them the advantage of a phenologically earlier start of return migration (Both et al. 2006).

Changes on the wintering grounds

On the open grasslands of the Sahel, which serve as winter quarters for so many migrant birds, it is the decline in rainfall that has influenced their numbers, rainfall which at times ceased altogether in the 1970s and 1980s. This resulted in severe population declines in a range of Afro-Palaearctic migratory species, a classic example being that of Sedge Warblers. A considerable proportion of the European population overwinters in the Sahel wetlands, such as the floodplains of the Inner Niger Delta or the Sénégal Delta. Large expanses of these floodplains dry up in periods of drought, reducing their carrying capacity and increasing competition for food among the Sedge Warblers, which ultimately results in a steep rise in density-dependent mortality. Peach et al. (1991) and Foppen et al. (1999) were able to show that rainfall in west Africa is the main factor that determines breeding numbers in both Britain and the Netherlands (fig. 13.4). The entire breeding population is likely to be affected when hydrological conditions in west Africa are particularly severe. In 1983-1984, the Great Sahel Drought alone caused declines in various European Sedge Warbler populations ranging from 51 to 79%. The survival of adults reaching southern England was estimated to be lower than 4% between 1983 and 1984. In turn, wet years in west Africa enable rapid recoveries (Zwarts et al. 2009). The problem of the Sahel region is that reduced rainfall, together with agricultural intensification, have a negetative effect on the vegetation, leading to widespread desertification. This combination of climate change and habitat loss is

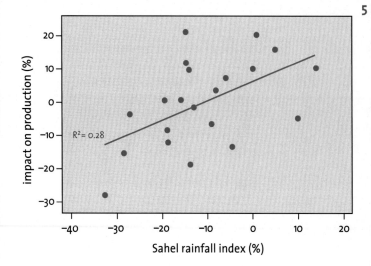

Fig. 13.5 Productivity (number of juveniles relative to adults captured on Constant Effort Sites) of British Sedge Warblers as a function of Sahel rainfall (Zwarts *et al.* 2009).

probably the biggest threat of all to Sahelian biodiversity (Sanderson *et al.* 2006). In contrast to Sedge Warblers, those Eurasian Reed Warbler populations that winter in the Sahel region are not so dependent on rainfall because they prefer drier habitats. Moreover substantial numbers spend the winter in intertidal mangroves along the west African coast (Altenburg & van Spanje 1989, Zwarts *et al.* 2009).

Sadly that is not the end of an ominous story. Following increased winter mortality, the central European Sedge Warbler populations have so far been unable to return to their previous numbers. The fatal effects of habitat loss and fragmentation in their summer quarters are devastatingly amplified by mortality during the winter months (Peach *et al.* 1991, Foppen *et al.* 1999). Any individuals which survive the drought period in their overwintering areas retain carry-over effects into the breeding season. For British Sedge Warblers, a significant correlation between productivity (ratio of juveniles to adults on British Constant Effort Sites) and Sahel rainfall in the preceding year has been demonstrated (fig. 13.5). The diminished fitness of the survivors of Sahel droughts can be explained by their poor condition on arrival on their breeding grounds (Zwarts *et al.* 2009).

5 Conservation matters

Of the 33 continental acrocephalids, six are classed as Vulnerable on the IUCN Red List (BirdLife International 2004-2010). For the majority of the remaining species long-term population declines have been noted more and more frequently, at least at regional levels. A serious additional factor is that many species suffer impacts not only on their breeding sites but also on their migration stopover areas and winter quarters. The priority aim of the conservation management of threatened species is the protection and restoration of their habitats; those of acrocephalids have to be protected against over-exploitation, water loss, eutrophication, and vegetational succession. In the case of the island species, that list must also be augmented by the eradication of introduced predators or food competitors.

Restoration of habitats

A precondition of effective conservation is that the biology and habitat requirements of the species to be protected must be thoroughly studied and understood. Once again, a good example would be Seychelles Warblers, whose habitat on the tiny island of Cousin was almost completely smothered by coconut plantations in 1968. The removal of coconut palms and systematic replanting with morinda (Indian mulberry or noni) trees were the first important steps towards effective habitat restoration. As the most insect-rich tree, morinda is preferred by the warblers for foraging. The greater the morinda cover and insect abundance in warbler territories, the higher the reproductive success and survival rate of the warblers. Once the population had returned to a situation of habitat saturation following these measures in 1987, the food supply continued to guarantee high habitat quality and high reproductive success (Komdeur 1994b & Pels 2005). Habitat restoration also has positive effects for a variety of co-inhabiting organisms (Samways *et al.* 2010).

In the case of European acrocephalids, prevention of eutrophication and vegetational succession are among the most important conservation measures. Very many reedswamps are increasingly overaged, matted and tangled on their landward side as a result of succession processes, and hence of little use as habitat for many reed-dwelling species. Sometimes the reason for this lies in the fact that human utilization of the reeds has become uneconomical and the beds are therefore abandoned. The traditional winter cutting of reed protected many reedbeds from matting and scrub invasion, keeping them in a stable balance. Unfortunately, the removal and transport of biomass in the course of conservation management is today barely affordable, given the labour costs. Winter reed cutting is an effective conservation measure provided it is carefully carried out. In recently cut, newly sprouting young reeds, the population densities of both arthropods and reed-dwelling birds will clearly be lower than before; Sedge Warblers may be reduced by a factor of between 6 and 50, while the numbers of Eurasian Reed Warblers may be reduced by up to 2.5 times (Baldi & Moskát 1995, Graveland 1999, Schmidt *et al.* 2005). It is only two to three years after cutting that the advantages become clear, when the densities of a host of reedswamp species in now optimal habitat conditions can rise again, often to levels exceeding those before cutting (Wüst-Graf 1992, Graveland 1999, Poulin & Lefebvre 2002). To avoid too much disturbance it is advisable not to cut the entire reedbed, but to adopt a rotational practice, cutting in a mosaic or strip pattern at defined time intervals from the reappearance of reed degradation. In the modelling of high quality reed warbler habitats the aim should be to keep reedbed fragmentation to a minimum (see below), while at the same time creating sufficiently tall vegetation edges to increase arthropod diversity. However, it should be noted that predation risk at the edges of reedbeds is higher than inside them (Foppen *et al.* 1999, Baldi & Batary 2005; section 8.4).

Although Eurasian Reed Warblers are not yet counted among the threatened species, their western and central European populations are subject to increasing fragmentation of their habitats, which are mostly embedded in a landscape dominated by intensive agriculture (fig. 13.6). If we wish to halt this fragmentation process, then large areas of reedswamp must be protected and conserved as source habitats. In addition, a number of small strips and corridors of reed vegetation in the neighbourhood of the larger reedbeds need to be created to function as buffer zones

Photos 13.9 and 13.10 An important task of conservation biology is to improve our knowledge of the landscapes and species to be protected so that the information gained can be turned into successful conservation measures. To this end, Aquatic Warbler populations are monitored in both their breeding (Biebrza, Poland) and winter habitats (Djoudj, Senegal) (photos Lars Lachmann, Martin Flade).

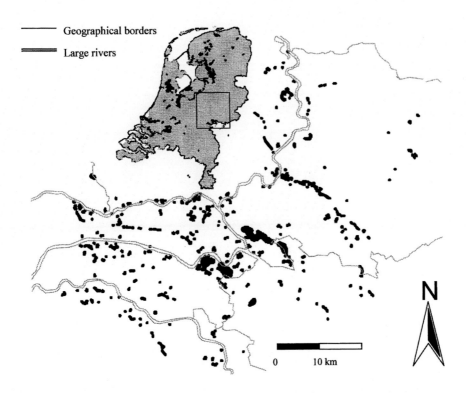

Fig. 13.6 Example of the fragmented distribution of marshlands (mainly reedbeds) in a 2000-km² study area in the Netherlands. Because most reedbeds are small and have a linear appearance, their presentation on the map is exaggerated. The insert shows the location of the study area and the location of large marshlands in the Netherlands (Foppen *et al.* 2000).

and even contribute, as sinks, to metapopulation stability. It is not only sink populations that profit from the densest possible habitat net. Mathematical models have shown that since dispersal flux from sink to source is greater than vice versa, sinks can also contribute to the stability of sources. Just how important a fine-meshed net of habitats can be is illustrated by the fact that Eurasian Reed Warblers are reluctant to cross gaps of as little as 50 m (Foppen *et al.* 2000, Bosschieter & Goedhart 2005).

The case of Aquatic Warblers, part two

The efforts surrounding the protection of Aquatic Warblers can serve as an exemplary model of international nature conservation at its best. Rare species such as Aquatic Warblers nearly always hold a particular fascination, but in their case a series of lucky circumstances contributed greatly to their survival (Flade 2008). In 1990, on the birds' breeding area in the Biebrza river valley in Poland, both Polish and German ornithologists were discussing how Aquatic Warblers and their habitat might be saved from complete eradication. To begin with, basic

research into the biology of the birds and their habitat was necessary, as well as a census of the entire population. First of all, Martin Flade privately organized search expeditions from Belarus to Siberia. The newly discovered large breeding populations in Ukraine, and especially in Belarus, were acutely threatened by gigantic land drainage projects that would quite possibly have threatened the very existence of the species (Kozulin & Flade 1999). As a result, action had to be taken quickly.

It was important to find financial sponsors and organizations willing to work in Belarus, despite an initially unfavourable political situation. With the courageous backing of the Michael Otto Foundation for Environmental Protection in Hamburg, early projects to secure the protection of the birds' habitat were devised. This enabled the Royal Society for the Protection of Birds (RSPB) to become involved, which first of all built up the Bird-Life Belarus (APB) organization, with the aim of making the project well known to the general public. Today Aquatic Warblers have become a flagship species for nature conservation, with their image appearing on postage stamps and their name known to every Belarusian schoolchild. The rate at which wetlands are being drained has been reduced, the remaining least degraded habitats placed under protection, several international conferences on the conservation of peatlands have been held, and management plans drawn up and put in place with support from the United Nations Development Programme (UNDP). Currently the restoration of 42 000 ha of drained peatland is planned, while further areas have been earmarked for future restoration within the framework of greenhouse gas emissions trading.

Parallel to this range of activities, the Aquatic Warbler Conservation Team (AWCT), first founded in 1998 by Martin Flade and Norbert Schäffer, was taken under the umbrella of BirdLife International. This group has already drawn up a second global Species Action Plan based on precise data and with clearly defined tasks (Flade & Lachmann 2008). In addition, AWCT, in cooperation with the RSPB, has managed to bring together representatives of 15 countries where the species occurs in the course of its annual cycle to sign a legally binding 'Memorandum of Understanding Concerning Conservation Measures for the Aquatic Warbler'. This is the first such international agreement to be formulated for the protection of a single species of 'little brown birds' and it should ensure a sound foundation for national and international conservation projects and sponsoring appeals (Flade 2008).

The previous rapid population decline of Aquatic Warblers on their Belarusian breeding grounds has been slowed down, at least for the time being, simply by saving their habitat. Nevertheless, a gradual decrease in numbers continues, especially in the peripheral populations (western Poland, Lithuania, Hungary), which often have to make do with small breeding sites (Flade, *in litt.* 2011). The reason is habitat change due to vegetation matting and scrub invasion, because the traditional extensive farming methods of cutting, mowing, and grazing have been abandoned (Lachmann 2008). Within the framework of the European Union's LIFE project, which has been active since 2005, substantial sums of money have been made available for the study of habitat structure and the implementation of habitat management plans in both Poland and Germany (Tanneberger *et al.* 2008, 2009). Landowners receive a financial contribution for their maintenance work. On the Biebrza wetlands and in

western Poland, several thousand hectares of fen mire were cut after the breeding season and invading scrub removed using innovative techniques, such as a tracked snowfield vehicle rejigged to function as a mowing machine that would avoid damage to the sensitive peat substrate. The cuttings thus collected were turned into pellets for industrial combined heat and power plants. This model project is currently on the brink of commercial viability. What is more, it is not only the Aquatic Warblers that profit from all these measures. Since the conservation projects began, their population on the Biebrza is certainly growing, but so also are those of many other animals and plants of the area, including a range of waders (Lachmann 2008, Lachmann *et al.* in press).

It is not only in the breeding quarters that help is needed; after all, Aquatic Warblers do spend eight months of the year outside their breeding areas. In the western European passage regions, as well as in the overwintering sites in Sahelian Africa, several projects are currently in progress to improve the habitats of Aquatic Warblers and associated wildlife in order to protect the populations throughout their entire annual cycle (Flade 2008, Lachmann & Flade 2008, Flade *et al.* 2011).

As these examples show, habitat management should not only be practised locally but should be embedded in large-scale regional projects and always follow tried and tested formulae. The greater the area covered by networks of appropriate habitat, then the greater will be the benefit. There is good evidence that intelligent habitat management and restoration are not only able to increase the reproductive output of a species, but also to improve its resilience to catastrophic events (Komdeur & Pels 2005). However the main problem remains the financing of such labour-intensive and complex projects. The difficulties faced by nature conservation are huge and the challenges we must meet are indeed immense.

6 Aspects of population genetics

The smaller and more isolated animal populations are, the greater the risk that genetic variation will be reduced through inbreeding. Molecular studies have confirmed that the founding of a new population involves a loss of genetic diversity indirectly proportional to the size of the founder group.

Table 13.3 Global and national population estimates of Aquatic Warblers (singing males with northeast German and northwest Polish populations unified as 'Pomeranian population') and population trends according to the current 'best knowledge' (Lachmann & Tanneberger in prep.).

Country/ region	Population estimate 2009		Population trend, comments
	total or geometric mean	range	
Belarus	5 489	3 959 - 7 609	stable/fluctuating since 1996, possibly underlying slow decline
Ukraine	4 556	4 215 - 4 925	possible increase since 2003 due to high water levels, but data quality questionable
Eastern Poland	3 115	3 115 - 3 115	stable/fluctuating since 1996
Hungary	195	190 - 200	steep decline since 2001. < 10 males in 2011
Lithuania	150		strong decline since 2004
Pomerania	54	54 - 55	strong continuous decline since mid-1990s
Russia incl. western Siberia		0 - 500	last breeding season records from 2000 (western Siberia), possibly now extinct
Latvia	0		last breeding season records 2000-2002
Global population	**13 809**	**11 443 - 16 369**	stable/fluctuating since 1996 (annual minimum 10 160 in 2001, max. 13 811 in 2005)

For instance, in the newly founded and isolated Great Reed Warbler population on Lake Kvismaren in southern Sweden, an increased rate of partial albinism was discovered. Albinism results from the expression of recessive alleles due to inbreeding. The rate of albinism only increased during the initial years, gradually declining as the population grew, and the lifetime reproductive success of albinistic warblers was not reduced (Bensch et al. 2000). Neither could the lower hatching success of a few females in this population be assigned to inbreeding, but rather to other, maternal effects (Knape et al. 2008; section 8.5). In the same population, no evidence was found for inbreeding avoidance via kin discrimination, hence even in small populations dispersal can be regarded as a succesful mechanism to counteract inbreeding (Hansson et al. 2007).

Since island populations are subject to genetic drift, and may have passed through several bottlenecks (during colonization and subsequent stochastic population collapses), their genetic diversity (allele frequencies) can be greatly reduced under certain conditions. In the bottlenecked and endangered island endemics, Nihoa Millerbirds and Seychelles Warblers, a substantially reduced genetic diversity has been found in their small populations compared with Laysan Millerbirds and the much more widespread Great Reed Warblers respectively. In contrast, a continental endangered species with an extremely small breeding range, Basra Reed Warblers, contained as much variation as Great Reed Warblers (Hansson & Richardson 2005, Fleischer et al. 2007). Nihoa Millerbirds have even less genetic variation than other insular acrocephalids or other Hawaiian songbirds. Given their tiny population size (the genetically effective number of breeders is between 5 and 13 individuals; Addison & Diamond 2011) they should lose 25 to 75% of the genetic variation of their starting population within 50 generations, according to calculations made by Robert Fleischer and colleagues (as cited in Addison & Diamond 2011). Despite this, there are at present no detectable symptoms of inbreeding depression (Conant & Morin 2001).

For bird populations, island life commonly leads to a higher susceptibility to introduced diseases. In this situation the immune functions of insular birds are apparently not attenuated but are rather 'reorganized' in general (Matson 2006). Results from two reed warbler species on Polynesian islands also point in this direction.

They showed completely different responses to an experimental immune challenge (Beadell *et al.* 2007; section 12.3). Investigating the role played in this by the genetic structures of island species is a completely new field of research. In this regard, studies on a number of reed warblers could be just as significant as those on Seychelles Warblers for a better understanding of the genetic basis of inbreeding depression (section 8.5). The precondition for such studies is that they must be long-term, since the disadvantages suffered by populations with reduced genetic diversity – such as an impaired ability to adapt to environmental changes or a higher susceptibility to disease – often reveal themselves only in times of crisis. In the face of increasing threats to ever more species, a major aim in conservation must be to maintain and restore sufficient genetic variation in small and endangered populations.

The world in which the acrocephalids must make their way is one of constant change. Whether it is man-made habitat loss through intensive agriculture or through manipulation of the water table on the breeding grounds, whether it is drought and desertification in the winter quarters and stopover sites, or whether it is the multifarious effects of global climate change, every one of these alterations represents an enormous disturbance to their mode of life. The list of reasons why acrocephalid species are in decline is becoming longer and longer. If several of these challenges coincide, then some species may no longer be able to cope with the burden. It is high time that we recognize that we have to protect Nature from ourselves.

Summary

In recent decades many acrocephalid species have undergone significant declines in their numbers. The reasons include the intense man-made pressures on their habitats, prominent among them a whole range of manipulations of wetland water regimes and agricultural intensification. As long as their habitats remain intact, the breeding densities of reed-dwelling acrocephalids can reach levels that are among the highest of all bird species. The two commonest European species are Sedge and Icterine Warblers, with their greatly extended distributional ranges.

In the course of the general eutrophication of wetlands in the 19th century, the populations of many reed warblers at first underwent significant increases; among the most impressive of these was the spread of Eurasian Reed Warblers to Scandinavia and eastern Europe. However, with increasing pressure from an expanding human population, and its attendant accelerated intensification in agriculture, many acrocephalids began to show declining population trends. In Europe there has been a large-scale decline in reedbeds, though the principal cause of this – the process known as reed die-back – has yet to be fully explained. Fifty per cent of all wetlands have been destroyed on the global scale, particularly in the tropical regions.

The various effects of global climate change are becoming increasingly significant. In their central European breeding areas some acrocephalids are among the climate change winners, but others have suffered dramatic losses. Eurasian Reed Warblers now arrive up to 21 days earlier than they did 40 years ago, and show an increase in the number of viable young, thanks to earlier vegetation growth and food abundance throughout their entire breeding period. In contrast, forest-dwelling Icterine Warblers are among the climate change losers. As extreme long-distance migrants they cannot bring their arrival on the breeding grounds forward and probably suffer from a lack of synchronicity with their prey, which now reaches an ever earlier maximum in deciduous forests.

On the wintering or staging grounds, for instance in Sahelian Africa, climate change manifests itself in a decline in precipitation and resulting periods of drought, made worse by overgrazing and agricultural intensification. In long dry periods the mortality of birds such as Sedge Warblers rises substantially. Such winter mortality together with habitat loss in the summer quarters fatally reinforce each other.

Aquatic Warblers are perhaps the continental species most seriously affected by population decline arising from habitat loss in both breeding and wintering quarters. Currently their drastically fragmented breeding area covers less than 500 km². Given this alarming state of affairs, the Aquatic Warbler Conservation Team has developed concepts and programmes to save the species, which are already beginning to bear fruit and are now regarded as a model for the conservation of other threatened species.

14

Ecological equivalents and convergence

Curve-billed Reedhaunter.

The Otamendi Nature Reserve, Argentina, 70 km northwest of Buenos Aires, is where the level pampa meets the broad floodplains of the Río Parana and Río Uruguay. There are temporary shallow pans, long water-filled depressions, lagoons fringed with emergents, backwaters and bayous, meanders with riverside vegetation, and frequently inundated shoals near the banks. In contrast to the wet marshes of the Old World, where *Phragmites* is the dominant emergent, here we mainly find stands of *Typha* (totorales) and the grass *Zizaniopsis bonariensis* (espadanales), which, with its broad, stiffly upright leaves, looks like a cross between reed and reedmace and can reach up to three metres high (fig. 14.1: hydrosere of a pampa lagoon). This is a paradise for the various species of wetland birds that occur here in large numbers. In the reedswamp we first spot two Curve-billed Reedhaunters, rather plump-looking passerines with plain brownish upperparts, whitish supercilium, and white underside, moving boldly through the tangle of blades and stalks; they allow us just a brief view as they fly up to a song-perch. In both appearance and behaviour, their resemblance to a large reed warbler is remarkable. They belong to the ovenbird family, Furnariidae. Other members of the family soon appear, including a confiding Yellow-chinned Spinetail, whose cinnamon upperparts are so reminiscent of a small plain-coloured acrocephalid, followed by Sulphur-bearded Spinetails and Straight-billed Reedhaunters, which skulk so deeply within the vegetation that they are best seen when lured into the open using tape playback.

These reedswamp birds attain their greatest diversity in the New World in the southern temperate and subtropical marshes of South America and are impressive both for their species richness and for the density of individuals. (Orians 1980). Many of the commonest species belong to the Icteridae and Tyrannidae families. The observer is immediately struck by the large number of New World blackbirds around the marsh edges; the males in particular catch our eye with their intensely coloured plumage patches or epaulets. Now, in the austral spring, the early-breeding Brown-and-yellow Marshbirds are feeding their fledged young; flocks of Yellow-winged Blackbirds fly between their breeding sites and feeding grounds, while Scarlet-headed Blackbirds, or Federales, with their brilliantly red heads and velvet-black plumage, are conspicuously perched on the tops of high reeds to signal their territories. Warbling Doraditos, tyrannids with yellow underparts and long slender legs, roam in pairs through the reedbeds. At the deep-water edge of the swamp, where the emergent vegetation consists only of pure stands of Californian bulrush, Many-colored Rush-tyrants and Wren-like Rushbirds, another member of the Furnariidae, are commonly to be seen.

We observe how the gaudily coloured rush-tyrants clamber around among the stems even more skilfully than the doraditos, gleaning tiny insects from the aquatic vegeta-tion with their slender bills. Nearby a rushbird is working on its half-built nest. This nest-building method is unique within the ovenbird family, resembling the technique used by reed warblers on vertical stems (section 8.1). The bird weaves wet plant fibres tightly around five supporting stems before cementing them together with mud. Its industrious activity is so striking that it is called *trabajador* (the worker) in some South American countries. However, in contrast to the open cup of the Old World acrocephalids, the end product will be an untidy domed nest in the form of an inverted pear.

We find it quite remarkable how much so many of these New World reedswamp inhabitants remind us of reed warblers, whose distribution is restricted to the Old World. Furthermore, both New and Old World reed-dwellers could hardly be further apart in the systematic classification of the songbirds. An entire series of ecological surrogates for the acrocephalids lives in the reedbeds of the western hemisphere (the 'blank area' on the map in fig. 1.1). In North America it is chiefly the members of the wren family (Troglodytidae) and the *Ammodramus* New World sparrows (Emberizidae) that have occupied the position of the reed warblers; some examples would be Marsh Wrens in deep-water marshes, Sedge Wrens at marsh edges, and Saltmarsh Sparrows in the coastal marshes. The birds that most remind us of the acrocephalids in South America are principally representatives of two suboscine families, the ovenbirds (Furnariidae) and tyrant-flycatchers (Tyrannidae). In the 60-million-year-long 'splendid isolation' of the south American continent, the ovenbirds and tyrant-flycatchers have undergone a remarkable radiation, which has produced species that are astonishingly similar to Old World birds both in ecology and appearance.

One ovenbird species in particular, Wren-like Rushbirds, has long been recognized as an ecological equivalent of both the Old World reed warblers and the North American Marsh Wrens (Wetmore 1926, Koepcke 1974). Two further marsh-dwellers traditionally seen as replacements for the acrocephalids are the Curve-billed and Straight-billed Reedhaunters. Both were collected by Darwin in Uruguay during his voyage on HMS *Beagle*, and placed in a single genus by John Gould. Are such counterparts really similar to the reed warblers, and, if so, in which characteristics in particular?

1

Morphological convergences

A comparison of the similarities between sets of unrelated species living in the same environments on different continents offers an attractive opportunity to look more closely at the convergent evolution we encountered in our expedition to the Argentinian wetlands. Just like researchers before us, we too were fascinated by the similar way in which South American suboscines and totally unrelated Old World birds utilize the same habitat and show such a remarkable parallel development in form

and plumage markings. The two ovenbirds, Curve-billed Reedhaunters and Wren-like Rushbirds, really do appear to reflect the spectrum seen in the reed-dwelling acrocephalids, from the large plain-coloured species that frequent the upper strata of the reed stand to the striped species living lower down near the base of the plants. In appearance and behaviour, Wren-like Rushbirds are not only a genuine ecological equivalent to Moustached Warblers, but also bear a striking resemblance to the Marsh Wrens of North America. All three have a dark crown, a prominent whitish supercilium contrasting with dark ear-coverts, and a boldly streaked mantle. They are therefore impressive examples of how a similar plumage pattern evolves convergently when individual species of three widely separate lineages live in the same environmental conditions, since the disruptive head patterns and streaked upperparts perfectly match the flecks of light and shadow of the reedswamp understorey and help to camouflage the birds against the background.

Riegner (2008) outlined possible paths to such independent correspondences, but concluded that plumage configurations subject to selection may be developmentally constrained. This means that functional explanations alone are insufficient to understand just why certain chromatic and morphological features covary so consistently (see chapter 2 for parallel evolution of head patterns within the acrocephalids). J. V. Remsen, Jr., who provided a comprehensive treatment of the ovenbirds in volume 8 of the *Handbook of the Birds of the World*, pointed out that while their many obvious convergences have often been described, they have never actually been studied within a quantitative framework (Remsen 2003).

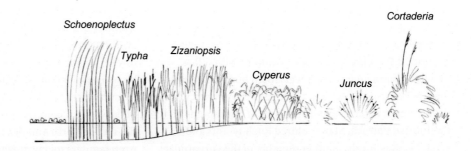

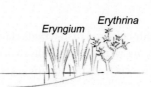

Fig. 14.1 Transect through a pampas lagoon showing the habitat-shaping plant genera (drawing by Bernd Leisler).

Photos 14.1 to 14.3 There are no acrocephalids in the New World. The reedswamps here have been settled by other groups which could hardly be further away in the taxonomic system of the songbirds. For instance, Curve-billed Reedhaunters, Yellow-chinned Spinetails, and Sulphur-bearded Spinetails (from left to right) are very reminiscent of reed warblers in their appearance, their inconspicuous brown plumage, their perching posture, and their movements through the vertically structured vegetation, and yet they belong to the ovenbirds, a family in the suboscines. Such parallels in the morphology of unrelated taxa arise via convergent evolution as adaptations to the same habitat requirements (near Buenos Aires, Argentina; photos Alec Earnshaw).

How can such supposed similarities in the morphology of two different taxa be explained? First of all, a definition might help us: convergences are resemblances in certain characteristics of two taxa which belong to separate lineages, in which the corresponding evolutionary novelties have been independently acquired as adaptations to similar environmental conditions (Schluter 1986, Mayr 2001). The resemblances can refer to a close overall similarity or they can be limited to only a few characters or a subset of structures, such as the slightly curved bill of various unrelated nectarivorous birds on different continents or archipelagos, for example.

Whether convergence has taken place can be demonstrated by comparing the morphology of individuals from a particular habitat with that of congeneric – or the most closely related – species from other habitats, since members of a genus have evolved from a common ancestor, and the morphological differences between congeners indicate the direction in which the character has changed (Niemi 1985). Leisler & Winkler (2001) used this method as a starting point for both an intercontinental comparison of South American and Old World reedbed birds, and for a study involving various species living in African papyrus swamps, a habitat that can be regarded as a

Photo 14.4 Not only the plain-coloured but also the striped reed warblers have their New World 'doubles', as the plumage pattern of this Wren-like Rushbird shows (near Buenos Aires, Argentina; photo Alec Earnshaw).

'mega-reedswamp' (section 3.1). In addition to the general hypothesis that such specific habitats produce overall similarities, they tested the hypothesis that convergence affects characters related only to bipedal locomotion, since reeds pose the most stringent demands in this regard because of their tall, vertical structure (section 5.1). A further significant characteristic of reedbeds is their cage-like structure, which allows their inhabitants only brief flights and forces them to employ gleaning from perches as their main foraging strategy.

Marsh dwellers in South America

Only when seen in the light of the recent resolution of the molecular phylogenies – and hence exact relationships – of the ovenbirds and tyrant-flycatchers can a meaningful comparison be made between the external characters in the Old World reed-dwelling Acrocephalidae family and South American Furnariidae and Tyrannidae families (see www.knnvuitgeverij.nl/EN/appendix-the-reedwarblers) and those of congeneric or most closely related species (within the three lineages) from other habitats (Olson *et al.* 2005, Irestedt *et al.* 2006, Ohlson *et al.* 2008, Tello *et al.* 2009). These phylogenies made it possible for the

first time to find the actual nearest relatives of the reed-dwellers concerned in order to make the comparison, which involved 14 ovenbird and 10 tyrant-flycatcher species together with 37 acrocephalids; see www.knnvuitgeverij.nl/EN/appendix-the-reedwarblers). Previously, the assumed affinities were uncertain and often erroneous. For instance, the two reedhaunters are now known to be totally unrelated to each other, since molecular genetics have indicated that Curve-billed Reedhaunters are most closely related to Wren-like Rushbirds. Surprisingly, the sister taxon of both, the Sharp-tailed Streamcreepers, are a species that had formerly been positioned in a completely different ovenbird clade. The Icteridae family are excluded from this comparative analysis because most of this family's members forage away from reedbeds, using feeding techniques other than gleaning.

As in other morphological studies, the variables in external shape (19 in all) were analysed using various approaches. Results indicated that the reed-dwellers in the three families converge in their overall morphology, that is, their shapes came to resemble each other independently on both continents fig. 14.2). All reedswamp-dwellers have the following features in common: large feet, small wings, and long narrow bills. However, a more detailed analysis of the three function-complexes (hind limb, flight apparatus, and bill) shows that the evolutionary path to reed-dwelling is a different one for each group, because they differ in the respective magnitude of the necessary changes in these function-complexes. The wings became smaller mainly in the two South American groups; the reed-dwelling tyrannids developed clearly longer and narrower bills than their hawking and striking relatives in other habitats (see also Fitzpatrick 2004). The feet and legs of both ovenbirds and reed warblers underwent considerable changes as they became more suited for clinging to vertical elements.

A further method of visualizing convergent differences in reed-dwellers as opposed to non-reed-dwellers in various lineages was carried out in two stages. In the first we used discriminant analysis to see how the three families are morphologically differentiated in general. As expected, the differences discovered between the groups of related species reflect adaptations for their overall mode of life and their phylogenetic characteristics. For example, the ovenbirds have strongly graduated tails and relatively short legs. They also lack rictal bristles. These traits separate them clearly from the other two families, which have greater morphological congruences (x-axis in fig. 14.2). In the second stage, we formed two groups from reed-dwell-

Photos 14.5 and 14.6 Warbling Doraditos and Many-colored Rush-tyrants, which are also specialists at living in reed-swamps, belong to the tyrannids. Their methods of moving and foraging immediately remind the observer of reed warblers (near Buenos Aires, Argentina; photos Alec Earnshaw).

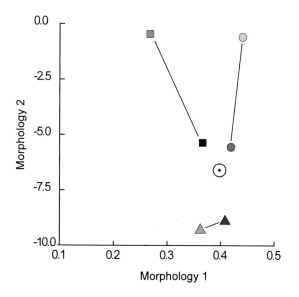

Fig. 14.2 Discriminant analysis of convergent development in the external morphology of reed-dwelling tyrants (squares), ovenbirds (triangles), and reed warblers (circles). The target symbol represents the central point of the reed-dweller space (group means, dark symbols). The group means of the non-reed-dwelling members of the three groups (families) are indicated by light symbols. The reed-dwellers in all three families approach each other on the discriminant axes.

ers (24 species) and non-reed-dwellers (taking no account of their phylogenetic relationships) and searched for possible morphological differences between them. This analysis revealed that the reed-dwellers differ significantly from the other species in the possession of relatively longer toes and claws, and also larger feet, stronger legs, and longer bills (fig. 14.2).

Papyrus birds

In our second example of the evolution of convergent morphological traits, we shall consider papyrus swamp – a habitat so special that only a few passerine species have entered it and are now restricted and endemic to it. According to Britton (1978), on the basis of their permanent occurrence in the habitat, eight species from six lineages can be regarded as papyrus specialists: Papyrus Gonoleks (bushshrikes from the Malaconotidae family); Carruthers's Cisticolas (Cisticolidae); White-winged Warblers (Locustellidae); two reed warbler species, Greater Swamp- and Papyrus Yellow Warblers (Acrocephalidae); two weaverbird species, Slender-billed and Northern Brown-throated Weavers (Ploceidae), and Papyrus Canaries (Fringillidae). Again, Leisler & Winkler (2001) compared their overall morphology with that of their nearest relatives from other habitats and were able to separate phylogenetic differences from those arising from adaptation to the papyrus habitat. Convergence affected only a few traits of the foot and only one of the wing, but did not affect the overall phenotype. Thus all papyrus birds have developed larger feet and broader wings but have otherwise retained the overall morphology of their different ancestries.

The results of both studies make clear that in all groups hind limb change is the most dominant common element that alters in a convergent manner with the change to a reedbed mode of life. Long, powerful toes and long claws are obviously necessary for all reed-dwellers, regardless of their preferred climbing technique, as they are essential for strong perching, clinging in the reed warbler fashion, or standing or hanging 'straddled' between two erect stems, with the legs widely spread as each foot grasps a separate culm (section 5.1).

Small changes – great consequences

Other investigations of morphological convergence have shown that it is first and foremost the locomotor apparatus, especially the hind limb, (and also body size) that are affected by convergent developments; in other words body traits that directly interact with habitat structures. For example, Niemi (1985) compared patterns of morphological evolution of passerines in the peatlands of Minnesota and Finland. In both continents, North America and Eurasia, shrub-dwellers of different lineages or families (among them Sedge Wrens and Sedge Warblers as the corresponding commonest species) developed striking

Photos 14.7 and 14.8 It is not only the species richness of southern temperate marshes in South America that is remarkable, but also the density of individuals. Many of the commonest species belong to the New World blackbirds. They signal their territories with their intensely coloured plumage patches, known as epaulets; their song is less elaborate than that of reed warblers (Brown-and-yellow Marshbird; Yellow-winged Blackbird near Buenos Aires, Argentina; photos Alec Earnshaw).

Photo 14.9 In North America, the ecological position of the reed warblers has been occupied by representatives of the wrens and New World sparrows. Marshlands of the Old and of the New World have rich food supplies in common and hence high population densities of reedswamp-dwelling birds. This leads to pronounced interspecific territoriality and interspecific hostility. Marsh Wrens, for instance, are well known for destroying the eggs of both unrelated species and conspecifics (Utah; photo Paul Higgins).

Fig. 14.3 Plot resulting from two discriminant analyses of the morphological characters separating the three families of tyrannids (species C & D), furnariids (species A & B), and acrocephalids (species E & F) (x-axis) and reedbed (dark symbols, B, D, F) vs non-reedbed (light symbols, A, C, E) species (y-axis). The x-axis shows that morphologically the tyrants more strongly resemble the Old World reed warblers than they do the ovenbirds, to which they are much more closely related. In all three groups the reed-dwelling species are differentiated from the non-reedbed-dwellers by a score derived from differences in toe and claw length, tarsus thickness, and bill length.

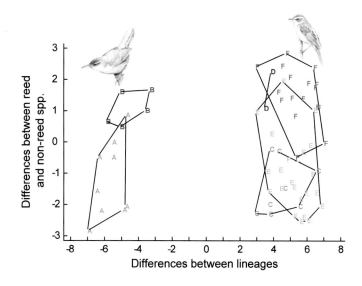

morphological convergences in the peatland habitat, associated with movement through the vegetation, when compared with their respective closest relatives from coniferous forests.

Convergence is a predictable, directional change in a character (or set of characters) towards a common phenotype that is adapted to a particular habitat. However, the necessary changes do vary depending on the evolutionary point of departure in the various groups. Dawkins (1986) gives accounts of many examples where independent lineages appear to have converged on what looks very much like the same endpoint from widely different starting points. Nevertheless he makes an important reservation: 'When we look in detail we find – *it would be worrying if we didn't* – that the convergence is *not total*' (our italics). The members of the lineages concerned can only develop a solution to the common ecological problem that is the best possible for themselves; evolution can only work with what is already there. An exact correspondence cannot be expected. If the same ecological problem (e.g. access to nectar secreted in the calyx of a flower) can be solved in a way that does not involve the use of a slightly curved bill, then convergences towards this character state are not to be expected in the first place. The flowerpiercers in the tanager family (Thraupidae) do not introduce their bill into the calyx to reach the

Photo 14.10 New World deep-water marshes like this one in Utah (taken in November) are characterized by high richness of invertebrate prey, just like Old World wetlands (photo Paul Higgins).

nectar, but have developed the habit of pecking at the base of the flower to break into the nectar store. Thus, in this instance, nectar is gained with a completely different bill form: straight, with a substantial hook at the tip, and the lower mandible curved slightly upwards. Another general conclusion we can draw from our convergence investigations, and other ecomorphological studies, is that small morphological changes can have large ecological consequences (chapter 5).

2 Behavioural convergences

Adaptation to similar ecological roles causes unrelated species of birds to converge in aspects of both appearance and behaviour. Just as in morphology, in comparative studies of behaviour we must be absolutely clear exactly which selection pressure is being responded to in the same way by different species.

Interspecific aggression

One ecological factor in particular is shared by the marshlands of the Old and New Worlds: the high population densities of their inhabitants as a result of the rich food supplies (chapters 6 and 13). Marsh-nesting blackbirds (Icteridae) achieve higher densities than their close relatives in upland habitats, and Marsh Wrens can breed in tiny territories of only 60 m² in their high-density populations (Orians 1980, Kroodsma & Brewer 1997, Brewer 2001). One consequence of this is that an interspecific spatial segregation has arisen among the marsh-nesting songbirds of North America, very like the interspecific territoriality of the Old World reed warblers (Orians & Willson 1964, Verner 1971). As a result, mutual interspecific hostility between icterids and Marsh Wrens is the rule. Blackbirds exclude Marsh Wrens from available space, while Marsh Wrens in turn seek to destroy the eggs of other birds, both conspecific or even unrelated (Picman 1988).

This peculiar and extraordinary behaviour of destroying nests, pecking eggs, or killing small young was not newly invented by wetland-dwelling Marsh Wrens, but exists in many wren species of other genera and in totally different habitats. It fulfils a variety of functions: in cavity-nesting species it is directed at competitors for holes, while other species prey on eggs as food, such as the two large species Giant and Bicolored Wrens of central and northern South

sexes puncture eggs in their case. By destroying nearby nests wrens presumably reduce competition for food and breeding space, both intra- and interspecifically. 'Aggressive' interactions with other species are probably a consequence both of this competition and of the egg-destroying habits of wrens. Unfortunately, we still lack good data that document cause-and-effect relationships among these behaviours (Picman 1988, Kroodsma & Verner 1997).

In the same way that the Old World Great Reed Warblers dominate their smaller congeners, the larger Red-winged and Yellow-headed Blackbirds react aggressively towards Marsh Wrens and exclude them by driving them away and destroying their nests. If the blackbirds move away or are experimentally removed from parts of a marsh, the Marsh Wrens will immediately expand into the vacated areas (Leonard & Picman 1986). Since Marsh Wrens are classed as predators on account of their egg-destroying habits in North American wetlands, the resulting interspecific spatial segregation is not absolutely identical to that of reed warblers, which probably developed mainly as a reaction against common enemies and also perhaps via apparent competition with other species (section 6.2). On the other hand, the important role of learning in the development of interspecific aggressive interactions is very similar in both Old World reed warblers and New World marshland birds.

Photos 14.11 and 14.12 Red-winged and Yellow-headed Blackbirds are common North American marsh dwellers, and they breed in polygynous systems. In their marsh habitats, the patchy distribution of important resources for females (i.e. nest sites and food) allows the males of these icterids – just as in reed warblers – to control high-quality territories, leading to the evolution of polygyny with female-biased parental care (Utah; photos Paul Higgins).

Mating systems

A rich and patchy distribution of food is characteristic of both North American and Old World marshlands, whose low or 'monolayer' vegetation offers abundant prey. Even more significantly, safe nest sites are also patchily distributed, especially in deep-water marshes (Picman *et al.* 1993; see also chapters 6 and 10).

Polygyny in deep-water marshes

For female birds, important resources such as nest sites or food are irregular or even clumped in distribution, whereas for males they are easier to monopolize and defend. Control of high quality territories then leads to the evolution of resource-defence or harem-defence polygyny with female-biased parental care (section 10.5). A strong pointer to convergence from different ancestries arises from the fact that, in the two continents of America and Eurasia, many marsh-dwellers breed in polygynous systems, even though they belong to completely different groups, such as the wrens, icterids or acrocephalids,

America. They even have the reputation of entering chicken coops to eat the eggs, earning them the local name of *chupahuevo*, or egg-sucker. Yet this egg-destroying behaviour is absent in other wren species (Kroodsma & Brewer 2005). In several members of the Troglodytidae family, the egg-puncturing habit can apparently be easily switched 'on and off' in the course of evolution, and can also be less or more pronounced. It is obligate in the polygynous Marsh Wrens, in which the males build multiple nests and help to raise the young if necessary; both

Photos 14.13 to 14.15 Another striking example of morphological and behavioural convergences is between Aquatic Warblers (top right) and two North American sharp-tailed sparrows, the Saltmarsh Sparrows and the closely related Nelson's Sparrows. The similarity of their habitats – open, wet meadows subject to frequent changes in their water level – is remarkable. The birds resemble each other, not only in plumage and in their method of locomotion in the vegetation, but both warblers and sparrows have a promiscuous mating system with no territoriality, no pair bonds, brood care being carried out only by the females, and very high levels of multiple paternity (Texas and Belarus; photos Greg Lasley, Alexei Kozulin, Mark Bartosik).

(Dyrcz 1988, Greenlaw 1988). Recent phylogenetic analyses also show that polygynous breeding systems have developed independently within icterids and acrocephalids under almost identical ecological conditions, and that they have evolved separately, not once, but several times from monogamous ancestors. Interesting parallels have emerged even when only the resource-based polygyny seen in Great Reed Warblers and icterids have been compared. Because of the risk that males might not assist at the nest, females base their choices less on indicators of male quality than on indicators of territory quality (e.g. resources such as safe nest sites or high food availability). An additional factor that promotes the aggregation of blackbird females is the advantage they gain through communal nest defence (Orians 1985).

Because of the high population density on these marshes, for all the above reasons the competition of males for high-quality resources or female groups appears to be increased, resulting in a higher intensity of intrasexual selection. This male-male competition then drives the evolution of secondary sexual traits, though on each continent the male characters affected are quite different. In the Americas the males in icterid lineages that settled in marsh habitats developed carotenoid-coloured patches and epaulets for signalling purposes when they confront their rivals (Johnson & Lanyon 2000). These striking plumage ornamentations in the extant marsh breeders arose several times from all-black ancestors living in non-marsh habitats. In the polygynous large reed warblers of Eurasia, the same task is performed by their complex songs (Catchpole 1980, 2000; chapter 7).

Table 14.1 Comparison of the common ecological factors of the habitats (in italics) of Saltmarsh Sparrows and Aquatic Warblers and the correspondences in their breeding systems.

	Saltmarsh Sparrow	Aquatic Warbler
Habitat	*tidal marshes*	*wet grasslands*
Fluctuations of nest sites	*high, predictable*	*moderate, unpredictable*
Food supply	*high*	*high*
Care	maternal	maternal
Territorial defence	absent	absent
Social bonding	absent	absent
Social mating system	promiscuity	promiscuity
Multiple paternity (in % of broods)	extreme (95%)	very high (60 - 75%)
Paternity guard	none	prolonged copulation

Promiscuity on marsh edges

In Europe and North America a further outstanding example of behavioural convergence has evolved in another type of marsh habitat, that of frequently flooded marsh edges or coastal marshes. Both the habitat and breeding system of Aquatic Warblers show several remarkable parallels to two North American sharp-tailed sparrows, primarily Saltmarsh Sparrows, but also the closely related Nelson's Sparrows (formerly considered to be a single species; Leisler 1985, Schulze-Hagen *et al.* 1995, Hill *et al.* 2010). Nelson's Sparrows mainly inhabit the edges of freshwater marshes in interior North America, while Saltmarsh Sparrows – true to their name – occur on the salt marshes of the Atlantic coast from southern Maine to North Carolina/Virginia. Like Aquatic Warblers, these New World sparrows inhabit flat, open, wet meadows with grass cover less than one metre tall. Another factor such habitats have in common is that their frequent changes in water level – floods, high tides, or drying-out – can lead to breeding losses and, above all, to a rapid alteration in the dispersion and availability of suitable nest sites. Saltmarsh Sparrows are especially vulnerable to predictable tidally-influenced inundations, which can cause a high level of nest losses. The high spring tides occur every four weeks, corresponding to the lunar cycle; this is the exact length of time it takes for a Saltmarsh Sparrow to raise a brood. The sparrows have optimally adapted to the situation by initiating renesting immediately after nest flooding (Shriver *et al.* 20007). Such rap-

idly changing conditions in their habitats, together with the extremely dense vegetation, mean that localization of fertile females can be very difficult for males of all three species. At the same time, the rich food supply provided by these habitats allows the females of both Aquatic Warblers and Saltmarsh Sparrows to rear their young completely unaided (section 10.6).

Saltmarsh Sparrow females raising a brood on their own have been found to be just as productive as sympatric Seaside Sparrow pairs with both parents doing the same job, yet with no apparent survival costs to the Saltmarsh Sparrow females (Post & Greenlaw 1982, Greenlaw & Post 1985). Neither male Aquatic Warblers nor male Saltmarsh Sparrows play any role in caring for chicks, neither do they defend territory or demonstrate any distinct antagonistic behaviour. In neither species do the sexes bond together to form pairs, which is unusual in songbirds. Both species are promiscuous and individuals of both sexes have been observed mating with more than one member of the opposite sex (table 14.1). This mating pattern results in a very high level of multiple paternities in both Aquatic Warblers and Saltmarsh Sparrows. In the latter, nearly every clutch was the product of more than one father and the average brood had been sired by more than 2.5 fathers – making this sparrow the most promiscuous bird species studied to date (Dyrcz 1993, Schulze-Hagen *et al.* 1993, 1995, Leisler & Wink 2000, Dyrcz *et al.* 2002, Hill *et al.* 2010; table 14.1; see also section 10.6).

Unfortunately, our present knowledge of the mechanics of how this unusual breeding system evolved is only fragmentry; because of the great difficulty involved in actually observing the birds in the dense grass vegetation of their preferred habitats, neither do we fully understand the role both sexes play in it. Saltmarsh Sparrow males appear to pursue a form of lekking or a competitive mate-searching, whereby they survey successive areas from 'lookout' perches and attempt to intercept and copulate with females, occasionally using forced copulations (Greenlaw & Post unpubl. ms). Likewise, Aquatic Warbler males also patrol large areas seeking fertile females. In neither species do males follow or guard mates. Females of both species are apparently either unable to choose or are uninterested in choosing the best possible male to sire their young (Hill *et al.* 2010). One limiting factor in the search for the best possible male seems to be the high pressure of time under which nesting or renesting Saltmarsh Sparrow females must operate. They appear to make up for the possible disadvantage of a copulation with a poor-quality partner by mating with several males.

The promiscuous systems of Aquatic Warblers and Saltmarsh Sparrows are derived states within clades of mostly monogamous species (Leisler *et al.* 2002, Hill *et al.* 2010, Greenlaw & Post unpubl. ms). The mating system of Nelson's Sparrows provides us with an idea of what an intermediate stage in the evolution of such pronounced promiscuity might have looked like. In this *Ammodramus* species some cases of bonding behaviour have been recorded, in the form of pair bonds and occasional biparental care (Greenlaw & Post unpubl. ms). While it is true that in all three species males have been observed forming temporary aggregations around a fertile female, only in Nelson's Sparrows do dominant males fight and drive off competitors from a female, which they guard, at least temporarily. In this fashion a few males attain a disproportionate number of copulations (Greenlaw & Rising 1994, Greenlaw & Post unpubl. ms). In communication between the sexes in all three species, the 'tuc' calls of the females seem to play a role in that they signal their lack of interest in the males. Only with clarification of the problem of how active or passive the females are in mating shall

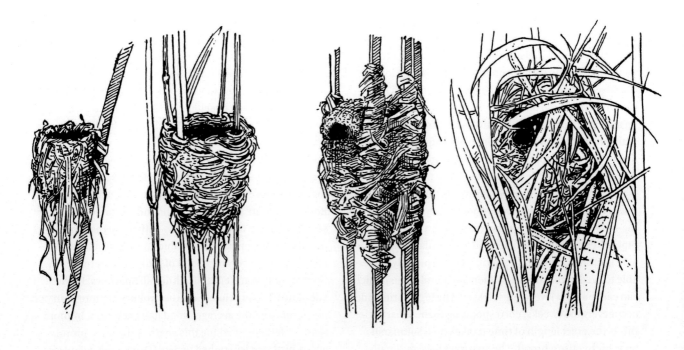

Fig. 14.4 Nests of inhabitants of deep-water marshes. From left to right: Many-colored Rush-tyrant, Eurasian Reed Warbler, Wren-like Rushbird, and Marsh Wren (from Koepcke 1974).

we hope to fully understand the breeding systems of these birds Nevertheless, the examples discussed above are impressive documentation of just how strongly the evolution of mating systems is determined by ecological conditions, how flexible and easily altered they are, and consequently how convergences can evolve.

Until now, very little, if any, attention has been paid to possible convergences in song or nest building (fig 14.4). Remsen (2003) speculates on a bioacoustic convergence between Marsh Wrens and Wren-like Rushbirds, which resemble each other in their use of harsh mechanical notes and high rate of song delivery. It is conceivable that there could be a growing congruence in the utilization of short elements in the songs of New and Old World reed-swamp species. However that may be, what struck us in Argentina was that all five wetland ovenbird species sang remarkably simple song-phrases. In chapter 7 we reported that acrocephalids in low vegetation (i.e.grassland- and reedbed-dwellers) also construct their songs from short elements. Although the songs of the suboscine ovenbirds are on the whole rather simple in structure – consisting of a series of similar elements that may accelerate or decelerate, and rise or fall in pitch (Remsen 2003) – the reedbed-dwellers appear to sing even more simply than their closest relatives from other habitats, with the length of the elements especially shortened.

Convergencies graphically illustrate how selection makes use of already existing variation in organisms to produce forms that are adaptive to extreme ecological niches. We are delighted to observe that there is still plenty of excellent material for further studies on convergent characteristics in the acrocephalids and their companion taxa in the Americas.

Summary

In the reedswamps of the New World there is a whole series of passeriform birds that bears an almost uncanny resemblance to the reed warblers of the Old World. In North America they are representatives of the wrens, in South America of the ovenbirds and tyrant-flycatchers, two suboscine families that have evolved there in isolation for over 60 million years. This remarkable convergent development in sets of morphological traits between unrelated taxa is evolution's answer to the same ecological conditions in similar environments on different continents.

Only by resolving the molecular phylogenies of the families under comparison were we able to discover their closest relative in each case among the non-reed-dwelling species, so that a comparison with the reed specialists could be made. This revealed that the reed-dwellers of the Old and the New Worlds have large feet, small wings, and long narrow bills in common. A further example of convergent evolution can be found in the species belonging to the passerine guild in the papyrus swamps of east Africa, which are members of six different songbird lineages. A comparison of their overall morphology with that of their respective nearest relatives from other habitats enabled us to isolate the adaptations that papyrus-dwellers have evolved to cope with their habitat, namely larger feet and broader wings. Through direct interaction with a common habitat structure, it is principally the locomotor apparatus, especially the hindlimb, that is subject to convergent developments.

Alongside the morphological convergences there is also a variety of behavioural convergences between marsh-dwelling songbirds in the Old and New Worlds. One characteristic that these habitats have in common on all continents is their high productivity and resulting abundant food supplies. On the one hand these allow high population densities, to which the marsh inhabitants – independently on different continents – respond with intensified interspecific territoriality, and on the other, the abandonment of biparental brood care.

This last phenomenon, together with an unequal distribution of critical resources, stimulated the evolution of polygynous breeding systems in the deep-water marshes of both the Palaearctic and the Nearctic. At the same time, frequent water-level changes in flat, wet meadows allowed the abundant and constantly replenishing food supplies which gave rise to the development of promiscuity and the ability for uniparental rearing of young. The promiscuous breeding systems of Aquatic Warblers in the Old World and Saltmarsh Sparrows in the New are strikingly similar and admirably demonstrate the extent to which the evolution of mating systems can be determined by ecological conditions. Yet no matter how astonishing might be the similarities between unrelated taxa resulting from the same environmental pressures, a key feature of convergence is that it is never total.

15

Postscript

Clamorous Reed Warbler.

In our studies of reed warblers over 50 years or so, we have been fortunate to have travelled the path of both science and natural history, to observe our fellow travellers and even to alter the route a little. Our journey began in the 1960s as curious young observers, who, on entering the dark recesses of a reedbed came suddenly and unexpectedly on the sight of a reed warbler's nest. This was the start of a long relationship – and one that has not dimmed with time. We could never have imagined how much we, and others, would discover about the lives of these unobtrusive little brown birds.

In the 1960s ornithology was largely a descriptive science, focussing on breeding biology, taxonomy based on morphology and biogeography; the study of animal behaviour was still in its infancy.

The reed warblers – the acrocephalids – comprise a wonderful group for research, as they are uniform, yet sufficiently diverse to provide the opportunity for meaningful comparisons between life history strategies. This book is a summary of what we currently know. In ornithological terms fifty years is a long time – as evident in the extraordinary explosion of knowledge relating to the behaviour and ecology of reed warblers. Inevitably, much more remains to be discovered. Thus, in a sense, this is a progress report on the current state of our knowledge. Scientists always seek 'the truth', but because science is far from static, what we know at the present time is much better described as 'the truth for now'.

Photo 15.1 This Aquatic Warbler, which carries a geolocator, is a fine example of how natural history knowledge, field experience, high-tech, and science can be combined. While much field experience and many skills are needed in order to find and trap this nondescript little bird (without harming it) in the vast flooded grasslands it inhabits, at the same time scientific know-how is necessary to create the prerequisites for such a project and to be able to analyse the results. Aquatic Warblers are the smallest migratory species so far to be fitted with geolocators. For this to be possible, a miniaturized version of the data logger had to be developed. Several of the birds were able to carry this 'rucksack' for an entire year without any hindrance, reproducing during that period and undertaking the complete round trip to their wintering grounds in Africa and back again (Belarus; photo Volker Salewski).

Photos 15.2 and 15.3 Molecular genetic studies have revolutionized behavioural ecology. The main work now takes place in the laboratory (Great Reed Warbler nestling; Staffan Bensch at his work in the lab; photos Dennis Hasselquist).

Our own research has been influenced by three factors, all of which belong to the long tradition of the Max Planck Institute for Behavioural Physiology – today's Max Planck Institute for Ornithology. First, the combination of field, aviary, and laboratory studies that facilitate standardization, quantification, and experimentation. Secondly, the application of Niko Tinbergen's 'four questions' (Tinbergen 1963), which are essential for the correct interpretation of biological findings: (i) how does a given feature function, (ii) how does it develop during an individual's life? (iii) how has it evolved and (iv) what is its adaptive value? The last two questions concerning evolutionary history and survival value and history have been particularly close to our hearts and constitute the main emphasis of this book. Thirdly, the fruitful cooperation with a host of amateur ornithologists, whose wealth of experience and commitment made it possible to carry out long-term, large-scale network-based projects. Cooperative ventures like these are often at the interface of natural history and science, where they mutually enrich each other – confirming Aristotle's dictum that 'the whole is greater than the sum of its parts'.

Reflecting on our own development and careers, it is clear that over the past 50 years there have been major changes in both society and education. From our own perspective, perhaps the most alarming is the divide that now exists between natural history and science. The two were once considered to be one and the same (Birkhead 2008), but as science has become wider, more technical, and more expensive, so it has also become more removed from the natural world. *'Decades of emphatic promotion of experimental ecology, at the cost of natural history education, have robbed many young biologists of exposure to organisms in their environment. Our universities are training biologists strong in theory, but weak in the ability to apply these theories to organisms and the natural world'* (McGlynn 2008).

Much has been written about this division and its consequences (Noss 1996, Grant 2000, Dayton 2003, Greene 2005, Schmidly 2005, Willson & Armesto 2006, Beehler 2010). Reed Noss fears that the application of *'new technologies and new statistical methods in the absence of a firm foundation of field experience, void of the "naturalist's intuition" that is gained only by many years of immersion in raw Nature and through a ceaseless hunger of knowledge about living things, will lead us astray. A scientist who lacks familiarity with Nature will have difficulty interpreting any kind of result realistically. We have no shortage of fabulous models and supercomputers. What we lack in many cases are good field data to plug into the models'* (Noss 1996).

We, and many of our contemporaries, have been privileged to have enjoyed a thorough education in natural history. We are absolutely convinced that this has both enhanced and informed the quality of our science. The often ground-breaking insights achieved by our co-workers through the intelligent and skilful combination of organismal, evolutionary, environmental, and molecular biology are the highest rewards of such a synthesis.

Harry Greene gives three reasons why we still need descriptive ecological and ethological research:

'(i) Until recently, one might have at least claimed that an understanding of diverse organisms was irrelevant to biochemistry and molecular biology, whereas the most exciting frontiers in these disciplines now involve integration and reciprocal illumination. (ii) New facts about organisms often reset the research cycles of hypothesis testing and theory refinement that underlie good progressive science. (iii) Organisms themselves (what they look like, where they live, and what they do) provoke widespread human curiosity, including that of conceptually focused biologists and of the public, who financially support scientific research' (Greene 2005).

Photos 15.4 to 15.6 Long-term projects, and those in remote or unspoilt regions of the world, contribute to raising the value of scientific studies, such as those in conservation biology (Jan Komdeur at work in a Dutch reedbed, top left; Dennis Hasselquist carrying out long-term monitoring of Great Reed Warbler territories in southern Sweden, below left; Phil Round in Thailand, top right; David Pearson and a colleague ringer in Kenya, below right; photos Jan Komdeur, Staffan Bensch, Phil Round, Colin Jackson).

Photo 15.7 Field excursions offer great opportunities to develop new ideas for scientific projects that arise from simply watching birds and discussions with colleagues (Kazakhstan, from left to right: Tim R. Birkhead, Karl Schulze-Hagen, Bernd Leisler, Bård G. Stokke; photo Klaus Nigge).

Mary Willson and Juan Armesto find that very often *'it is simple natural history observations, planned or unplanned, that tweak the imagination into challenging existing dogma, asking novel questions, and seeing natural phenomena from a different perspective. So it seems that natural history is an integral part of the process of scientific inquiry in ecology and evolution'* (Willson & Armesto 2006).

Nowhere is the combination of natural history and science more relevant than in conservation biology. How is conservation supposed to work if not by the integration of descriptive and theory-based approaches? We wish therefore to make a plea for a renaissance of natural

history education as an important grounding for later scientific activities. 'Organisms *are the heart of biology education and research, and to prepare students as holistic thinkers we must show them how to see the organism for the cells as well as the forest for the trees. Infusing curiosity and knowledge about Nature into undergraduates is best achieved with creative pedagogical approaches'* (McGlynn 2008). For this to be realized, governments need to provide funds and to rethink the balance between administration and politics.

The sheer volume of biodiversity lost in the last 50 years is staggering. It ought therefore to be an obligation of all naturalists and scientists to contribute in some way to

Photo 15.8 The wonder in the face of this child might perhaps light the fire of a later passion for Nature. At a time when most people are further removed from the natural world than any previous generation, it is of crucial importance that young people are given the chance to get to know Nature and gradually build a close relationship with her (Great Reed Warbler, Sweden; photo Johan Stenlund).

the reversal of this process by bringing people closer to Nature. We have to communicate our passion for, wonder at and commitment to Nature to the general public. We also need to find ways of getting as many people as possible to participate in natural history and conservation.

The Max Planck Institute for Ornithology recently announced novel projects whose aim is to introduce children and teenagers to Nature, to get to know animals and plants, and to take part in scientific studies – an offer that has been enthusiastically taken up by the youngsters themselves. For university students the MPG (Max Planck Gesellschaft) has established an International Research

School. In a similar vein, the Department of Animal and Plant Sciences at the University of Sheffield, UK, offers interactive lectures on natural history for schoolchildren. The magazine *GEO*, the European equivalent of the *National Geographic*, has an annual Biodiversity Day, in which thousands of people take part. The non-profit KNNV Publishing (started by the Royal Dutch Natural History Society or Koninklijke Nederlandse Natuurhistorische Vereniging) is dedicated to increasing public involvement with wildlife, Nature, and conservation.

These are some of the first promising steps towards getting Nature back into people's awareness and their lives. The need is greater now than at any other time.

This Great Reed Warbler casually shows us just what it has to be able to do while clinging to a vertical stalk, while simultaneously scratching its head. This plain-coloured, nondescript, vertical-artist can do this directly or with its leg *under* its wing (underwing or direct method). Its striped relatives, and most other songbirds, scratch their head by raising the foot *over* the dropped wing (overwing or indirect method). The difference is that they do not have to perform this while always standing on vertical structures! These behavioural differences were thought to be a useful taxonomic character until well into the 1970s. Today we are in a position to enter or plot such (behavioural) traits into the phylogenetic tree (derived from genetic data) of an avian group and thereby test whether they dependably reflect true relationships (photo Kai Rösler).

Is it perhaps the inquisitive look in the eyes of this Great Reed Warbler that confronts us with the question of whether the things we think we know about it are actually true?

Acknowledgements

This book has a long history, stretching back to our work on the reed warbler chapters in the *Handbuch der Vögel Mitteleuropas* (Glutz & Bauer 1991). Inspired by the achievement of Douglass H. Morse in his *American Warblers – An ecological and behavioral perspective* (1989), we could envisage even then how attractive and exciting a similar comparative presentation of the reed warbler family might be. However, at that time we had no conception of just how much time and labour would be necessary to produce such a survey. The actual work of gathering material, researching the literature and writing the first drafts of the book began in 2006, though it would always be interrupted by the professional commitments of both authors. The long and stony path to its publication would have been too great a challenge for us without the invaluable help of a whole host of friends and colleagues, as well as experts in a variety of fields. Many of them assisted us as selflessly as if the project had been their own. The friendship and commitment we were privileged to experience on the journey to this book have been an unexpected gift that has helped us through rough patches and periods of frustration. This has been the true reward for our efforts.

We wish to express our deep gratitude to the Max Planck Institute for Ornithology in Radolfzell, in particular its Director Martin Wikelski. He has always regarded the reed warbler book as a priority project. The outstanding working conditions at the Institute, together with the on-going logistic and financial backing for the work of Bernd Leisler, ensured a solid foundation on which the project could be built. The entire team at the Institute identified themselves with the book and enthusiastically helped us in a multitude of ways. We thank Wolfgang Fiedler for his logistic support from the early beginning. Claudia Engele, Edith Sonnenschein and Ursula Tschakert provided us with much secretarial help. In addition, Monika Krome, Evi Fricke and Karl Heinz Siebenrock created the figures and tables, and also made sure that the bibliography was in perfect order. Kristina Schlegel, Kami Safi, and Bart Kranstauber helped with additional complex illustrations and maps. Ruth von Tauffkirchen, Rose Klein and Christine Fathelbab graciously gave their time for library research over many years, while for Alex Krikellis no trouble was too much in his search for the most obscure literature. Franz Bairlein, Peter Berthold, Tim Birkhead, Henrik Brumm, Clive Catchpole, Alice Cibois, Vladimir Ivanitzkii, Barbara Helm, Wolfgang Forstmeier, Sjouke Kingma, Stefan Leitner, Willy Ley (†), Gerd Nikolaus, Volker Salewski, Bård Stokke and Hans Winkler were kind enough to read various chapters for us. Their constructive comments and discussions played a crucial role in ensuring the success of the book. A further circle of colleagues and experts willingly and generously answered our calls for information, put unpublished material at our disposal, discussed controversial points, and generally assisted us in many ways. Every piece of data they supplied us with has contributd to the final structure. Among them are Per Alström, Anton Antonov, Franz Bairlein, Staffan Bensch, Ben Bell, Marta Borowiec, Peter Castell, Nikita Chernetsov, Alice Cibois, Janske van de Crommenacker, Francoise Dowsett-Lemaire, Renate van den Elzen, Alec Earnshaw, Bärbel Fiedler, Wolfgang Fiedler, Martin Flade, Wolfgang Forstmeier, Silke Fregin, Shoji Hamao, Dennis Hasselquist, Barbara Helm, Jens Hering, Jan Komdeur, Lars Lachmann, Manfred Lieser, Uladimir Malashevic, Oldrich Mikulica, Stephen Mosher, Andrey Mukhin, Csaba Moskát, Eulalia Moreno, Klaus Nigge, Gert Niklaus, David Pearson, Douglas Pratt, Phil Round, Frank Steinheimer, Bård Stokke, Lars Svensson, Franziska Tanneberger, Jean-Claude Thibault, and Michael Wink.

We are indebted to all the museum curators who enabled us to make use of their collections.

We humans are primarily visual animals and enjoy depictions of the natural world around us. However, the purpose of the photographs is not simply to portray the acrocephalids but rather to afford us an insight into their way of life and habitats. That this has succeeded is solely due to the skills and passion of a great many photographers. We are indebted to all of them who generously offered their photos. Some of them allowed us to use their images without payment, or even went into the field to shoot some photos just for this book. We are particularly grateful to Oldrich Mikulica, Jens Hering, and Mike Pope. We also wish to thank Abdulrahman Al-Sirhan, Anton Antonov, Franz Bairlein, Dave Bartlett, Mark Bartosik, Staffan Bensch, Jiri Bohdal, Nik Borrow, Lyanne Brouwer, Phil Chapman, Peter Castell, Owen Chiang, Alice Cibois, Mark Coates, Keith Cochrane, David Cookson, Elizabeth Crapo, Janske van de Crommenacker, Greg & Yvonne Dean, Brendan Doe, Alec Earnshaw, Cas Eikenaar, Dany Ellinger, Elio della Ferrara, Karsten Gärtner, Jan van der Greef, Mike Grimes, Shoji Hamao, Dennis Hasselquist, Paul Higgins, Joseph Hlasek, Colin Jackson, Renatas Jakaitis, Jack Jeffries, Bob Johnson, Marten van Kempen, Grzegorz & Tomasz Klosowscy, Jan Komdeur, Alexej Kozulin, Manfred Krauß, Jan Kube, Greg Lasley, Arnaud LeNevé, Alfred Limbrunner, Helen Lowe, Wolfgang Mädlow, Irina Marova-Kleinbub, Clemens Martin, Jonathan Martinez, Roland Mayr, John Mittermaier, Greg Morgan, Czaba Moskát, Klaus Nigge, Gert Nikolaus, Mark Nollet, Daniele Occhiato, Romy Ocon, Per Olsen, Wolfgang Ostendorp, David Pearson, Yoav Perlman, Garry Prescott, Milan Radisics, David Richardson, Heinz Ritter, Kai Rösler, Phil Round, Volker Salewski, Manuel San Miguel Bento, Norbert Schäffer, Wolfram Scheffler, Karl Seddon, Yuri Shibnev, Sunil Singhal, Albert Stehen-Hansen, Johan Stenlund, Bård Stokke, Jean-Claude Thibault, Huib Tom, Zdenek Tunka, Markus Varesvuo, Jan Vermeer, William Vivarelli, Simon White, Robert Wienand, Hans Winkler, Rene Winter, Winfried Wisniewski, Sergey Yeliseev, Toshiyuki Yoshino, and Carlos Zumalacarregui Martinez. Their photographs have brought colour into our sometimes dry text.

Special thanks are due to Brian Hillcoat, who has translated our texts from the very beginning, in many small steps following critical remarks from our reviewers or the insertion of yet more new literature. His patience, precision, and passion for ornithology appear to be inexhaustible. Jack Folkers, the editor responsible at KNNV Publishing, has made our project his own, helping the authors to navigate safely among the rocks of book production. Marie-Anne Martin has, as copy-editor, immersed herself in reed warbler biology, employing her practical way of seeing things to make complex facts readable. We will fondly remember her British sense of humour and constant encouragement during the stressful final straight. The lively and delicate drawings by David Quinn give the book that extra something. Together with him we had many enjoyable hours watching various reed warbler species.

We are more than grateful to two special friends, Tim Birkhead und Hans Winkler, who not only always encouraged and supported us but also read the entire manuscript. Hans Winkler carried out all the data and statistical analyses. Finally, the greatest thanks must go to our wives Gerti Leisler and Mechthild Schulze-Hagen, who, in all the years of our absences (physical and mental), stress and agitation, have shown extraordinary patience. To them we dedicate this book.

References

Addison, J.A. & Diamond, A.W. 2011. Population genetics and effective population size of the critically endangered Nihoa Millerbird (*Acrocephalus familiaris kingi*). *Auk* **128**: 265-272.

Airey, D.C., Buchanan, K.L., Szekely, T., Catchpole, C.K. & DeVoogd, T.J. 2000. Song, sexual selection, and a song control nucleus (HVc) in the brains of European sedge warblers. *J. Neurobiol.* **44**: 1-6.

Akcay, E. & Roughgarden, J. 2007. Extra-pair paternity in birds: review of the genetic benefits. *Evol. Ecol. Res.* **9**: 855-868.

Akesson, M., Bensch, S., Hasselquist, D., Tarka, M. & Hansson, B. 2008. Estimating heritabilities and genetic correlations: comparing the ‚animal model‘ with parent-offspring regression using data from a natural population. *PLoS ONE* **3**: e1739.

Alatalo, R.V., Gustafsson, L. & Lundberg, A. 1984. Why have young passerine birds shorter wings than older ones? *Ibis* **126**: 410-415.

Albat, T. 1996. Der Einfluß verschiedener ökologischer Faktoren auf Vorkommen, Polygynie und Bruterfolg des Drosselrohrsängers (*Acrocephalus arundinaceus* L.,1758) im Fränkischen Weihergebiet. Kiel: Christian-Albrechts-Universität, pp. 1-81.

Alerstam, T. 2009. Flight by night or day? Optimal daily timing of bird migration. *J. Theor. Biol.* **258**: 530-536.

Alström, P., Olssen, U. & Round, P.D. 1991. The taxonomic status of *Acrocephalus agricola tangorum*. *Forktail* **6**: 3-13.

Alström, P., Ericson, P.G.P., Olsson, U. & Sundberg, P. 2006. Phylogeny and classification of the avian superfamily Sylvioidea. *Mol. Phylogenet. Evol.* **38**: 381-397.

Altenburg, W. & van Spanje, T. 1989. Utilization of mangroves by birds in Guinea-Bissau. *Ardea* **77**: 57-74.

Amezian, M., Cortes, J., Thompson, I., Bensusan, K., Perez, C., Louah, A., El Agbani, M.A. & Qninba, A. 2010. Complete moult of an undescribed resident taxon of the Reed Warbler *Acrocephalus scirpaceus/baeticatus* complex in the Smir marshes, Northern Morocco. *Ardea* **98**: 225-234.

Andersson, M. 1994. *Sexual selection*. Princeton,NJ: Princeton University Press.

Anon. 1996. 40 Jahre Österreichisch-Ungarische Gewässerkommission, Jubiläumsschrift, Vienna

Antonov, A. 2004. Smaller Eastern Olivaceous Warbler *Hippolais pallida elaeica* nests suffer less predation than larger ones. *Acta Ornithol.* **39**: 87-92.

Antonov, A., Stokke, B.G., Moksnes, A., Kleven, O., Honza, M. & Røskaft, E. 2006a. Eggshell strength of an obligate brood parasite: a test of the puncture resistance hypothesis. *Behav. Ecol. Sociobiol.* **60**: 11-18.

Antonov, A., Stokke, B.G., Moksnes, A. & Røskaft, E. 2006b. Egg rejection in Marsh Warblers (*Acrocephalus palustris*) heavily parasitized by Common Cuckoos (*Cuculus canorus*). *Auk* **123**: 419-430.

Antonov, A., Stokke, B.G., Moksnes, A. & Røskaft, E. 2007a. Factors influencing the risk of common cuckoo *Cuculus canorus* parasitism on marsh warblers *Acrocephalus palustris*. *J. Avian Biol.* **38**: 390-393.

Antonov, A., Stokke, B.G., Moksnes, A. & Røskaft, E. 2007b. First evidence of regular common cuckoo, *Cuculus canorus*, parasitism on eastern olivaceous warblers, *Hippolais pallida elaeica*. *Naturwissenschaften* **94**: 307-312.

Antonov, A., Stokke, B.G., Moksnes, A. & Røskaft, E. 2008. Getting rid of the cuckoo *Cuculus canorus* egg: why do hosts delay rejection? *Behav. Ecol.* **19**: 100-107.

Antonov, A., Stokke, B.G., Moksnes, A. & Røskaft, E. 2009. Evidence for egg discrimination preceding failed rejection attempts in a small cuckoo host. *Biol. Lett.* **5**: 169-171.

Antonov, A., Stokke, B.G., Vikan, J.R., Fossøy, F., Ranke, P.S., Røskaft, E., Moksnes, A., Møller, A.P. & Shykoff, J.A. 2010. Egg phenotype differentiation in sympatric cuckoo *Cuculus canorus* gentes. *J. Evol. Biol.* **23**:1170-1182.

Aquatic Warbler Conservation Team. 1999. World population, trends and conservation status of the Aquatic Warbler *Acrocephalus paludicola*. *Vogelwelt* **120**:65-85.

Arnold, K.E. & Owens, I.P.F. 1999. Cooperative breeding in birds: the role of ecology. *Behav. Ecol.* **10**: 465-471.

Arnold, K.E. & Owens, I.P.F. 2002. Extra-pair paternity and egg dumping in birds: life history, parental care and the risk of retaliation. *Proc. Roy. Soc. B-Biol. Sci.* **269**: 1263-1269.

Arnqvist, G. & Kirkpatrick, M. 2005. The evolution of infidelity in socially monogamous passerines: the strength of direct and indirect selection on extrapair copulation behavior in females. *Am. Nat.* **165**: S26-S37.

Aschoff, J. 1955. Jahresperiodik der Fortpflanzung bei Warmblütern. *Studium Generale* **8**: 742-776.

Ash, J.S., Pearson, D.J. & Bensch, S. 2005. A new race of Olivaceous Warbler (*Hippolais pallida*) in Somalia. *Ibis* **147**: 841-843.

Aviles, J.M., Stokke, B.G., Moksnes, A., Røskaft, E., Asmul, M. & Møller, A.P. 2006. Rapid increase in cuckoo egg matching in a recently parasitized reed warbler population. *J. Evol. Biol.* **19**: 1901-1910.

Aviles, J.M., Stokke, B.G., Moksnes, A., Røskaft, E. & Møller, A.P. 2007. Environmental conditions influence egg color of reed warblers *Acrocephalus scirpaceus* and their parasite, the common cuckoo *Cuculus canorus*. *Behav. Ecol. Sociobiol.* **61**: 475-485.

Aviles, J.M., Vikan, J.R., Fossøy, F., Antonov, A., Moksnes, A., Røskaft, E., Shykoff, J.A., Møller, A.P., Jensen, H., Procházka, P. & Stokke, B.G. 2011. The common cuckoo *Cuculus canorus* is not locally adapted to its reed warbler *Acrocephalus scirpaceus* host. *J. Evol. Biol.* **24**: 314-325.

Ayé, R., Hertwig, S.T. & Schweizer, M. 2010. Discovery of a breeding area of the enigmatic large-billed reed warbler *Acrocephalus orinus*. *J. Avian Biol.* **41**: 452-459.

Baccetti, N. 1985. The vertical distribution of three passerine birds in a marshland of Central Italy. *Ring. Migr.* **6**: 93-96.

Badyaev, A.V. & Leaf, E.S. 1997. Habitat associations of song characteristic in *Phylloscopus* and *Hippolais* warblers. *Auk* **114**: 40-46.

Bairlein, F. 1981. Ökosystemanalyse der Rastplätze von Zugvögeln: Beschreibung und Deutung der Verteilungsmuster von ziehenden Kleinvögeln in verschiedenen Biotopen der Stationen des „Mettnau-Reit-Illmitz-Programmes". *Ökologie Vögel* **3**: 7-137.

Bairlein, F. 1996. *Ökologie der Vögel.* Stuttgart: Gustav Fischer.

Bairlein, F. 1998. The European-African songbird migration network: new challenges for large-scale study of bird migration. *Biol. Conserv. Fauna* **102**: 13-27.

Bairlein, F. 2000. Photoperiode und Nahrungsangebot beeinflussen zugzeitliche Fettdeposition. *Jber.Institut Vogelforschung* **4**: 5.

Bairlein, F. 2002. How to get fat: nutritional mechanisms of seasonal fat accumulation in migratory songbirds. *Naturwissenschaften* **89**: 1-10.

Bairlein, F. 2006. Family Sylviidae (Old World Warblers). In: del Hoyo, J., Elliott, A. & Christie, D.A. eds. *Handbook of the birds of the world.* Vol. 11. Barcelona: Lynx Edicions, pp. 492-575.

Baker, R.H. 1951. The avifauna of Micronesia, its origin, evolution and distribution. *Univ.Kansas Publ. Mus. Nat. Hist.* **3**: 1-359.

Baker, R.R. 1993. The function of post-fledging exploration: a pilot study of three species of passerines ringed in Britain. *Ornis Scand.* **24**: 71-79.

Baker, M.C. 1996. Depauperate meme pool of vocal signals in an island population of singing honeyeaters. *Anim. Behav.* **51**: 853-858.

Baker, K. 1997. *Warblers of Europe, Asia and North Africa.* London: Christopher Helm.

Balanca, G. & Schaub, M. 2005. Post-breeding migration ecology of Reed *Acrocephalus scirpaceus*, Moustached *A. melanopogon* and Cetti's Warblers *Cettia cetti* at a Mediterranean stopover site. *Ardea* **93**: 245-257.

Baldi, A. & Batáry, P. 2005. Nest predation in European reedbeds: different losses in edges but similar losses in interiors. *Folia Zool.* **54**: 285-292.

Báldi, A. & Batáry, P. 2007. The effect of nest predation on the survival of real and artificial Great Reed Warbler's nests. In: Hengsberger, R., Hofmann, H., Wagner, R. & Winkler, H. eds. *Abstracts of the 6th European Ornithologists' Union Conference, 24-29 August 2007, Vienna, Austria.* Vienna: Austrian Academy of Sciences, p. 100.

Baldi, A.& Moskát, C. 1995: Effect of reed burning and cutting on breeding bird communities. In: Bissonette, J.A. & Krausman, P.R. eds. *Integrating people and wildlife for a sustainable future.* Bethesda, Md: The Wildlife Society, pp. 637-642.

Barabás, L., Gilicze, B., Takasu, F. & Moskát, C. 2004. Survival and anti-parasite defense in a host metapopulation under heavy brood parasitism: a source-sink dynamic model. *J. Ethol.* **22**: 143-151.

Barta, Z., McNamara, J.M., Houston, A.I., Weber, T.P., Hedenström, A. & Fero, O. 2008. Optimal moult strategies in migratory birds. *Phil. Trans. Soc. B-Biol. Sci.* **363**: 211-229.

Basciutti, P., Negra, O. & Spina, F. 1997. Autumn migration strategies of the sedge warbler *Acrocephalus schoenobaenus* in northern Italy. *Ring. Migr.* **18**: 59-67.

Batáry, P. & Baldi, A. 2005. Factors affecting the survival of real and artificial great reed warbler's nests. *Biologia* **60**: 215-219.

Batáry, P., Winkler, H. & Baldi, A. 2004. Experiments with artificial nests on predation in reed habitats. *J. Ornithol.* **145**: 59-63.

Batáry, P., Báldi, A. & Trnka, A. 2007. Edge effect on avian nest predation in European scale - a meta analysis. In: Hengsberger, R., Hofmann, H., Wagner, R. & Winkler, H. eds. *Abstracts of the 6th European Ornithologists' Union Conference, 24-29 August 2007, Vienna, Austria.* Vienna: Austrian Academy of Sciences, p 41.

Bayly, N.J. 2006. Optimality in avian migratory fuelling behaviour: a study of a trans-Saharan migrant. *Anim. Behav.* **71**: 173-182.

Bayly, N.J. 2007. Extreme fattening by sedge warblers, *Acrocephalus schoenobaenus*, is not triggered by food availability alone. *Anim. Behav.* **74**: 471-479.

Beadell, J.S., Atkins, C., Cashion, E., Jonker, M. & Fleischer, R.C. 2007. Immunological change in a parasite-impoverished environment: divergent signals from four island taxa. *PLoS ONE* **2**(9): e896.

Becker, P. & Lütgens, H. 1976. Sumpfrohrsänger (*Acrocephalus palustris*) in Südwestafrika. *Madoqua* **9**: 41-44.

Beehler, B.M. 2010. The forgotten science: a role for natural history in the twenty-first century? *J. Field Ornithol.* **81**: 1-4.

Begon, M., Harper, J.L. & Townsend, C.R. 1990. *Ecology: individuals, populations and communities.* Oxford: Blackwell.

Beier, J. 1993. Bestandsentwicklung und Habitatkonkurrenz von Drossel-und Teichrohrsänger im Fränkischen Weihergebiet. *Beih. Veröff. Natursch. Landschaftspfl. Bad.-Württ.* **68**: 137-139.

Beier, J., Leisler, B. & Wink, M. 1997. Ein Drossel- x Teichrohrsänger-Hybride *Acrocephalus arundinaceus* x *A. scirpaceus* und der Nachweis seiner Elternschaft. *J. Ornithol.* **138**: 51-60.

Belgers, J. & Arts, G. 2003. Moerasvogels op peil. Deelrapport 1: Peilen op riet; literatuurstudie naar de sturende processen en factoren voor de achteruitgang en herstel van jonge verlandingspopulaties van Riet (*Phragmites australis*) in laagveenmoerassen en rivierkleigebieden. Wageningen: Alterra-rapport 828.1.

Bell, B.D. 2001. Endemic island Acrocephalus warblers of the Pacific: how do they differ in morphology and song from their continental counterparts, and why? Paper presented at *The ecology of insular biotas conference,* 12-16th February ,Victoria University of Wellington, New Zealand.

Bell, B.D., Borowiec, M., McConkey, K.R. & Ranoszek, E. 1997. Settlement, breeding success and song repertoires of monogamous and polygynous Sedge Warblers (*Acrocephalus schoenobaenus*). *Vogelwarte* **39**: 87-94.

Bennett, P.M. & Harvey, P.H. 1987. Active and resting metabolism in birds: allometry, phylogeny and ecology. *J. Zool.* **213**: 327-363.

Bensch, S. 1993. Costs, benefits and strategies for females in a polygynous mating system: a study on the great reed warbler. Lund, Sweden: [Dissertation] Lund University.

Bensch, S. 1996. Female mating status and reproductive success in the great reed warbler: is there a potential cost of polygyny that requires compensation? *J. Anim. Ecol.* **65**: 283-296.

Bensch, S. 1997. The cost of polygyny: definitions and applications. *J. Avian Biol.* **28**: 345-352.

Bensch, S. & Hasselquist, D. 1991. Territory infidelity in the polygynous great reed warbler (*Acrocephalus arundinaceus*): the effect of variation in territory attractiveness. *J.Anim. Ecol.* **60**: 857-871.

Bensch, S. & Hasselquist, D. 1992. Evidence for active female choice in a polygynous warbler. *Anim. Behav.* **44**: 301-311.

Bensch, S. & Hasselquist, D. 1994. Higher rate of nest loss among primary than secondary females: infanticide in the great reed warbler. *Behav. Ecol.Sociobiol.* **35**: 309-317.

Bensch, S. & Pearson, D. 2002. The Large-billed Reed Warbler *Acrocephalus orinus* revisited. *Ibis* **144**: 259-267.

Bensch, S., Hasselquist, D., Hedenström, A. & Ottosson, U. 1991. Rapid moult among palaearctic passerines in West Africa: an adaptation to the oncoming dry season? *Ibis* **133**: 47-52.

Bensch, S., Hasselquist, D. & von Schantz, T. 1994. Genetic similarity between parents predicts hatching failure: nonincestuous inbreeding in the Great Reed Warbler. *Evolution* **48**: 317-326.

Bensch, S., Hasselquist, D., Nielsen, B. & Hansson, B. 1998. Higher fitness for philopatric than for immigrant males in a semi-isolated population of great reed warblers. *Evolution* **52**: 877-883.

Bensch, S., Hansson, B., Hasselquist, D. & Nielsen, B. 2000. Partial albinism in a semi-isolated population of great reed warblers. *Hereditas* **133**: 167-170.

Bensch, S., Hasselquist, D., Nielsen, B., Nihl, N.C. & Frodin, P. 2001. Food resources and territory quality in the polygynous Great Reed Warbler. *Biosyst. Ecol.* **18**: 49-71.

Bensch, S., Waldenström, J., Jonzen, N., Westerdahl, H., Hansson, B., Sejberg, D. & Hasselquist, D. 2007. Temporal dynamics and diversity of avian malaria parasites in a single host species. *J. Anim. Ecol.* **76**: 112-122.

Beresford, P., Barker, F.K., Ryan, P.G. & Crowe, T.M. 2005. African endemics span the tree of songbirds (Passeri): molecular systematics of several evolutionary 'enigmas'. *Proc. Roy. Soc. B-Biol. Sci.* **272**: 849-858.

Berg, M.L., Beintema, N.H., Welbergen, J.A. & Komdeur, J. 2006. The functional significance of multiple nest-building in the Australian Reed Warbler *Acrocephalus australis*. *Ibis* **148**: 395-404.

Berndt, R.K. & Sternberg, H. 1968. Terms, studies and experiments on the problems of bird dispersion. *Ibis* **110**: 256-269.

Berndt, R.K. & Struwe-Juhl, B. 2004. Why has the breeding population of the great reed warbler (*Acrocephalus arundinaceus*) decreased in Schleswig-Holstein? *Corax* **19**: 281-301.

Berthold, P. 2001. *Bird migration: a general survey.* 2nd ed. Oxford: Oxford University Press.

Berthold, P. & Leisler, B. 1980. Migratory restlessness of the Marsh Warbler *Acrocephalus palustris*: a reflection of its unusual migration. *Naturwissenschaften* **67**: 472.

Berthold, P., Fliege, G., Heine, G., Querner, U. & Schlenker, R. 1991. Wegzug, Rastverhalten, Biometrie und Mauser von Kleinvögeln in Mittereuropa. *Vogelwarte* **36** (Sonderheft):1-221.

Bibby, C.J. 1978. Some breeding statistics of Reed and Sedge Warblers. *Bird Study* **25**: 207-222.

Bibby, C.J. 1982. Studies of West Palearctic birds,184: Moustached Warbler. *Brit. Birds* **75**: 346-359.

Bibby, C.J. & Green, R.E. 1981. Autumn migration strategies of Reed and Sedge Warblers. *Ornis Scand.* **12**: 1-12.

Bibby, C.J. & Thomas, D.K. 1985. Breeding and diets of the Reed Warbler: at a rich and a poor site. *Bird Study* **32**: 19-31.

Biebach, H. 1990. Strategies of Trans-Sahara Migrants. In: Gwinner, E. ed. *Bird migration, physiology and ecophysiology.* Berlin,Heidelberg: Springer, pp. 352-367.

BirdLife International. 2004. *Birds in the European Union: a status assessment.* Wageningen, The Netherlands: BirdLife International.

BirdLife International 2004-2010. Species factsheets (Acrocephalidae species). www.birdlife.org

Birkhead, T.R. 1987. Sperm competition in birds. *Trend Ecol. Evol.* **2**: 268-272.

Birkhead, T.R. 1995. Sperm competition: evolutionary causes and consequences. *Reprod. Fert. Develop.* **7**: 755-775.

Birkhead, T.R. 2008. *The wisdom of birds.* London: Bloomsbury.

Birkhead, T.R. & Møller, A.P. 1992. *Sperm competition in birds.* London: Academic Press.

Birkhead, T.R. & Møller, A.P. 1996. Monogamy and sperm competition in birds. In: Black, J.M. ed. *Partnerships in birds: the study of monogamy.* Oxford: Oxford University Press, pp. 323-343.

Birkhead, T.R., Briskie, J.V. & Møller, A.P. 1993. Male sperm reserves and copulation frequency in birds. *Behav. Ecol. Sociobiol.* **32**: 85-93.

Birkhead, T.R., Hemmings, N., Spottiswoode, C.N., Mikulica, O., Moskát, C., Ban, M. & Schulze-Hagen, K. 2011. Internal incubation and early hatching in brood parasitic birds. *Proc. Roy. Soc. B-Biol. Sci* **278**: 1019-1024.

Blomberg, S.P., Garland, T.J. & Ives, A.R. 2003. Testing for phylogenetic signal in comparative data: behavioral traits are more labile. *Evolution* **57**: 717-745.

Blomqvist, D., Fessl, B., Hoi, H. & Kleindorfer, S. 2005. High frequency of extra-pair fertilisations in the moustached warbler, a songbird with a variable breeding system. *Behaviour* **142**: 1133-1148.

Blomqvist, D., Hoi, H. & Weinberger, I. 2006. To see or not to see: the role of habitat density in the occurrence of extra-pair paternity and assurance behaviors. *Acta Zool. Sinica* **52** (Suppl.): 229-231.

Blondel, J. 2000. Evolution and ecology of birds on islands: trends and prospects. *Vie Milieu* **50**: 205-220.

Blondel, J. & Aronson, J. 1999. *Biology and wildlife of the Mediterranean region.* Oxford: Oxford University Press.

Bochenski, Z. & Kusnierczyk, P. 2003. Nesting of the Acrocephalus warblers. *Acta Zool. Cracov* **46**: 97-195.

Bock, W.J. 1980. The definition and recognition of biological adaptation. *Am. Zool.* **20**: 217-227.

Bolshakov, C.V. 2003. Nocturnal migration of passerines in the desert-highland zone of Western Central Asia: selected aspects. In: Berthold, P., Gwinner, E. & Sonnenschein, E. eds. *Avian migration.* Berlin, Heidelberg: Springer, pp. 225-236.

Bolshakov, C.V., Bulyuk, V. & Chernetsov, N. 2003. Spring nocturnal migratory of Reed Warbler (*Acrocephalus scirpaceus*): departure, landing and body condition. *Ibis* **145**: 106-112.

Boncoraglio, G. & Saino, N. 2007. Habitat structure and the evolution of bird song: a meta-analysis of the evidence for the acoustic adaptation hypothesis. *Funct. Ecol.* **21**: 134-142.

Borowiec, M. 1992. Breeding ethology and ecology of the Reed Warbler, (*Acrocephalus scirpaceus*, Hermann,1804) at Milicz, SW Poland. *Acta Zool. Cracov.* **35**: 315-350.

Borowiec, M. & Lontkowski, J. 1988. Polygyny in the Sedge Warbler *Acrocephalus schoenobaenus*. *Vogelwelt* **109**: 222-226.

Bosschieter, L. & Goedhart, P.W. 2005. Gap crossing decisions by reed warblers (*Acrocephalus scirpaceus*) in agricultural landscapes. *Landscape Ecol.* **20**: 455-468.

Both, C. & te Marvelde, L. 2007. Climate change and timing of avian breeding and migration throughout Europe. *Climate Res.* **35**: 93-105.

Both, C., Bouwhuis, S., Lessells, C.M. & Visser, M.E. 2006. Climate change and population declines in a long-distance migratory bird. *Nature* **441**: 81-83.

Both, C., van Asch, M., Bijlsma, R.G., van den Burg, A.B. & Visser, M.E. 2009. Climate change and unequal phenological changes across four trophic levels: constraints or adaptations. *J. Anim. Ecol.* **78**: 73-83.

Both, C. 2010a. Why migrants fail to adapt to climate change: phenotypic plasticity masked by environmental constraints. *Curr. Biol.* **20**: 243-248.

Both, C. 2010b. Food availability, mistiming and climatic change. In: Møller, A.P., Fiedler, W. & Berthold, P. eds. *Effects of climate change on birds*. Oxford: Oxford University Press, pp.129-148.

Both, C., Van Turnhout, C.A.M., Bijlsma, R.G., Siepel, H., Van Strien, A.J. & Foppen, R.P.B. 2010c. Avian population consequences of climate change are most severe for long-distance migrants in seasonal habitats. *Proc. Roy. Soc. B-Biol. Sci* **277**: 1259-1266.

Bowlin, M.S., Wikelski, M.C. & Cochran, W.W. 2004. The relationship between individual morphology, atmospheric conditions, and inter-individual variation in heart rate and wingbeat frequency during natural migration in the Swainsons Thrush (*Catharus ustulatus*). *Integr. Comp. Biol.* **44**: 529-529.

Bowman, R.I. 1983. The evolution of song in Darwin's Finches. In: Bowman, R.I., Berson, M. & Leviton, A.E. eds. *Patterns of evolution in Galapagos organisms*. San Francisco, Ca: American Association for the Advancement of Science. Special Publication No.1, pp. 237-537.

Brawn, J. 2006. What do we really know about the demography of tropical birds? *J. Ornithol.* **147**: 39-40.

Brensing, D. 1985. Alterskennzeichen bei Sumpf- und Teichrohrsänger (*Acrocephalus palustris, A.scirpaceus*): Quantitative Untersuchung. *J. Ornithol.* **126**: 125-153.

Brensing, D. 1989. Ökophysiologische Untersuchungen der Tagesperiodik von Kleinvögeln. *Ökol. Vögel* **11**: 1-148.

Brewer, D. 2001. *Wrens, dippers and thrashers*. London: Christopher Helm.

British Trust for Ornithology. 2010. *Birdfacts*. www.bto.org

Britton, P.L. 1978. Seasonality, density and diversity of birds of a papyrus swamp in western Kenya. *Ibis* **120**: 450-466.

Brommer, J.E., Alho, J.S., Biard, C., Chapman, J.R., Charmantier, A., Dreiss, A., Hartley, I.R., Hjernquist, M.B., Kempenaers, B., Komdeur, J., Laaksonen, T., Lehtonen, P.K., Lubjuhn, T., Patrick, S.C., Rosivall, B., Tinbergen, J.M., van der Velde, M., van Oers, K., Wilk, T. & Winkel, W. 2010. Passerine extrapair mating dynamics: a Bayesian modeling approach comparing four species. *Am. Nat.* **176**: 178-187.

Brooke, M. de L. & Davies, N.B. 1987. Recent changes in host usage by Cuckoos *Cuculus canorus* in Britain. *J. Anim. Ecol.* **56**: 873-883.

Brooke, M. de L. & Davies, N.B. 1989. Provisioning of nestling Cuckoos *Cuculus canorus* by Reed Warbler *Acrocephalus scirpaceus* hosts. *Ibis* **131**: 250-256.

Brooke, M. de L. & Hartley, I.R. 1995. Nesting Henderson Reed-Warblers (*Acrocephalus vaughani taiti*) studied by DNA fingerprinting: unrelated coalitions in a stable habitat? *Auk* **112**: 77-86.

Brooke, M. de L., Davies, N.B. & Noble, D.G. 1998. Rapid decline of host defences in response to reduced cuckoo parasitism: behavioural flexibility of reed warblers in a changing world. *Proc. Roy. Soc. B-Biol. Sci.* **265**: 1277-1282.

Brouwer, L., Komdeur, J. & Richardson, D.S. 2007. Heterozygosity-fitness correlations in a bottlenecked island species: a case study on the Seychelles warbler. *Mol. Ecol.* **16**: 3134-3144.

Brouwer, L., Barr, I., van de Pol, M., Burke, T., Komdeur, J. & Richardson, D.S. 2010. MHC-dependent survival in a wild population: evidence for hidden genetic benefits gained through extra-pair fertilizations. *Mol. Ecol.* **19**: 3444-3455.

Brown, P.E. & Davies, M.G. 1949. *Reed-warblers: an introduction to their breeding biology and behaviour*. East Molesey, Surrey: Foy.

Brumm, H. & Naguib, M. 2009. Environmental acoustics and the evolution of bird song. *Adv.Stud. Behav.* **40**: 1-33.

Brumm, H., Robertson, K.A. & Nemeth, E. 2011. Singing direction as a tool to investigate the function of birdsong: an experiment on sedge warblers. *Anim. Behav.* **81**: 653-659.

Bruner, P.L. 1974. Behavior, ecology, and taxonomic status of three southeastern Pacific Warblers of the genus *Acrocephalus*. [MSc thesis] Louisiana State University.

Buchanan, K.L. & Catchpole, C.K. 1997. Female choice in the sedge warbler, *Acrocephalus schoenobaenus*: multiple cues from song and territory quality *Proc. Roy. Soc. B-Biol. Sci.* **264**: 521-526.

Buchanan, K.L. & Catchpole, C.K. 2000a. Song as an indicator of male parental effort in the sedge warbler. *Proc. Roy. Soc. B-Biol. Sci.* **267**: 321-326.

Buchanan, K.L. & Catchpole, C.K. 2000b. Extra-pair paternity in the socially monogamous Sedge Warbler *Acrocephalus schoenobaenus* as revealed by multilocus DNA fingerprinting. *Ibis* **142**: 44-66.

Buchanan, K.L., Catchpole, C.K., Lewis, J.W. & Lodge, A. 1999. Song as an indicator of parasitism in the sedge warbler. *Anim. Behav.* **57**: 307-314.

Bulyuk, V. 2003. Relationship between realization of juvenile dispersal in the reed warbler (*Acrocephalus scirpaceus*) and weather conditions. *Vogelwarte* **42**: 85.

Bussmann, C. 1979. Ökologische Sonderung der Rohrsänger Südfrankreichs aufgrund von Nahrungsstudien. *Vogelwarte* **30**: 84-101.

Butyev, V.T., Shitikov, D.A. & Fedotova, S.E. 2007. Nesting biology of the booted warbler (*Hippolais caligata*, Aves, Passeriformes) at the northern boundary of its range. *Zool. Zh.* **86**: 81-89.

Camp, R.J., Pratt, T.K., Marshall, A.P., Amidon, F. & Williams, L.L. 2009. Recent status and trends of the land bird avifauna on Saipan, Mariana Islands, with emphasis on the endangered Nightingale Reed-warbler *Acrocephalus luscinia*. *Bird Conserv. Int.* **19**: 323-337.

Campobello, D. & Sealy, S.G. 2011. Use of social over personal information enhances nest defense against avian brood parasitism. *Behav. Ecol.* **22**: 422-428.

Cantos, F.J. & Telleria, J.L. 1994. Stopover site fidelity of 4 migrant warblers in the Iberian Peninsula. *J. Avian Biol.* **25**: 131-134.

Capek, M. & Kloubec, B. 2002. Seasonal and diel patterns of song output by great reed warblers. *Biologia* **57**: 267-276.

Carey, C. 2009. The impacts of climate change on the annual cycles of birds. *Phil. Trans. Soc. B-Biol. Sci.* **364**: 3321-3330.

Castell, P. 2001. Clamorous reed warblers *Acrocephalus stentoreus* nesting in maize. *Bull. Afr. Bird Club* **8**: 56.

Castell, P. & Kirwan, G.M. 2005. Will the real Sykes's Warbler please stand up? Breeding data support specific status for *Hippolais rama* and *H. caligata*, with comments on the Arabian population of `booted warbler`. *Sandgrouse* **27**: 30-36.

Catchpole, C.K. 1972. A comparative study of territory in the Reed warbler (*Acrocephalus scirpaceus*) and Sedge warbler (*A. schoenobaenus*). *J. Zool.* **166**: 213-231.

Catchpole, C.K. 1973. Conditions of co-existence in sympatric breeding populations of Acrocephalus warblers. *J. Anim. Ecol.* **42**: 623-635.

Catchpole, C.K. 1974. Habitat selection and breeding success in the reed warbler (*Acrocephalus scirpaceus*). *J. Anim. Ecol.* **43**: 363-380.

Catchpole, C.K. 1976. Temporal and sequential organisation of song in the sedge warbler (*Acrocephalus schoenobaenus*). *Behaviour* **59**: 226-246.

Catchpole, C.K. 1977. Aggressive responses of male sedge warblers (*Acrocephalus schoenobaenus*) to playback of species song and sympatric species song, before and after pairing. *Anim. Behav.* **25**: 489-496.

Catchpole, C.K. 1978. Interspecific territorialism and competition in Acrocephalus warblers as revealed by playback experiments in areas of sympatry and allopatry. *Anim. Behav.* **26**: 1072-1080.

Catchpole, C.K. 1980. Sexual selection and the evolution of complex songs among European warblers of the genus *Acrocephalus*. *Behaviour* **74**: 149-166.

Catchpole, C.K. 1982. The evolution of bird sounds in relation to mating and spacing behavior. In: Kroodsma, D. & Miller, E.H. eds. *Acoustic communication in birds*. New York: Academic Press, pp. 297-317.

Catchpole, C.K. 1983. Variation in the song of the great reed warbler *Acrocephalus arundinaceus* in relation to mate attraction and territorial defence. *Anim. Behav.* **31**: 1217-1225.

Catchpole, C.K. 2000. Sexual selection and the evolution of song and brain structure in *Acrocephalus* warblers. *Adv. Stud. Behav.* **29**: 45-97.

Catchpole, C.K. & Leisler, B. 1986. Interspecific territorialism in Reed Warblers: a local effect revealed by playback experiments. *Anim. Behav.* **34**: 299-300.

Catchpole, C.K. & Leisler, B. 1989. Variation in the song of the aquatic warbler (*Acrocephalus paludicola)* in response to playback of different song structures. *Behaviour* **108**: 125-138.

Catchpole, C.K. & Slater, P.J.-B. 2008. *Bird song*. Cambridge: Cambridge University Press.

Catchpole, C.K., Dittami, J. & Leisler, B. 1984. Differential responses to male song repertoires in female songbirds implanted with oestradiol. *Nature* **312**: 563-564.

Catchpole, C.K., Leisler, B. & Winkler, H. 1985. Polygyny in the great reed warbler (*Acrocephalus arundinaceus*): a possible case of deception. *Behav. Ecol. Sociobiol.* **16**: 285-291.

Catchpole, C.K., Leisler, B. & Dittami, J. 1986. Sexual differences in the responses of captive Great Reed Warblers (*Acrocephalus arundinaceus*) to variation in song structure and repertoire size. *Ethology* **73**: 69-77.

Chappuis, C. 1971. Un exemple de l`inflence du milieu sur les émissions vocales des oiseaux: l`évolution des chants en forêt équatoriale. *Terre* Vie **118**: 183-202.

Chernetsov, N. 1998. Habitat distribution during the post-breeding and post-fledging period in the reed warbler *Acrocephalus scirpaceus* and sedge warbler *A. schoenobaenus* depends on food abundance. *Ornis Svecica* **8**: 77-82.

Chernetsov, N. 2005. Spatial behavior of medium and long-distance migrants at stopovers studied by radio tracking. *Ann. NY Acad. Sci.* **1046**: 242-252.

Chernetsov, N. & Manukyan, A. 1999. Feeding strategy of Reed Warblers (*Acrocephalus scirpaceus*) on migration. *Avian Ecol. Behav.* **3**: 59-68.

Chernetsov, N. & Manukyan, A. 2000. Foraging strategy of the Sedge Warbler (*Acrocephalus schoenobaenus*) on migration. *Vogelwarte* **40**: 189-197.

Chernetsov, N. & Titov, N. 2001. Movement patterns of European Reed Warblers *Acrocephalus scirpaceus* and Sedge Warblers *A. schoenobaenus* before and during autumn migration. *Ardea* **89**: 509-515.

Chernetsov, N., Mukhin, A. & Ktitorov, P. 2004. Contrasting spatial behaviour of two long-distance passerine migrants at spring stopovers. *Avian Ecol. Behav.* **12**: 53-61.

Chernetsov, N., Bulyuk, V.N. & Ktitorov, P. 2007. Migratory stopovers of passerines in an oasis at the crossroads of the African and Indian flyways. *Ring. Migr.* **23**: 243-251.

Cherry, M.I., Bennett, A.T.D. & Moskát, C. 2007a. Do cuckoos choose nests of great reed warblers on the basis of host egg appearance? *J. Evol. Biol.* **20**: 1218-1222.

Cherry, M.I., Bennett, A.T.D. & Moskát, C. 2007b. Cuckoo egg matching, host intra-clutch variation and egg rejection by great reed warblers. *Naturwissenschaften* **94**: 441-447.

Cibois, A., Pasquet, E. & Schulenberg, T.S. 1999. Molecular systematics of the Malagasy babblers (Passeriformes: Timaliidae) and warblers (Passeriformes: Sylviidae), based on cytochrome b and 16S rRNA sequences. *Mol. Phylogenet. Evol.* **13**: 581-595.

Cibois, A., Slikas, B., Schulenberg, T.S. & Pasquet, E. 2001. An endemic radiation of Malagasy songbirds is revealed by mitochondrial DNA sequence data. *Evolution* **55**: 1198-1206.

Cibois, A., Thibault, J.C. & Pasquet, E. 2004. Biogeography of eastern polynesian monarchs (Pomarea): an endemic genus close to extinction. *Condor* **106**: 837-851.

Cibois, A., Thibault, J.C. & Pasquet, E. 2007. Uniform phenotype conceals double colonization by reed-warblers of a remote Pacific archipelago. *J. Biogeogr.* **34**: 1150-1166.

Cibois, A., Thibault, J.C. & Pasquet, E. 2008. Systematics of the extinct reed warblers *Acrocephalus* of the Society Islands of eastern Polynesia. *Ibis* **150**: 365-376.

Cibois, A., David, N., Gregory, S.M.S. & Pasquet, E. 2010a. Bernieridae (Aves: Passeriformes): a family-group name for the Malagasy sylvioid radiation. *Zootaxa* **2554**: 65-68.

Cibois, A., Thibault, J.C. & Pasquet, E. 2010b. Influence of quaternary sea-level variations on land bird endemic to Pacific atolls *Proc. Roy. Soc. B-Biol. Sci.* **277**: 3445-3451.

Cibois, A., Beadell, J.S., Graves, G.R., Pasquet, E., Slikas, B., Sonsthagen,

S.A., Thibault, J.-C. & Fleischer, R.C. 2011a. Charting the course of reed-warblers across the Pacific islands. *J.Biogeogr.* **38**: 1963-1975.

Cibois, A., Thibault, J.-C. & Pasquet, E. 2011b. Molecular and morphological analysis of Pacific reed warbler specimens of dubious origin, including *Acrocephalus luscinius astrolabii. Bull. Brit. Ornithol. Club* **131**: 32-40.

Claramunt, S. 2010. Discovering exceptional diversifications at continental scales: the case of the endemic families of neotropical suboscine passerines. *Evolution* **64**: 2004-2019.

Clarke, A.L., Øien, I.J., Honza, M., Moksnes, A. & Røskaft, E. 2001. Factors effecting Reed Warbler risk of brood parasitism by the Common Cuckoo. *Auk* **118**: 534-538.

Clegg, S.M. & Owens, I.P.F. 2002. The 'island rule' in birds: medium body size and its ecological explantion. *Proc. Roy. Soc. B-Biol. Sci.* **269**: 1359-1365.

Cody, M.L. 1974. *Competition and the structure of bird communities.* Princeton, NJ: Princeton University Press.

Comtesse, H. 1974. Elementtypen und Sequenzbeziehungen im Reviergesang des Sumpfrohrsängers (*Acrocephalus palustris Bechstein*). [Diploma thesis] Kaiserslautern.

Conant, S., and Morin, M. 2001: Why isn't the Nihoa Millerbird extinct? *Stud. Avian Biol.* **22**: 338-346.

Constantine, M. & The Sound Approach. 2006. *The Sound Approach to birding.* Dorset: The Sound Approach.

Copete, J.L., Bigas, D., Marine, R. & Martinez-Vilalta, A. 1998. Frequency of complete moult in adult and juvenile Great Reed Warblers (*Acrocephalus arundinaceus*) in Spain. *J. Ornithol.* **139**: 421-424.

Cosens, S.E. & Falls, B. 1984. A comparison of sound propagation and song frequency in temperate marsh and grassland habitats. *Behav. Ecol. Sociobiol.* **15**: 161-170.

Craig, R.J. 1992. Territoriality, habitat use and ecological distinctness of an endangered Pacific Island Reed-Warbler. *J. Field Ornithol.* **63**: 436-444.

Cramp, S. ed. 1992. *Handbook of the birds of Europe, the Middle East, and North Africa: the birds of the Western Palearctic.* Vol. 6: *Warblers.* Oxford: Oxford University Press.

Crespo, J., Alba, E. & Garrido, M. 1988. Tongue spots of nestling Olivaceous Warblers. *Brit. Birds* **81**: 470-471.

Crook, J.H. 1964. The evolution of social organization and visual communication in the weaver birds (Ploceinae). *Behaviour* Suppl. **10**: 1-178.

Cunningham, C.W., Omland, K.E. & Oakley, T.H. 1998. Reconstructing ancestral character states: a critical reappraisal. *Trends Ecol. Evol.* **13**: 361-366.

Cunningham, S.J., Alley, M.R. & Castro, I. 2011. Facial bristle feather histology and morphology in New Zealand birds: implications for function. *J. Morphol.* **272**: 118-128.

Daan, S., Dijkstra, C., Drent, R. & Meijer, T. 1988. Food supply and the annual timing of avian reproduction. In: Ouellet, H. ed. *Acta XIX Congressus Internationalis Ornithologici.* Ottawa: University of Ottawa Press, pp. 392-407.

Darwin, C. 1871. *The descent of man, and selection in relation to sex.* London: Murray,.

Davies, N.B. 2000. *Cuckoos, cowbirds and other cheats.* London: T. & A. D. Poyser.

Davies, N.B. 2011. Cuckoo adaptations: trickery and tuning. *J. Zool.* **284**: 1-14.

Davies, N.B. & Brooke, M. de L 1988. Cuckoos versus Reed Warblers:-adaptations and counteradaptations. *Anim. Behav.* **36**: 262-284.

Davies, N.B. & Brooke, M. de L. 1989a. An experimental-study of co-evolution between the Cuckoo, *Cuculus canorus*, and its hosts,1. Host egg discrimination. *J. Anim. Ecol.* **58**: 207-224.

Davies, N.B. & Brooke, M. de L. 1989b. An experimental-study of co-evolution between the Cuckoo, *Cuculus canorus*, and its hosts,2: Host egg markings, chick discrimination and general discussion. *J. Anim. Ecol.* **58**: 225-236.

Davies, N.B. & Green, R.E. 1976. The development and ecological significance of feeding techniques in the reed warbler (*Acrocephalus scirpaceus*). *Anim.Behav.* **24**: 213-229.

Davies, N.B. & Welbergen, J.A. 2009. Social transmission of a host defense against Cuckoo parasitism. *Science* **324**: 1318-1320.

Davies, N.B., Brooke, M. de L. & Kacelnik, A. 1996. Recognition errors and probability of parasitism determine whether reed warblers should accept or reject mimetic cuckoo eggs. *Proc. Roy. Soc. B-Biol. Sci.* **263**: 925-931.

Davies, N.B., Kilner, R.M. & Noble, D.G. 1998. Nestling cuckoos, *Cuculus canorus*, exploit hosts with begging calls that mimic a brood. *Proc. Roy. Soc. B-Biol. Sci.* **265**: 673-678.

Davies, N.B., Butchart, S.H.M., Burke, T.A., Chaline, N. & Stewart, I.R.K. 2003. Reed warblers guard against cuckoos and cuckoldry. *Anim. Behav.* **65**: 285-295.

Davies, N.B., Madden, J.R., Butchart, S.H.M. & Rutila, J. 2006. A host-race of the cuckoo *Cuculus canorus* with nestlings attuned to the parental alarm calls of the host species. *Proc. Roy. Soc. B-Biol. Sci.* **273**: 693-699.

Davis, J.M. & Stamps, J.A. 2004. The effect of natal experience on habitat preferences. *Trends Ecol. Evol.* **19**: 411-416.

Dawkins, R. 1986. *The blind watchmaker.* Harlow: Longman Scientific & Technical.

Dayton, P.K. 2003. The importance of the natural sciences to conservation. *Am. Nat.* **162**: 1-13.

de la Hera, I., Perez-Tris, J. & Telleria, J.L. 2010. Relationships among timing of moult, moult duration and feather mass in long-distance migratory passerines. *J. Avian Biol.* **41**: 609-614.

Dean, W.R.J. 2005. Dark-capped Yellow Warbler, *Chloropeta natalensis.* In: Hockey, P.A.R., Dean, W.R.J. & Ryan, P.G.. eds. *Roberts' birds of Southern Africa.* 7th ed. Cape Town: John Voelcker Bird Book Fund, pp. 805-806.

del Hoyo, J., Elliott, A. & Christie, D.A. eds. 2006. *Handbook of the birds of the world.* Vol. 11: *Old World Flycatchers to Old World Warblers.* Barcelona: Lynx Edicions.

Dolnik, V.R. 1990. Bird migration across arid and mountainous regions of Middle Asia and Kazakhstan. In: Gwinner, E., ed. *Bird migration: physiology and ecophysiology.* Berlin Heidelberg New York: Springer, pp. 368-386.

Dorsch, H. 1979. Möglichkeiten der Unterscheidung von Teich- und Sumpf-rohrsänger anhand morphologischer Merkmale. *Falke* **26**: 405-419.

Dowsett-Lemaire, F. 1978. Annual turnover in a Belgian population of marsh warblers, *Acrocephalus palustris*. *Gerfaut* **68**: 519-532.

Dowsett-Lemaire, F. 1979a. The imitative range of the song of the marsh warbler *Acrocephalus palustris*, with special reference to imitations of African birds. *Ibis* **121**: 453-468.

Dowsett-Lemaire, F. 1979b. Vocal behaviour of the marsh warbler. *Gerfaut* **69**: 475-502.

Dowsett-Lemaire, F. 1980. La territorialité chez la rousserolle verderolle, *Acrocephalus palustris*. *Terre Vie* **34**: 45-67.

Dowsett-Lemaire, F. 1981a. Eco-ethological aspects of breeding in the marsh warbler, *Acrocephalus palustris*. *Terre Vie* **35**: 437-491.

Dowsett-Lemaire, F. 1981b. The transition period from juvenile to adult song in the European Marsh Warbler. *Ostrich* **52**: 253-255.

Dowsett-Lemaire, F. 1994. The song of the Seychelles Warbler (*Acrocephalus sechellensis*) and its African relatives. *Ibis* **136**: 489-491.

Dowsett-Lemaire, F. & Dowsett, R.J. 1987b. European reed and marsh warblers in Africa: migration patterns, moult and habitat. *Ostrich* **58**: 65-85.

Duckworth, J.W. 1991. Responses of breeding Reed Warblers *Acrocephalus scirpaceus* to mounts of Sparrowhawk *Accipiter nisus*, Cuckoo *Cuculus canorus* and Jay *Garrulus glandarius*. *Ibis* **133**: 68-74.

Duckworth, J.W. 1992. Effects of mate removal on the behaviour and reproductive success of Reed Warbler (*Acrocephalus scirpaceus*). *Ibis* **134**: 164-170.

Dugatkin, L.A. 2009. *Principles of animal behavior*. 2nd ed. New York: W. W. Norton.

Dykyjovà, D., Ondok, J.P. & Priban, K. 1970. Seasonal changes In productivity and vertical structure of reed-stands (*Phragmites communis trin*). *Photosynthetica* **4**: 280-287.

Dynesius, M. & Jansson, R. 2000. Evolutionary consequences of changes in species' geographical distributions driven by Milankovitch climate oscillations. *Proc. Natl Acad. Sci. USA* **97**: 9115-9120.

Dyrcz, A. 1979. Die Nestlingsnahrung bei Drosselrohrsänger (*Acrocephalus arundinaceus*) und Teichrohrsänger (*Acrocephalus scirpaceus*) an den Teichen bei Milicz in Polen und zwei Seen in der Westschweiz. *Ornithol. Beobacht.* **76**: 305-316.

Dyrcz, A. 1981. Breeding ecology of great reed warbler (*Acrocephalus arundinaceus*) and reed warbler (*Acrocephalus scirpaceus*) at fish-ponds in SW Poland and lakes in Switzerland. *Acta Ornithol.* **18**: 307-334.

Dyrcz, A. 1986. Factors affecting facultative polygyny and breeding results in the Great Reed Warbler (*Acrocephalus arundinaceus*). *J.Ornithol.* **127**: 447-461.

Dyrcz, A. 1988. Mating systems in European marsh-nesting Passeriformes. In: Ouellet, H. ed. *Acta XIX Congressus Internationalis Ornithologici*. Ottawa, Ont: For National Museum of Natural Sciences by University of Ottawa Press, pp. 2613-2623.

Dyrcz, A. 1993. Nesting biology of the Aquatic Warbler *Acrocephalus paludicola* in the Biebrza Marshes (NE Poland). *Vogelwelt* **114**: 2-15

Dyrcz, A. 1995. Breeding biology and ecology of different European and Asiatic populations of the Great Reed Wabler *Acrocephalus arundinaceus*. *Jpn. J. Ornithol.* **44**:123-142.

Dyrcz, A. & Flinks, H. 2000. Potential food resources and nestling food in the Great Reed Warbler (*Acrocephalus arundinaceus arundinaceus*) and Eastern Great Reed Warbler (*Acrocephalus arundinaceus orientalis*). *J. Ornithol.* **141**: 351-360.

Dyrcz, A. & Halupka, L. 2006. Great reed warbler *Acrocephalus arundinaceus* and reed warbler *Acrocephalus scirpaceus* respond differently to cuckoo dummy at the nest. *J. Ornithol.* **147**: 649-652.

Dyrcz, A. & Halupka, L. 2009. The response of the Great Reed Warbler *Acrocephalus arundinaceus* to climate change. *J. Ornithol.* **150**: 39-44.

Dyrcz, A. & Nagata, H. 2002. Breeding ecology of the Eastern Great Reed Warbler *Acrocephalus arundinaceus orientalis* at Lake Kasumigaura, central Japan. *Bird Study* **49**: 166-171.

Dyrcz, A. & Zdunek, W. 1993. Breeding ecology of the Aquatic Warbler (*Acrocephalus paludicola*) on the Biebrza marshes, northeast Poland. *J. Ornithol.* **135**: 181-189.

Dyrcz, A. & Zdunek, W. 1996. Potential food resources and nestlings food in the Great Reed Warbler (*Acrocephalus arundinaceus*) and Reed Warbler (*Acrocephalus scirpaceus*) at Milicz Fish-Ponds. *Ptaki Śląska* **11**: 123-132.

Dyrcz, A., Borowiec, M. & Czapulak, A. 1994. Nestling growth and mating system in four *Acrocephalus* species. *Vogelwarte* **37**: 179-182.

Dyrcz, A., Wink, M., Backhaus, A., Zdunek, W., Leisler, B. & Schulze-Hagen, K. 2002. Correlates of multiple paternity in the Aquatic Warbler (*Acrocephalus paludicola*). *J .Ornithol.* **143**: 430-439.

Dyrcz, A., Wink, M., Kruszewicz, A. & Leisler, B. 2005. Male reproductive success is correlated with blood parasite levels and body condition in the promiscuous aquatic warbler (*Acrocephalus paludicola*). *Auk* **122**: 558-565.

Dyrcz, A., Zdunek, W. & Schulze-Hagen, K. 2011. Increased male singing in response to predator presence may represent reproductive investment in a promiscuous species, the Aquatic Warbler *Acrocephalus paludicola*. *Acta Ornithol.* **46**: 97-100.

Eck, S. 1994. Die geographisch-morphologische Vikarianz der großen palaearktischen Rohrsänger. *Zool. Abh. Mus.Tierk. Dresden* **48**: 161-168.

Edvardsen, E., Moksnes, A., Røskaft, E., Øien, I.J. & Honza, M. 2001. Egg mimicry in cuckoos parasitizing four sympatric species of *Acrocephalus* warblers. *Condor* **103**: 829-837.

Eikenaar, C., Berg, M.L. & Komdeur, J. 2003. Experimental evidence for the influence of food availability on incubation attendance and hatching asynchrony in the Australian reed warblers (*Acrocephalus australis*). *J.Avian Biol.* **34**: 419-427.

Eikenaar, C., Richardson, D.S., Brouwer, L. & Komdeur, J. 2007. Parent presence, delayed dispersal, and territory acquisition in the Seychelles warbler. *Behav. Ecol.* **18**: 874-879.

Eikenaar, C., Richardson, D.S., Brouwer, L. & Komdeur, J. 2008a. Sex biased natal dispersal in a closed, saturated population of Seychelles warblers *Acrocephalus sechellensis*. *J. Avian Biol.* **39**: 73-80.

Eikenaar, C., Komdeur, J. & Richardson, D.S. 2008b. Natal dispersal patterns are not associated with inbreeding avoidance in the Seychelles Warbler. *J. Evol. Biol.* **21**: 1106-1116.

Eikenaar, C., Brouwer, L., Komdeur, J. & Richardson, D.S. 2010. Sex biased natal dispersal is not a fixed trait in a stable population of Seychelles warblers. *Behaviour* **147**: 1577-1590.

Eising, C.M., Komdeur, J., Buys, J., Reemer, M. & Richardson, D.S. 2001. Islands in a desert: breeding ecology of the African Reed Warbler *Acrocephalus baeticatus* in Namibia. *Ibis* **143**: 482-493.

Emlen, S.T. & Oring, L.W. 1977. Ecology, sexual selection, and the evolution of mating systems. *Science* 197: 215-223.

Engler, J., Elle, O. & Rödder, D. 2010. Der Eine geht, der Andere kommt: Untersuchung der Arealdynamik von Orpheus- und Gelbspötter mittels Artverbreitungsmodellen. *Vogelwarte* 48: 405-406.

Eriksen, A., Slagsvold, T. & Lampe, H.M. 2011. Vocal plasticity - are Pied Flycatchers, *Ficedula hypoleuca*, open-ended learners? *Ethology* 117: 188-198.

Ernst, S. 2001. Der Buschspötter. *Falke* 48: 114-117.

Ezaki, Y. 1984. Notes on the moult of the Easten Great Reed Warbler (*Acrocephalus arundinaceus orientalis*) in the breeding grounds. *J. Yamashina Inst. Ornithol.* 16: 88-91.

Ezaki, Y. 1987. Male time budgets and recovery of singing rate after pairing in polygamous Great Reed Warblers. *Jpn. J. Ornithol.* 36: 1-11.

Ezaki, Y. 1990. Female choice and the causes and adaptiveness of polygyny i n great reed warblers. *Anim. Ecol.* 59: 103-119.

Ezaki, Y. 1992. Importance of communal foraging grounds outside the reed marsh for breeding great reed warblers. *Ecol. Res.* 7: 63-70.

Ezaki, Y. & Urano, E. 1995. Intraspecific comparison of ecology and mating system of the Great Reed Wabler *Acrocephalus arundinaceus*: why different results from different populations? *Jpn. J.Ornithol.* 44: 107-122.

Faivre, B., Secondi, J., Ferry, C., Chastragnat, L. & Cezilly, F. 1999. Morphological variation and the recent evolution of wing length in the Icterine warbler: a case of unidirectional introgression? *J. Avian Biol.* 30: 152-158.

Ferry, C. & Deschaintre, D. 1974. Le chant, signal interspecifique chez *Hippolais icterina* et *polyglotta. Alauda* 42: 289-312.

Ferry, C. & Faivre, B. 1991a. *Hippolais icterina* – Gelbspötter. In: Glutz von Blotzheim, U.N. & Bauer, K.M. eds. *Handbuch der Vögel Mitteleuropas.* Wiesbaden: Aula, pp. 568-601.

Ferry, C. & Faivre, B. 1991b. *Hippolais polyglotta* – Orpheusspötter. In: Glutz von Blotzheim, U.N. & Bauer, K.M. eds. *Handbuch der Vögel Mitteleuropas.* Wiesbaden: Aula, pp. 601-626.

Fessl, B. & Hoi, H. 1996. The significance of a two part song in the moustached warbler (*Acrocephalus melanopogon*). *Ethol. Ecol. Evol.* 8: 265-278.

Fessl, B., Kleindorfer, S., Hoi, H. & Lorenz, K. 1996. Extra male parental behaviour: evidence for an alternative mating strategy in the Moustached Warbler *Acrocephalus melanopogon. J. Avian Biol.* 27: 88-91.

Fiedler, W. 2005. Ecomorphology of the external flight apparatus of blackcaps (*Sylvia atricapilla*) with different migration behaviour. *Ann. NY Acad. Sci.* 1046: 253-263.

Fiedler, B. 2011. Die Evolution des Gesanges der Acrocephalinae (*Hippolais, Acrocephalus* und *Chloropeta*) unter Einbeziehung der Phylogenie und morphologischer, ökologischer und sozialer Faktoren. [Dissertation] Universität Oldenburg.

Filardi, C.E. & Moyle, R.G. 2005. Single origin of a pan-Pacific bird group and upstream colonization of Australasia. *Nature* 438: 216.

Fischer, S., Frommolt, K.H. & Tembrock, G. 1996. Gesanggsvariabilität beim Drosselrohrsänger (*Acrocephalus arundinaceus*). *J. Ornithol.* 137: 503-513.

Fitzpatrick, J. 2004. Family Tyrannidae (Tyrant-Flycatchers). In: del Hoyo, J., Elliott, A. & Christie, D.A. eds. *Handbook of the birds of the world.* Vol. 9: *Cotingas to Pipits and Wagtails.* Barcelona: Lynx Edicions, pp. 170-257.

Fjeldså, J. 1980. The bird fauna of the swamp Regnemark Mose, Zealand, with estimates on the role of birds in the energy budget of marshland. *Dansk Orn. Foren. Tidsskr.* 74: 91-104.

Flade, M. 2008. [The aquatic warbler story: the *paludicola* operation.]. *Falke* 55: 90-99.

Flade, M. & Lachmann, L. 2008. *International species action plan for the Aquatic Warbler Acrocephalus paludicola.* Birdlife International on behalf of The European Commission.

Flade, M., Diop, I., Haase, M., Le Nevé, A., Oppel, S., Tegetmeyer, C., Vogel, A. & Salewski, V. 2011. Distribution, ecology and threat status of the Aquatic Warblers *Acrocephalus paludicola* wintering in West Africa. *J. Ornithol.* 152: S129-S140.

Fleischer, R.C., Slikas, B., Beadell, J., Atkins, C., McIntosh, C.E. & Conant, S. 2007. Genetic variability and taxonomic status of the Nihoa and Laysan Millerbirds. *Condor* 109: 954-962.

Foote, K. 2009. Groundbreaking research for the Nihoa millerbird. *USFWS Endangered Species Bull.* **Spring:** 20-21.

Foppen, R., ter Braak, C.J.F., Verboom, J. & Reijnen, R. 1999. Dutch sedge warblers (*Acrocephalus schoenobaenus*) and West-African rainfall: empirical data and simulation modelling show low population resilience in fragmented marshlands. *Ardea* 87: 113-127.

Foppen, R.P.B., Chardon, J.P. & Liefveld, W. 2000. Understanding the role of sink patches in source-sink metapopulations: Reed Warbler in an agricultural landscape. *Conserv. Biol.* 14: 1881-1892.

Forstmeier, W. & Leisler, B. 2004. Repertoire size, sexual selection, and offspring viability in the great reed warbler: changing patterns in space and time. *Behav. Ecol.* 15: 555-563.

Forstmeier, W., Leisler, B. & Kempenaers, B. 2001a. Bill morphology reflects female independence from male parental help. *Proc. Roy. Soc. B-Biol. Sci.* 268: 1583-1588.

Forstmeier, W., Kuijper, D.P.J. & Leisler, B. 2001b. Polygyny in the dusky warbler, *Phylloscopus fuscatus*: the importance of female qualities. *Anim. Behav.* 62: 1097-1108.

Forstmeier, W., Bourski, O.V. & Leisler, B. 2001c. Habitat choice in *Phylloscopus* warblers: the role of morphology, phylogeny and competition. *Oecologia* 128 :566-576.

Forstmeier, W., Hasselquist, D., Bensch, S. & Leisler, B. 2006. Does song reflect age and viability? A comparison between two populations of the great reed warbler *Acrocephalus arundinaceus. Behav. Ecol. Sociobiol.* 59: 634-643.

Forstmeier, W., Martin, K., Bolund, E., Schielzeth, H. & Kempenaers, B. 2011. Female extrapair mating behavior can evolve via indirect selection on males. *Proc. Natl Acad. Sci. USA* 108: 10608-10613.

Fossøy, F., Antonov, A., Moksnes, A., Røskaft , E., Vikan, J.R., Møller, A.P., Shykoff, J.A. & Stokke, B.G. 2011. Genetic differentiation among sympatric cuckoo host races: males matter. *Proc. Roy. Soc. B-Biol. Sci.* 278: 1639-1645.

Fransson, T. & Stolt, B-O. 2005. Migration routes of North European Reed Warblers (*Acrocephalus scirpaceus*). *Ornis Svecica* 15: 153-160.

Fregin, S., Haase, M., Olsson, U. & Alström, P. 2009. Multi-locus phylogeny of the family Acrocephalidae (Aves: Passeriformes): the traditional taxonomy overthrown. *Mol. Phylogenet. Evol.* 52: 866-878.

Fry, C.H., Ash, J.S. & Ferguson-Lees, I.J. 1970. Spring weights of some Palaearctic migrants at Lake Chad. *Ibis* **112**: 58-82.

Fuchs, J., Fjeldså, J., Bowie, R.C.K., Voelker, G. & Pasquet, E. 2006. The African warbler genus *Hyliota* as a lost lineage in the Oscine songbird tree: molecular support for an African origin of the Passerida. *Mol. Phylogenet. Evol.* **39**: 186-197.

Garamszegi, L.Z., Eens, M., Hurtrez-Boussès, S. & Møller, A.P. 2005. Testosterone, testes size, and mating success in birds: a comparative study. *Horm. Behav.* **47**: 389-409.

Garamszegi, L.Z., Eens, M., Pavlova, D.Z., Aviles, J. & Møller, A.P. 2007. A comparative study of the function of heterospecific vocal mimicry in European passerines. *Behav. Ecol.* **18**: 1001-1009.

Gärtner, K. 1982. Zur Ablehnung von Eiern und Jungen des Kuckucks *Cuculus canorus* durch die Wirtsvögel: Beobachtungen und experimentelle Untersuchungen am Sumpfrohrsänger *Acrocephalus palustris*. *Vogelwelt* **103**: 201-224.

Gaston, A.J. 1976. The moult of Blyth's reed warbler (*Acrocephalus dumetorum*) with notes on the moult of other Palaearctic warblers in India. *Ibis* **118**: 247-251.

Geffen, E. & Yom-Tov, Y. 2000. Are incubation and fledging periods longer in the tropics? *J. Anim. Ecol.* **69**: 59-73.

Gießing, B. 2002. Viele Väter für eine Brut - vorteilhaft oder unausweichlich für das Weibchen? [Dissertation] Universität zu Köln.

Gil, D. & Gahr, M. 2002. The honesty of bird song: multiple constraints for multiple traits. *Trends Ecol. Evol.* **17**: 133-141.

Gill, F.B. 2007. *Ornithology.* New York: W.H.Freeman and Company.

Ginn, H.B. & Melville, D.S. 1983. *Moult in birds*. BTO Guide No. 19.Tring, UK: British Trust for Ornithology.

Glutz von Blotzheim, U.N. & Bauer, K.M. eds. 1991. *Handbuch der Vögel Mitteleuropas*. Band 12/1: *Sylvidae*. Wiesbaden: Aula.

Goodman, S.M., Tello, G.T. & Langrand, O. 2000. Patterns of morphological and molecular variation in *Acrocephalus newtoni* on Madagascar. *Ostrich* **71**: 367-370.

Gowaty, P.A. 1996. Battles of the sexes and origins of monogamy. In: Black, J.M. ed. *Partnerships in birds*. Oxford: Oxford University Press, pp. 21-52.

Gowaty, P.A. 1999. Extra-pair paternity and paternal care: differential male fitness via exploitation of variation among females. In Adams, N.J. & Slotow, R.H. eds. *Proceedings of the 22nd International Ornithological Congress, Durban, South Africa, 1998*. Johannesburg: BirdLife South Africa, pp. 2639-2656.

Grant, P.R. 1998. Patterns on islands and microevolution. In: Grant, P.R. ed. *Evolution on islands*. Oxford: Oxford University Press, pp. 1-17.

Grant, P.R. 2000. What does it mean to be a naturalist at the end of the twentieth century? *Am. Nat.* **155**: 1-12.

Graveland, J. 1998. Reed die-back, water level management and the decline of the Great Reed Warbler *Acrocephalus arundinaceus* in The Netherlands. *Ardea* **86**: 187-201.

Graveland, J. 1999. Effects of reed cutting on density and breeding success of Reed Warbler *Acrocephalus scirpaceus* and Sedge Warbler *A. schoenobaenus*. *J. Avian Biol.* **30**: 469-482.

Graves, G.R. 1992. The endemic land birds of Henderson Island, Southeastern Polynesia: notes on natural history and conservation. *Wilson Bull.* **104**: 32-43.

Gray, E.M. 1997. Female red-winged blackbirds accrue material benefits from copulating with extra-pair males. *Anim. Behav.* **53**: 625-639.

Green, R.E. & Davies, N.B. 1972. Feeding ecology of Reed and Sedge Warblers. *Wicken Fen Group Rep.* **4**: 1-14.

Greenberg, R. & Mettke-Hoffman, C. 2001. Ecological aspects of neophobia and neophilia in birds. *Curr. Ornithol.* **16**: 119-178.

Greene, H.W. 2005. Organisms in nature as a central focus for biology. *Trends Ecol. Evol.* **20**: 23-27.

Greenlaw, J.S. 1988. On mating systems in passerine birds of American marshlands. In: Ouellet, H., ed. *Acta XIX Congressus Internationalis Ornithologici*. Ottawa, Ont: For National Museum of Natural Sciences by University of Ottawa Press, pp. 2645-2667.

Greenlaw, J.S. & Post, W. 1985. Evolution of monogamy in seaside sparrows, *Ammodramus maritimus*: tests of hypotheses. *Anim. Behav.* **33**: 373-383.

Greenlaw, J.S. & Rising, J.D. 1994. Sharp-Tailed Sparrow (*Ammodramus caudacutus*). In: Poole, A. & Gill, F. eds. *The birds of North America*. Philadelphia: Academy of Natural Sciences, pp. 1-28.

Griffith, S.C. 2000. High fidelity on islands: a comparative study of extrapair paternity in passerine birds. *Behav. Ecol.* **11**: 265-273.

Griffith, S.C. 2007. The evolution of infidelity in socially monogamous passerines: neglected components of direct and indirect selection. *Am. Nat.* **169**: 274-281.

Griffith, S.C., Owens, I.P.F. & Thuman, K.A. 2002. Extra pair paternity in birds: a review of interspecific variation and adaptive function. *Mol. Ecol.* **11**: 2195-2212.

Grim, T. 2002. Why is egg mimicry in cuckoo eggs sometimes so poor? *J. Avian Biol.* **33**: 302-305.

Grim, T. 2006. An exceptionally high diversity of hoverflies (Syrphidae) in the food of the reed warbler (*Acrocephalus scirpaceus*). *Biologia* **61**: 235-239.

Grim, T. 2007. Experimental evidence for chick discrimination without recognition in a brood parasite host. *Proc. Roy. Soc. B-Biol. Sci.* **274**: 373-381.

Grim, T., Kleven, O. & Mikulica, O. 2003. Nestling discrimination without recognition: a possible defence mechanism for hosts towards cuckoo parasitism? *Proc. Roy. Soc. B-Biol. Sci.* **270**: S73-S75.

Grimmett, R., Inskipp, C. & Inskipp, T. 1998. *Birds of the Indian Subcontinent*. London: Christopher Helm.

Grinkevich, V., Chernetsov, N. & Mukhin, A. 2007. Juvenile Reed Warblers *Acrocephalus scirpaceus* see the world but settle close to home. *Avian Ecol. Behav.* **16**: 3-10.

Grüll, A. & Zwicker, E. 1993. Zur Siedlungsdichte von Schilfsingvögeln (*Acrocephalus* und *Locustella*) am Neusiedlersee in Abhängigkeit vom Alter der Röhrichtbestände. *Beih.Veröff. Natursch. Landschaftspfl. Bad.-Wurtt.* **68**: 159-171.

Grünberger, S. 1992. Die Ausbildung von Habitatpräferenzen bei der Tannenmeise (*Parus ater*): Verschränkung angeborener und erfahrungsbedingter Mechanismen. [Dissertation] Universität Konstanz.

Gwinner, E. 1967. Circannuale Periodik der Mauser und der Zugunruhe bei einem Vogel. *Naturwissenschaften* **54**: 447.

Gwinner, E. 1986. *Circannual rhythms*. Heidelberg: Springer.

Gwinner, E. 2005. Avian circannual clocks: dependence on and interactions with photoperiods. In: Honma, K.H. & Honma, S. eds. *Biological rhythms*. Proceedings of the 10th Sapporo symposium on biological rhythms, September 8-10, 2003. Sapporo: Hokkaido University Press, pp. 19-40.

Haartman, L. von 1982. Two modes of clutch size determination in passerine birds. *J.Yamashina Inst.Ornithol.* **14**: 214-219.

Hagemeijer, W.J.M. & Blair, M.J. (eds) 1997. *The EBCC atlas of European breeding birds: their distribution and abundance.* London: T. & A.D. Poyser.

Hall, K.S.S. & Tullberg, B.S. 2004. Phylogenetic analyses of the diversity of moult strategies in Sylviidae in relation to migration. *Evol. Ecol.* **18**: 85-105.

Hall, K.S.S., Ryttman, H., Fransson, T. & Stolt, B-O. 2004. Stabilising selection on wing length in reed warblers *Acrocephalus scirpaceus. J. Avian Biol.* **35**: 7-12.

Halupka, K. 1998. Partial predation in an altricial bird selects for the accelerated development of young. *J. Avian Biol.* **29**: 129-133.

Halupka, L. 1999. Nest defence in an altricial bird with uniparental care: the influence of offspring age, brood size, stage of the breeding season and predator type. *Ornis Fennica* **76**: 97-105.

Halupka, L. 2003. Intersexual differences in nest defence in the genus *Acrocephalus. Vogelwarte* **42**:128.

Halupka, L., Dyrcz, A. & Borowiec, M. 2008. Climate change affects breeding of reed warblers *Acrocephalus scirpaceus. J. Avian Biol.* **39**: 95-100.

Halupka, L., Sztwiertnia, H., Tomasik, L., Leisler, B. & Borowiec, M. (in press) No polygamy and a single polygyny attempt in a long studied population of reed warblers *Acrocephalus scirpaceus. Acta Ornithol.*

Hamao, S. 2003. Reduction of cost of polygyny by nest predation in the Black-browed Reed Warbler. *Ornithol. Sci.* **2**: 113-118.

Hamao, S. 2005. Predation risk and nest-site characteristics of the Black-browed Reed Warbler(*Acrocephalus bistrigiceps*): the role of plant strength. *Ornithol. Sci.* **4**: 147-153.

Hamao, S. 2008. Singing strategies among male Black-browed Reed Warblers *Acrocephalus bistrigiceps* during the post-fertile period of their mates. *Ibis* **150**: 388-394.

Hamao, S. & Eda-Fujiwara, H. 2004. Vocal mimicry by the Black-browed Reed Warbler (*Acrocephalus bistrigiceps*): objective identification of mimetic sounds. *Ibis* **146**: 61-68.

Hamao, S. & Saito, D.S. 2005. Extrapair fertilization in the black-browed reed warbler (*Acrocephalus bistrigiceps*): effects on mating status and nesting cycle of cuckolded and cuckolder males. *Auk* **122**: 1086-1096.

Hamilton, W.D. 1964. The genetical evolution of social behaviour. *J. Theor. Biol.* **7**: 1-52.

Hansell, M. 2000. *Bird nests and construction behaviour.* Cambridge: Cambridge University Press.

Hansson, B. & Richardson, D.S. 2005. Genetic variation in two endangered *Acrocephalus* species compared to a widespread congener: estimates based on functional and random loci. *Anim. Conserv.* **8**: 83-90.

Hansson, B., Bensch, S. & Hasselquist, D. 1997. Infanticide in great reed warblers: secondary females destroy eggs of primary females. *Anim. Behav.* **54**: 297-304.

Hansson, B., Bensch, S. & Hasselquist, D. 2000. Patterns of nest predation contribute to polygyny in the Great Reed Warbler. *Ecology* **81**: 319-328.

Hansson, B., Bensch, S., Hasselquist, D. & Akesson, M. 2001. Microsatellite diversity predicts recruitment of sibling great reed warblers. *Proc. Roy. Soc. B-Biol. Sci.* **268**: 1287-1291.

Hansson, B., Bensch, S., Hasselquist, D. & Nielsen, B. 2002a. Restricted dispersal in a long-distance migrant bird with patchy distribution: the great reed warbler. *Oecologia* **130**: 536-542.

Hansson, B., Bensch, S. & Hasselquist, D. 2002b. Predictors of natal dispersal in great reed warblers: results from small and large census areas. *J. Avian Biol.* **33**: 311-314.

Hansson, B., Bensch, S. & Hasselquist, D. 2003a. A new approach to study dispersal: immigration of novel alleles reveals female-biased dispersal in great reed warblers. *Mol. Ecol.* **12**: 631-637.

Hansson, B., Bensch, S. & Hasselquist, D. 2003b. Heritability of dispersal in the great reed warbler. *Ecol. Lett.* **6**: 290-294.

Hansson, B., Gavrilov, E. & Gavrilov, A. 2003c. Hybridisation between great reed warblers (*Acrocephalus arundinaceus*) and clamorous reed warblers (*A.stentoreus*): morphological and molecular evidence. *Avian Sci.* **3**: 145-151.

Hansson, B., Hasselquist, D. & Bensch, S. 2004a. Do female great reed warblers seek extra-pair fertilizations to avoid inbreeding? *Proc. Roy. Soc. B-Biol. Sci.* **271**: S290-S292.

Hansson, B., Westerdahl, H., Hasselquist, D., Akesson, M. & Bensch, S. 2004b. Does linkage disequilibrium generate heterozygosity-fitness correlations in Great Reed Warblers? *Evolution* **58**: 870-879.

Hansson, B., Bensch, S. & Hasselquist, D. 2004c. Lifetime fitness of short- and long-distance dispersing great reed warblers. *Evolution* **58**: 2546-2557.

Hansson, B., Roggeman, W. & De Smet, G. 2004d. Molecular evidence of a reed warbler x great reed warbler hybrid (*Acrocephalus scirpaceus* x *A. arundinaceus*) in Belgium. *J. Ornithol.* **145**: 159-160.

Hansson, B., Jack, L., Christians, J.K., Pemberton, J.M., Akesson, M., Westerdahl, H., Bensch, S. & Hasselquist, D. 2007. No evidence for inbreeding avoidance in a great reed warbler population. *Behav. Ecol.* **18**: 157-164.

Hansson, B., Hasselquist, D., Tarka, M., Zehtindjiev, P. & Bensch, S. 2008. Postglacial colonisation patterns and the role of isolation and expansion in driving diversification in a passerine bird. *PLoS ONE* **3**: e2794.

Haslam, S.M. 1973. Some aspects of the life history and autecology of *Phragmites communis trin. Pol. Arch. Hydrobiol.* **20**: 79-100.

Hasselquist, D. 1994. Male attractiveness, mating tactics and realized fitness in the polygynous great reed warbler. [Ph.D thesis] Lund University.

Hasselquist, D. 1995. Demography and lifetime reproductive success in the polygynous great reed warbler. *Jpn. J. Ornithol.* **44**: 181-194.

Hasselquist, D. 1998. Polygyny in great reed warblers: a long-term study of factors contributing to male fitness. *Ecology* **79**: 2376-2390.

Hasselquist, D. & Bensch, S. 1991. Trade-off between mate guarding and mate attraction in the polygynous great reed warbler. *Behav. Ecol. Sociobiol.* **28**: 187-193.

Hasselquist, D. & Langefors, A. 1998. Variable social mating system in the sedge warbler, *Acrocephalus schoenobaenus*. *Ethology* **104**: 759-769.

Hasselquist, D. & Sherman, P.W. 2001. Social mating systems and extrapair fertilizations in passerine birds. *Behav. Ecol.* **12**: 457-466.

Hasselquist, D., Bensch, S. & von Schantz, T. 1994. Low frequency of extrapair paternity in the polygnous great reed warbler (*Acrocephalus arundinaceus*). *Behav. Ecol.* **6**: 27-38.

Hasselquist, D., Bensch, S. & von Schantz, T. 1996. Correlation between male song repertoire, extra-pair paternity and offspring survival in the great reed warbler. *Nature* **381**: 229-232.

Hasselquist, D., Östman, O., Waldenström, J. & Bensch, S. 2007. Temporal patterns of occurrence and transmission of the blood parasite *Haemoproteus payevskyi* in the great reed warbler *Acrocephalus arundinaceus*. *J. Ornithol.* **148**: 401-409.

Hatchwell, B.J. & Komdeur, J. 2000. Ecological constraints, life history traits and the evolution of cooperative breeding. *Anim.Behav.* **59**: 1079-1086.

Hau, M., Perfito, N. & Moore, I.T. 2008. Timing of breeding in tropical birds: mechanisms and evolutionary implications. *Ornitol. Neotrop.* **19**: 39-59.

Hedenström, A., Bensch, S., Hasselquist, D., Lockwood, M. & Ottosson, U. 1993. Migration, stopover and moult of the Great reed warbler (*Acrocephalus arundinaceus*) in Ghana, West Africa. *Ibis* **135**: 177-180.

Heinroth, O. 1903. Ornithologische Ergebnisse der I.Deutsche Südsee Expedition von Br.Mencke. Forts. *J. Ornithol.* **51**: 65-125.

Heise, G. 1970. Zur Brutbiologie des Seggenrohrsängers (*Acrocephalus paludicola* Vieillot). *J. Ornithol.* **111**:54-67.

Helb, H-W., Dowsett-Lemaire, F., Bergmann, H-H. & Conrads, K. 1985. Mixed singing in European songbirds: a review. *Z. Tierpsychol.* **69**: 27-41.

Helbig, A.J. 2000. Was ist eine Vogel-"Art"? Ein Beitrag zur aktuellen Diskussion um Artkonzepte in der Ornithologie. *Limicola* **14**: 58-79, 172-184, 220-247.

Helbig, A.J. 2003. Evolution of bird migration: a phylogenetic and biogeographic perspective. In Berthold, P., Gwinner, E. & Sonnenschein, E. eds. *Avian migration*. Berlin, Heidelberg: Springer, pp. 3-20.

Helbig, A.J. & Seibold, I. 1999. Molecular phylogeny of palearctic-African *Acrocephalus* and *Hippolais* warblers (Aves: Sylviidae). *Mol. Phylogenet Evol.* **11**: 246-260.

Helbig, A.J., Knox, A.G., Parkin, D.T., Sangster, G. & Collinson, M. 2002. Guidelines for assigning species rank. *Ibis* **144**: 518-525.

Helm, B. & Gwinner, E. 2006. Timing of molt as a buffer in the avian annual cycle. *Acta Zool. Sinica* **52** (Suppl.): 703-706.

Henry, C. 1979. Le concept de niche écologique Illustré par le cas de populations congénériques sympatriques du genre Acrocephalus. *Terr. Vie* **33**: 457-492.

Henshaw, I., Fransson, T., Jakobsson, S., Lind, J., Vallin, A. & Kullberg, C. 2008. Food intake and fuel deposition in a migratory bird is affected by multiple as well as single-step changes in the magnetic field. *J. Exp. Biol.* **211**: 649-653.

Hering, J. & Fuchs, E. 2009. Der Kapverdenrohrsängers *Acrocephalus brevipennis* auf Fogo (Kapverdische Inseln): Verbreitung, Dichte, Habitat und Brutbiologie. *Vogelwarte* **47**: 157-164.

Hering, J. & Hering, H. 2005. Discovery of Cape Verde warbler (*Acrocephalus brevipennis*) on Fogo, Cape Verde Islands. *Bull. Afr. Bird Club* **12**: 147-149.

Hering, J., Fuchs, E. & Winkler, H. 2011 Mangroverohrsänger *Acrocephalus scirpaceus avicenniae* als Baum- und Palmenbrüter in einer ägyptischen Saharaoase. *Limicola* **25**: 134-162.

Herremans, M. 1990. Body-moult and migration overlap in Reed Warblers (*Acrocephalus scirpaceus*) trapped during nocturnal migration. *Gerfaut* **80**: 149-158.

Herrera, C.M. 1978. Individual dietary differences associated with morphological variation in Robins (*Erithacus rubecula*). *Ibis* **120**: 542-545.

Heuwinkel, H. 1982. Schalldruckpegel und Frequenzspektren der Gesänge von *Acrocephalus arundinaceus, A. scirpaceus, A. schoenobaenus* und *A. palustris* und ihre Beziehung zur Biotopakustik. *Ökol.Vögel* **4**: 85-174.

Heuwinkel, H. 1990. The effect of vegetation on the transmission of songs of selected European Passeriformes. *Acta Biol. Benrodis* **2**: 133-150.

Higgins, P.J., Peter, J.M. & Cowling, S.J. 2006a. *Acrocephalus australis* - Australian Reed -Warbler. In: Higgins, P.J., Peter, J.M. & Cowling, S.J. eds. *Handbook of Australian, New Zealand and Antarctic Birds,* Vol. 7. Melbourne: Oxford Univ.Press, pp. 1605-1623.

Higgins, P.J., Peter, J.M. & Cowling, S.J. 2006b. *Acrocephalus orientalis* - Oriental Reed-Warbler. In: Higgins, P.J., Peter, J.M. & Cowling, S.J. eds. *Handbook of Australian, New Zealand and Antarctic Birds,* Vol. 7. Melbourne: Oxford Univ.Press, pp. 1623-1629.

Hill, C.E., Gjerdrum, C. & Elphick, C.S. 2010. Extreme levels of multiple mating characterize the mating system of the Saltmarsh Sparrow (*Ammodramus caudacutus*). *Auk* **127**: 300-307.

Hocke, H. 1903. Aus dem Leben der kleinen Rohrdommel. *Ornithol. Monatsschr.* **28**: 138-140.

Hockey, P.A.R. 2005. Predicting migratory behavior in landbirds. In: Greenberg, R. & Marra, P.P. eds. *Birds of two worlds: the ecology and evolution of migration*. Baltimore and London: Johns Hopkins University Press, pp. 53-62.

Hockey, P.A.R., Dean, W.R.J. & Ryan, P.G. eds. 2005. *Roberts - Birds of Southern Africa*. 7th ed. Cape Town: John Voelcker Bird Book Fund.

Hoi, H. & Ille, R. 1996. Trade-offs of territory choice in male and female Marsh Warblers. *Auk* **113**: 243-246.

Hoi, H. & Winkler, H. 1988. Feinddruck auf Schilfbrüter: eine experimentelle Untersuchung. *J. Ornithol.* **129**: 439-447.

Hoi, H. & Winkler, H. 1994. Predation on nests: a case of apparent competition. *Oecologia* **98**: 436-440.

Hoi, H., Eichler, T. & Dittami, J. 1991. Territorial spacing and interspecific competition in three species of reed warblers. *Oecologia* **87**: 443-448.

Hoi, H., Kleindorfer, S., Ille, R. & Dittami, J. 1995. Prey abundance and male parental behaviour in *Acrocephalus* warblers. *Ibis* **137**: 490-496.

Holmbring, J.A. 1973. The great reed warbler *Acrocephalus arundinaceus* in Sweden in 1971 and a review of its earlier status. *Vår Fågelvärld* **32**: 23-30.

Holmbring, J.A. 1979. The great reed warbler *Acrocephalus arundinaceus* in Sweden in 1972-76. *Vår Fågelvärld* **38**: 83-90.

Holmgren, N. & Hedenström, A. 1995. The scheduling of molt in migratory birds. *Evol. Ecol.* **9**: 354-368.

Holt, R.D. 1977. Predation, apparent competition, and the structure of prey communities. *Theor. Popul. Biol.* **12**: 197-229.

Holyoak, D.T. 1974. Undescribed land birds from the Cook Islands, Pacific Ocean. *Bull. Brit .Ornithol.Club* **94**: 145-150.

Holyoak, D.T. 1978. Variable albinism of the flight feathers as an adaptation for recognition of individual birds in some Polynesian populations of *Acrocephalus* Warblers. *Ardea* **66**: 112-117.

Holyoak, D.T. & Thibault, J-C. 1984. Contribution a l´étude des oiseaux de Polynésie orientale. *Memoir. Mus. Natl Hist. Nat.* **127**: 149-168.

Honza, M., Øien, I.J., Moksnes, A. & Røskaft, E. 1998. Survival of reed warbler *Acrocephalus scirpaceus* clutches in relation to nest position. *Bird Study* **45**: 104-108.

Honza, M., Moksnes, A., Røskaft, E. & Øien, I.J. 1999. Effect of Great Reed Warbler (*Acrocephalus arundinaceus*) on the reproductive tactics of the Reed Warbler *A.scirpaceus*. *Ibis* **141**: 489-506.

Honza, M., Moksnes, A., Røskaft, E. & Stokke, B.G. 2001. How are different common cockoo (*Cuculus canorus*) egg morphs maintained? An evaluation of different hypotheses. *Ardea* **89**: 341-352.

Howard, H.E. 1920. *Territory in bird life*. London: John Murray.

Howard, R.D. 1974. The influence of sexual selection and interspecific competition on Mockingbird song (*Mimus polyglottos*). *Evolution* **28**: 428-438.

Ille, R. & Hoi, H. 1995. Factors influencing fledgling survival in the Marsh Warbler (*Acrocephalus palustris*): food and vegetation density. *Ibis* **137**: 586-589.

Ille, R., Hoi, H. & Kleindorfer, S. 1996. Brood predation, habitat characteristics and nesting decisions in *Acrocephalus scirpaceus* and *A.palustris*. *Biologia* **51**: 219-225.

Imhof, G. 1972. Quantitative Aufsammlung schlüpfender Fluginsekten in einem semiterrestischen Lebensraum mittels flächenbezogener Elektoren. *Verh. Ber. Dtsch. Zool. Ges.* **65** Jahresvers: 120-123.

Impekoven, M. 1962. Die Jugendentwicklung des Teichrohrsängers (*Acrocephalus scirpaceus*): eine Verhaltensstudie. *Rev. Suisse Zool.* **69**: 77-191.

Irestedt, M., Fjeldsa, J. & Ericson, P.G.P. 2006. Evolution of the ovenbird-woodcreeper assemblage (Aves: Furnariidae) – major shifts in nest architecture and adaptive radiation. J. Avian Biol. **37**: 260-272.

Irwin, D.E. 2000. Song variation in an avian ring species. *Evolution* **54**: 998-1010.

Ivanitskii, V.V. 2001. The study of vocal and spatial interrelations between Blyth's reed warblers (*Acrocephalus dumetorum*) and marsh warblers (*A.palustris*). *Vest. Mosk. Univ. Ser. Biol.* **1**: 3-8

Ivanitskii, V.V. & Keseva, L. 1996. On the coexistence of the Garden *Acrocephalus dumetorum* and Swampy *A. palustris* in the Kostroma Region. *Vest. Mosk. Univ. Ser. Biol.* **0** (3): 60-66

Ivanitskii, V.V., Marova, I.M. & Kvartal'nov, P.V. 2002. Structure and

dynamics of multispecies reed warbler community *Acrocephalus* (Passeriformes, Sylviidae) in steppe lakes. *Zool. Zh.* **81**: 833-840.

Ivanitskii, V.V., Kalyakin, M.V., Marova, I.M. & Kvartal'nov, P.V. 2005. Ecological and geographical analysis of distribution of Reed Warblers (Acrocephalus, Sylviidae, Aves) and some problems of their evolution. *Zool. Zh.* **84**: 870-884.

Ivanitskii, V.V., Marova, I.M. & Kvartal'nov, P.V. 2006. The acoustic signal system and behavior of the paddy field warbler, *Acrocephalus agricola* (Passeriformes, Sylviidae). *Zool. Zh.* **85**: 971-982.

Janetos, A.C. 1980. Strategies of female mate choice: a theoretical analysis. *Behav. Ecol. Sociobiol.* **7**: 107-112.

Jenni, L. & Winkler, R. 1994. *Moult and ageing of European passerines*. London: Academic Press.

Jenni, L. & Schaub, M. 2003. Behavioural and physiological reactions to environmental variation in bird migration: a review. In: Berthold, P., Gwinner, E. & Sonnenschein, E. eds. *Avian migration*. Berlin, Heidelberg: Springer, pp. 155-171.

Jetz, W., Sekercioglu, C.H. & Böhning-Gaese, K. 2008. The worldwide variation in avian clutch size across species and space. *PLoS Biol.* **6**: e303.

Jilka, A. & Leisler, B. 1974. Die Einpassung dreier Rohrsängerarten (*Acrocephalus schoenobaenus, A. scirpaeus, A. arundinaceus*) in ihre Lebensräume in bezug auf das Frequenzspektrum ihrer Reviergesänge. *J. Ornthol.* **115**: 192-212.

Johansson, U.S., Fjeldså, J. & Bowie, R.C.K. 2008. Phylogenetic relationships within Passerida (Aves: Passeriformes): a review and a new molecular phylogeny based on three nuclear intron markers. *Mol. Phylogenet. Evol.* **48**: 858-876.

Johns, G.C. & Avise, J.C. 1998. A comparative summary of genetic distances in the vertebrates from the mitochondrial cytochrome b gene. *Mol. Biol. Evo.* **15**: 1481-1490.

Johnson, K.P. & Lanyon, S.M. 2000. Evolutionary changes in color patches of blackbirds are associated with marsh nesting. *Behav. Ecol.* **11**: 515-519.

Johnson, T.H. & Stattersfield, A.J. 1990. A global review of island endemic birds. *Ibis* **132**: 167-180.

Jones, P.J. 1995. Migration strategies of Palearctic passerines in Africa. *Israel J. Zool.* **41**: 393-406.

Jonsson, K.A. & Fjeldså, J. 2006. A phylogenetic supertree of oscine passerine birds (Aves: Passeri). *Zool. Scr.* **35**: 149-186.

Kagawa, T. 1989. Interspecific relationship between two sympatric warblers Great Reed Warbler (*Acrocphalus arundinaceus*) and Schrenck's Reed Warbler (*A. bistrigiceps*). *Jpn. J.Ornithol.* **37**: 129-144.

Kazlauskas, R., Pukas, A. & Meldazyte, R. 1986. On feeding of warblers (*Acrocephalus*) in reproductive period, Western Lithuania. In: Valius,

M. ed. *Ecology of birds of Lithuanian SSR. Part 3: Human impact on the avifauna and its conservation*. Vilnius: Akademija nauk Litovskoj SSR, pp. 130-149.

Kelsey, M.G. 1989. A comparison of the song and territorial behaviour of a long-distance migrant, the Marsh Warbler *Acrocephalus palustris*, in summer and winter. *Ibis* **131**: 403-414.

Kempenaers, B. 2007. Mate choice and genetic quality: a review of the heterozygosity theory. *Adv.Stud. Behav.* **37**: 189-278.

Kennerley, P. & Pearson, D. 2010. *Reed and bush warblers*. London: Christopher Helm.

Kerbiriou, C., Bargain, B., Le Viol, I. & Pavoine, S. 2011. Diet and fuelling of the globally threatened aquatic warbler at autumn migration stopover as compared with two congeners. *Anim. Conserv.* **14**: 261-270.

Kikkawa, J. & Blondel, J. 2002. Specialisation in island land birds. In: Schodde, R. ed. Proceedings of the 23rd International Ornithological Congress, Beijing, China, 11-17 August 2002. *Acta Zool. Sinica* **52** (Suppl.): 262-266

Kikuchi, H., Sakagami, S.F. & Konishi, M. 1957. Ethology of the Eastern Great Reed Warbler with special reference to the life history of a polygynous family: preliminary report. *Jpn .J. Ecol.* **7**: 155-160.

Kilner, R.M. 2010. Behavioural ecology: learn to beat an identity cheat. *Nature* **463**: 165-167.

Kilner, R.M. & Davies, N.B. 1999. How selfish is a cuckoo chick? *Anim. Behav.* **58**: 797-808.

Kilner, R.M., Noble, D.G. & Davies, N.B. 1999. Signals of need in parent-offspring communication and their exploitation by the common cuckoo. *Nature* **397**: 667-672.

Kleindorfer, S. & Hoi, H. 1997. Nest predation avoidance: an alternative explantion for male incubation *in Acrocephalus melanopogon*. *Ethology* **103**: 619-631.

Kleindorfer, S., Fessl, B. & Hoi, H. 1995. More is not always better: male incubation in two *Acrocephalus* warblers. *Behaviour* **132**: 607-625.

Kleindorfer, S., Hoi, H. & Ille, R. 1997. Nestling growth patterns and antipredator responses: a comparison between four *Acrocephalus* warblers. *Biologia* **52**: 677-685.

Kleindorfer, S., Fessl, B. & Hoi, H. 2003. The role of nest site cover for parental nest defence and fledging success in two *Acrocephalus* warblers. *Avian Sci.* **3**: 21-29.

Kleindorfer, S., Fessl, B. & Hoi, H. 2005. Avian nest defence behaviour: assessment in relation to predator distance and type, and nest height. *Anim. Behav.* **69**: 307-313.

Kleven, O., Moksnes, A., Røskaft, E. & Honza, M. 1999. Host species affects the growth rate of cuckoo (*Cuculus canorus*) chicks. *Behav. Ecol. Sociobiol.* **47**: 41-46.

Kleven, O., Moksnes, A., Røskaft, E., Rudolfsen, G., Stokke, R.G. & Honza, M. 2004. Breeding success of common cuckoos *Cuculus canorus* parasitising four sympatric species of *Acrocephalus* warblers. *J. Avian Biol.* **35**: 394-398.

Kloubec, B. & Capek, M. 2000. Diurnal, nocturnal, and seasonal patterns of singing activity in marsh warblers. *Biologia* **55**: 185-193.

Kluyver, H.N. 1955. Das Verhalten des Drosselrohrsängers am Brutplatz mit besonderer Berücksichtigung der Nestbautechnik und der Revierbehauptung. *Ardea* **43**: 1-50.

Knape, J., Skoeld, M., Jonzen, N., Akesson, M., Bensch, S., Hansson, B. & Hasselquist, D. 2008. An analysis of hatching success in the great reed warbler *Acrocephalus arundinaceus*. *Oikos* **117**: 430-438.

Koblik, E.A., Red'kin, Y.A., Meer, M.S., Derelle, R., Golenkina, S.A., Kondrashov, FA. & Arkhipov, V.Y. 2011. *Acrocephalus orinus*: a case of mistaken identity. PLoS ONE **6(4)**: e17716.

Koepcke, H-W. 1974. *Die Lebensformen*. Vol 2: *Die Arterhaltung. Die universelle Bedeutung der Lebensformen*. Krefeld: Goecke & Evers.

Kokko, H., Gunnarsson, T.G., Morrell, L.J. & Gill, J.A. 2006. Why do female migratory birds arrive later than males? *J. Anim. Ecol.* **75**: 1293-1303.

Komdeur, J. 1991. Cooperative breeding in the Seychelles warbler. [Ph.D thesis] University of Cambridge.

Komdeur, J. 1992. Importance of habitat saturation and territory quality for evolution of cooperative breeding in the Seychelles Warbler. *Nature* **358**: 493-495.

Komdeur, J. 1994a. Experimental evidence for helping and hindering by previous offspring in the cooperative-breeding Seychelles warbler *Acrocephalus sechellensis*. *Behav. Ecol. Sociobiol.* **34**: 175-186.

Komdeur, J. 1994b. Conserving the Seychelles Warbler (*Acrocephalus sechellensis*) by translocation from Cousin Island to the islands of Aride and Cousine. *Biol. Conserv.* **67**: 143-152.

Komdeur, J. 1996a. Seasonal timing of reproduction in a tropical bird, the Seychelles Warbler: a field experiment using translocation. *J. Biol. Rhythm.* **11**: 333-346.

Komdeur, J. 1996b. Influence of age on reproductive performance in the Seychelles warbler. *Behav. Ecol.* **7**: 417-425.

Komdeur, J. 2003a. Daughters on request: about helpers and egg sexes in the Seychelles Warbler. *Proc. Roy. Soc. B-Biol. Sci.* **270**: 3-11.

Komdeur, J. 2003b. Adaptations and maladaptations to island living in the Seychelles Warbler. *Ornithol. Sci.* **2**: 79-88.

Komdeur, J. 2006. Variation in individual investment strategies among social animals. *Ethology* **112**: 729-747.

Komdeur, J. & Daan, S. 2005. Breeding in the monsoon: semi-annual reproduction in the Seychelles warbler *Acrocephalus sechellensis*. *J. Ornithol.* **146**: 305-313.

Komdeur, J. & Edelaar, P. 2001a. Male Seychelles warblers use territory budding to maximize lifetime fitness in a saturated environment. *Behav. Ecol.* **12**: 706-715.

Komdeur, J. & Edelaar, P. 2001b. Evidence that helping at the nest does not result in territory inheritance in the Seychelles warbler. *Proc. Roy. Soc. B-Biol. Sci.* **268**: 2007-2012.

Komdeur, J. & Kats, R.K.H. 1999. Predation risk affects trade-off between nest guarding and foraging in Seychelles warblers. *Behav. Ecol.* **10**: 648-658.

Komdeur, J. & Pels, M.D. 2005. Rescue of the Seychelles Warbler on Cousin Island, Seychelles: the role of habitat restoration. *Biol. Conserv.* **124**: 15-26.

Komdeur, J. & Richardson, D.S. 2007. Molecular ecology reveals the hidden complexities of the Seychelles Warbler. *Adv. Stud. Behav.* **37**: 147-187.

Komdeur, J., Daan, S., Tinbergen, J. & Mateman, C. 1997. Extreme adaptive modification in sex ratio of the Seychelles Warbler's eggs. *Nature* **385**: 522-525.

Komdeur, J., Magrath, M.J.L. & Krackow, S. 2002. Pre-ovulation control of hatchling sex ratio in the Seychelles Warbler *Proc. Roy. Soc. B-Biol. Sci.* **269**: 1067-1072.

Komdeur, J., Piersma, T., Kraaijeveld, K., Kraaijeveld-Smit, F. & Richardson, D.S. 2004. Why Seychelles Warblers fail to recolonize nearby islands: unwilling or unable to fly there? *Ibis* **146**: 298-302.

Koskimies, P. 1991a. *Acrocephalus schoenobaenus* – Schilfrohrsänger. In: Glutz von Blotzheim, U.N. & Bauer, K.M. eds. *Handbuch der Vögel Mitteleuropas*. Wiesbaden: Aula, pp. 291-340.

Koskimies, P. 1991b. *Acrocephalus dumetorum* – Buschrohrsänger. In: Glutz von Blotzheim, U.N. & Bauer, K.M. eds. *Handbuch der Vögel Mitteleuropas*. Wiesbaden: Aula, pp. 352-376.

Kozulin, A. & Flade, M. 1999. Breeding habitat, abundance and conservation status of the aquatic warbler (*Acrocephalus paludicola*) in Belarus. *Vogelwelt* **120**: 97-111.

Kozulin, A.V., Flade, M. & Gritschik, W.W. 1999. Fen mires and the benefit of mobility: a hypothesis for the origin of promiscuity in Aquatic Warbler (*Acrocephalus paludicola*). *Subbuteo* **2**: 11-17.

Krebs, J.R. & Davies, N.B. 1993. *An introduction to behavioural ecology*. Oxford: Blackwell Scientific Publications.

Kroodsma, D. & Brewer, D. 2005. Family Troglodytidae (Wrens). In: del Hoyo, J., Elliott, A. & Christie, D.A. eds. *Handbook of the birds of the world*. Vol. 10: *Cuckoo Shrikes to Thrushes*. Barcelona: Lynx Edicions, pp. 356-401.

Kroodsma, D.E. & Verner, J. 1997. Marsh wren (*Cistothorus palustris*). In: Poole, A. & Gille, F. eds. *The birds of North America*, No. 308. Philadelphia Pa: The Academy of Natural Sciences, pp. 1-32.

Krüger, O. 2007. Cuckoos, cowbirds and hosts: adaptations, trade-offs and constraints. *Phil. Trans. Soc. B-Biol. Sci.* **362**: 1873-1886.

Krüger, O., Sorenson, M.D. & Davies, N.B. 2009. Does coevolution promote species richness in parasitic cuckoos? *Proc. Roy. Soc. B-Biol. Sci.* **276**: 3871-3879.

Kullberg, C., Jakobsson, S. & Fransson, T. 2000. High migratory fuel loads impair predator evasion in sedge warblers. *Auk* **117**: 1034-1038.

Kullberg, C., Henshaw, I., Jakobsson, S., Johansson, P. & Fransson, T. 2007. Fuelling decisions in migratory birds: geomagnetic cues override the seasonal effect. *Proc. Roy. Soc. B-Biol. Sci.* **274**: 2145-2151.

Lachmann, L. 2008. [Modern conservation: the EU Life project for the aquatic warbler.] *Falke* **55**: 430-435.

Lachmann, L. & Tanneberger, F. eds. (in prep). Aquatic Warbler conservation handbook.

Lack, D. 1968. *Ecological adaptations for breeding in birds*. London: Methuen.

Lack, P. 1985. The ecology of the land-birds of Tsavo East National Park, Kenya. *Scopus* **9**: 57-96.

Langefors, A., Hasselquist, D. & von Schantz, T. 1998. Extra-pair fertilizations in the sedge warbler. *J. Avian Biol.* **29**: 134-144.

Langmore, N.E., Hunt, S. & Kilner, R.M. 2003. Escalation of a coevolutionary arms race through host rejection of brood parasitic young. *Nature* **422**: 157-160.

Langmore, N.E., Stevens, M., Maurer, G., Heinsohn, R., Hall, M.L., Peters, A. & Kilner, R.M. 2011. Visual mimicry of host nestlings by cuckoos. *Proc. Roy. Soc. B-Biol. Sci.* **278**: 2455-2463.

Latchininsky, A.V. 2008. Grasshopper outbreak challenges conservation status of a small Hawaiian Island. *J. Insect Conserv.* **12**: 343-357.

Laußmann, H. & Leisler, B. 2001. The function of inter- and intraspecitic territoriality in warblers of the genus *Acrocephalus*. *Biosyst. Ecol.* **18**: 87-110.

Lee, P.L.M., Clayton, D.H., Griffiths, R. & Page, R.D.M. 1996. Does behaviour reflect phylogeny in swiftlets (Aves: Apodidae)? A test using cytochrome b mitochondrial DNA sequences. *Proc. Natl Acad. Sci. USA* **93**: 7091-7096.

Leisler, B. 1970. Vergleichende Untersuchungen zur ökologischen und systematischen Stellung des Mariskensängers (*Acrocephalus (Lusciniola) melanopogon*, Sylviidae), ausgeführt am Neusiedler See. [Dissertation] Universität Wien.

Leisler, B. 1972a. Die Mauser des Mariskensängers (*Acrocephalus melanopogon*) als ökologisches Problem. *J. Ornithol.* **113**: 191-206.

Leisler, B. 1972b. Artmerkmale am Fuß adulter Teich- und Sumpfrohrsänger (*Acrocephalus scirpaceus, A. palustris*) und ihre Funktion. *J. Ornithol.* **113**: 366-373.

Leisler, B. 1975. Die Bedeutung der Fußmorphologie für die ökologische Sonderung mitteleuropäischer Rohrsänger (*Acrocephalus*) und Schwirle (*Locustella*). *J. Ornithol.* **116**: 117-153.

Leisler, B. 1980. Morphological aspects of ecological specialization in bird genera. *Ökol. Vögel* **2**: 199-220.

Leisler, B. 1981. Die ökologische Einnischung der mitteleuropäischen Rohrsänger (*Acrocephalus*, Sylviinae). 1: Habitattrennung. *Vogelwarte* **31**: 45-74.

Leisler, B. 1985. Öko-ethologische Voraussetzungen für die Entwicklung von Polygamie bei Rohrsängern (*Acrocephalus*). *J. Ornithol.* **126**: 357-381.

Leisler, B. 1988a. Interspecific interactions among European marsh-nesting passerines. In: Ouellet, H. ed. *Acta XIX Congressus Internationalis Ornithologici*. Ottawa, Ont: For National Museum of Natural Sciences by University of Ottawa Press, pp. 2635-2644.

Leisler, B. 1988b. Intra- und interspezifische Aggression bei Schilf- und Seggenrohrsänger (*Acrocephalus schoenobaenus, A. paludicola*): ein Fall von akustischer Verwechslung? *Vogelwarte* **34**: 281-290.

Leisler, B. 1991a. *Acrocephalus melanopogon* – Mariskensänger. In: Glutz von Blotzheim, U.N. & Bauer, K.M. eds. *Handbuch der Vögel Mitteleuropas*. Wiesbaden: Aula, pp. 217-252.

Leisler, B. 1991b. *Acrocephalus arundinaceus* – Drosselrohrsänger. In: Glutz von Blotzheim, U.N. & Bauer, K.M. eds. *Handbuch der Vögel Mitteleuropas*. Wiesbaden: Aula, pp. 486-539.

Leisler, B. & Catchpole, C.K. 1992. The evolution of polygamy in European reed warblers of the genus *Acrocephalus*: a comparative approach. *Ethol. Ecol. Evol.* **4**: 225-243.

Leisler, B. & Dyrcz, A. 1988. Introduction to Symposium 47: Adaptations of marsh-nesting passerines. In: Ouellet, H. ed. *Acta XIX Congressus Internationalis Ornithologici*. Ottawa, Ont: For National Museum of Natural Sciences by University of Ottawa Press, pp. 2594-2596.

Leisler, B. & Wink, M. 2000. Frequencies of multiple paternity in three *Acrocephalus* species (Aves Sylviidae) with different mating systems (*A. palustris, A. arundinaceus, A. paludicola*). *Ethol. Ecol. Evol.* **12**: 237-249.

Leisler, B. & Winkler, H. 1979: Zur Unterscheidung von Teich- und Sumpfrohrsänger. *Vogelwarte* **30**: 44-48.

Leisler, B. & Winkler, H. 1985. Ecomorphology. In: Johnston, R.F. ed. *Current ornithology*. Vol. 2. New York: Plenum Press, pp. 155-186.

Leisler, B. & Winkler, H. 2001. Morphological convergence in papyrus dwelling passerines. *Ostrich* Suppl. **15**: 24-29.

Leisler, B. & Winkler, H. 2003. Morphological consequences of migration in passerines. In: Berthold, P., Gwinner, E. & Sonnenschein, E. eds. *Avian migration*. Berlin: Springer-Verlag, pp. 175-186.

Leisler, B. & Winkler, H. 2009. On the evolution of characters associated with migration. *Contrib. Nat. Hist.* **12**: 875-892.

Leisler, B., Ley, H.W. & Winkler, H. 1989. Habitat, behavior and morphology of *Acrocephalus* warblers: an integrated analysis. *Ornis Scand.* **20**: 181-186.

Leisler, B., Beier, J., Heine, G. & Siebenrock, K.H. 1995. Age and other factors influencing mating status in German Great Reed Wablers *Acrocephalus arundinaceus*. *Jpn. J. Ornithol.* **44**: 169-180.

Leisler, B., Heidrich, P., Schulze-Hagen, K. & Wink, M. 1997. Taxonomy and phylogeny of reed warblers (genus *Acrocephalus*) based on mtDNA sequences and morphology. *J. Ornithol.* **138**: 469-496.

Leisler, B., Beier, J., Staudter, H. & Wink, M. 2000. Variation in extra-pair paternity in the polygynous Great Reed Warbler (*Acrocephalus arundinaceus*). *J. Ornithol.* **141**: 77-84.

Leisler, B., Winkler, H. & Wink, M. 2002. Evolution of breeding systems in acrocephaline warblers. *Auk* **119**: 335-348.

Leitner, S., Nicholson, J., Leisler, B., DeVoogd, T.J. & Catchpole, C.K. 2002. Song and the song control pathway in the brain can develop independently of exposure to song in the sedge warbler. *Proc. Roy. Soc. B-Biol. Sci.* **269**: 2519-2524.

Lemaire, F. 1974. Le chant de la Rousserolle verderolle (*Acrocephalus palustris*) : étendue du répertoire imitatif, construction rythmique et musicalité. *Gerfaut* **64**: 3-28.

Lemaire, F. 1975a. Dialectical variations in the imitative song of the marsh warbler (*Acrocephalus palustris*) in western and eastern Belgium. *Gerfaut* **65**: 95-106.

Lemaire, F. 1975b. Le chant de la Rousserolle verderolle (*Acrocephalus palustris*): fidélité des imitations et relations avec les espèces limitées et avec les congenères. *Gerfaut* **65**: 95-106.

Lemaire, F. 1977. Mixed song, interspecific competition and hybridisation in the reed and marsh warblers (*Acrocephalus scirpaceus* and *palustris*). *Behaviour* **63**: 215-240.

Leonard, M.L. & Picman, J. 1986. Why are nesting marsh wrens and yellow-headed blackbirds spatially segregated? *Auk* **103**: 135-140.

Ley, H-W. 1988. Verhaltensontogenese der Habitatwahl beim Teichrohrsänger (*Acrocephalus scirpaceus*). *J. Ornithol.* **129**: 287-297.

Leyequien, E., de Boer, W.F. & Cleef, A. 2007. Influence of body size on coexistence of bird species. *Ecol. Res.* **22**: 735-741.

Lifjeld, J.T., Marthinsen, G., Myklebust, M., Dawson, D.A. & Johnsen, A. 2010. A wild Marsh Warbler x Sedge Warbler hybrid (*Acrocephalus palustris* x *A. schoenobaenus*) in Norway documented with molecular markers. *J. Ornithol.* **151**: 513-517.

Lindholm, A.K. 1999. Brood parasitism by the cuckoo on patchy reed warbler populations in Britain. *J. Anim. Ecol.* **68**: 293-309.

Lindholm, A. & Aalto, T. 2005. The calls of Sykes's and booted warblers. *Birding World* **18**: 395-396.

Lindholm, A., Bensch, S., Dowsett-Lemaire, F., Forsten, A. & Kärkkäinen, H. 2007. Hybrid Marsh x Blyth´s Reed Warbler with mixed song in Finland in June 2003. *Dutch Birding* **29**: 223-231.

Lockwood, R., Swaddle, J.P. & Rayner, J.M.V. 1998. Avian wingtip shape reconsidered: wingtip shape indexes and morphological adaptations to migration. *J. Avian Biol.* **29**: 273-292.

Loffredo, C.A. & Borgia, G. 1986. Sexual selection, mating systems, and the evolution of avian acoustical displays. *Am.Nat.* **128**: 773-794.

Lomolino, M.V. 2005. Body size evolution in insular vertebrates: generality of the island rule. *J. Biogeogr.* **32**: 1683-1699.

Long, R. 1975. Mortality of Reed Warblers in Jersey. *Ring. Migr.* **1**: 28-32.

Lotem, A. 1993. Learning To recognize nestlings Is maladaptive for Cuckoo *Cuculus canorus* hosts. *Nature* **362**: 743-745.

Lotem, A., Nakamura, H. & Zahavi, A. 1992. Rejection of Cuckoo eggs in relation to host age: a possible evolutionary equilibrium. *Behav. Ecol.* **3**: 128-132.

Louette, M., Herremans, M., Bijnens, L. & Janssens, L. 1988. Taxonomy and evolution in the brush warblers *Nesillas* on the Comoro Islands. *Tauraco* **1**: 110-129.

MacLean, I., Musina, J., Nalianya, N., Mahood, S., Martin, R. & Byuaruhanga, A. 2003. Systematics, distribution and vocalisation of Papyrus Yellow Warbler (*Chloropeta gracilirostris*). *Bull. Afr. Bird Club* **10**: 94-100.

Madden, J.R. & Davies, N.B. 2006. A host-race difference in begging calls of nestling cuckoos *Cuculus canorus* develops through experience and increases host provisioning. *Proc. Roy. Soc. B-Biol. Sci.* **273**: 2343-2351.

Mädlow, W. 1992. Zur Habitatwahl auf dem Wegzug rastender Kleinvögel in einer norddeutschen Uferzone. [Diplom] Berlin: Freie Universität.

Magrath, M.J.L., Vedder, O., van der Velde, M. & Komdeur, J. 2009. Maternal effects contribute to the superior performance of extra-pair offspring. *Curr. Biol.* **19**: 792-797.

Manne, L.L., Brooks, T.M. & Pimm, S.L. 1999. Relative risk of extinction of passerine birds on continents and islands. *Nature* **399**: 258-261.

Maragna, P. & Pesente, M. 1997. Complete moult confirmed in a Great Reed Warbler (*Acrocephalus arundinaceus*) population breeding in northern Italy. *Ring. Migr.* **18**: 57-58.

Marler, P. 1960. Bird songs and mate selection. In: Lanyon, W.E. & Tavolga, W.N. eds. *Animal sounds and communication*. Washington: American Institute of Biological Sciences, pp. 348-367.

Marova, I.M., Valchuk, O.P., Kvartalnov, P. & Ivanitskii, V. 2005. On the taxonomic position and evolutionary interrelations of the Thick-Billed Warbler (*Phragmaticola aeedon*). *Alauda* **73**: 308.

Marshall, R.C., Buchanan, K.L. & Catchpole, C.K. 2003. Sexual selection and individual genetic diversity in a songbird. *Proc. Roy. Soc. B-Biol. Sci.* **270**: S248-S250.

Marshall, R.C., Buchanan, K.L. & Catchpole, C.K. 2007. Song and female choice for extrapair copulations in the sedge warbler, *Acrocephalus schoenobaenus*. *Anim. Behav.* **73**: 629-635.

Martens, J. 1996. Vocalizations and speciation of Palearctic birds. In: Kroodsma, D.E. & Miller, E.H. eds. *Ecology and evolution of acoustic communication in birds*. Ithaca, NY: Cornell University Press, pp. 221-240.

Martin, T.E. 1993. Nest predation and nest sites. *Bioscience* **43**: 523-532.

Martin, P.R. & Martin, T.E. 2001a. Ecological and fitness consequences of species coexistence: a removal experiment with Wood Warblers. *Ecology* **82**: 189-206.

Martin, P.R. & Martin, T.E. 2001b. Behavioral interactions between coexisting species: song playback experiments with Wood Warblers. *Ecology* **82**: 207-218.

Martin, T.E., Auer, S.K., Bassar, R.D., Niklison, A.M. & Lloyd, P. 2007. Geographic variation in avian incubation periods and parental influences on embryonic temperature. *Evolution* **61** :2558-2569.

Martins, E.P. & Hansen, T.F. 1997. Phylogenies and the comparative method: a general approach to incorporating phylogenetic information into the analysis of interspecific data. *Am. Nat.* **149**: 646-667.

Mason, E.A. 1938. Determining sex in living birds. *Bird Banding* **9**: 46-48.

Matson, K.D. 2006. Are there differences in immune function between continental and insular birds? *Proc. Roy. Soc. B-Biol. Sci.* **273**: 2267-2274.

Matyjasiak, P. 2005. Birds associate species-specific acoustic and visual cues: recognition of heterospecific rivals by male blackcaps. *Behav. Ecol.* **16**: 467-471.

Mayr, E. 1941. The origin and the history of the bird fauna of Polynesia. *In: Proceedings of the 6th Pacific Science Congress of the Pacific Science Association, University of California, Berkeley, Stanford University, and San Francisco, 24 July-12 August 1939.* Vol 4, pp. 197-216.

Mayr, E. 1989. A new classification of the living birds of the world. *Auk* **106**: 508-516.

Mayr, E. 1990. Plattentektionik und die Geschichte der Vogelfaunen. In: van den Elzen, R., Schuchmann, K-L. & Schmidt-Koenig, K. eds., *Current topics in avian biology. Proceedings of the International Centenial Meeting of the Deutsche Ornithologen-Gesellschaft, Bonn, 1988.* Bonn: Verlag der D.O.G., pp. 1-17.

Mayr, E. 2001. What evolution is. New York: Basic Books.

Mayr, E. & Diamond, J. 2001. *The birds of Northern Melanesia.* New York: Oxford University Press.

Mayr, E. & Cottrell, G.W. (eds) 1986. *Check-list of birds of the world: a continuation of the work of James L. Peters.* Vol. XI. Cambridge Mass: Museum of Comparative Zoology.

McGlynn, T.P. 2008. Natural history and evolution as common themes in biology. *Nat. Hist. Educ.* **70**: 109-111.

McNab, B.K. 2002. Minimizing energy expenditure facilitates vertebrate persistence on oceanic islands. *Ecol. Lett.* **5**: 693-704.

Meise, W. 1976. Die Bedeutung der Oologie für die Systematik. In: Frith, H.J. & Calaby, J.H. eds. *Proceedings of the 16th International Ornithological Congress, Canberra, Australia, 12-17 August, 1974.* Canberra: Australian Academy of Sciences, pp. 208-216.

Merom, K., McCleery, R. & Yom-Tov, Y. 1999. Age-related changes in wing-length and body mass in the Reed Warbler (*Acrocephalus scirpaceus*) and Clamorous Reed Warbler (*A. stentoreus*). *Bird Study* **46**: 249-255.

Mettke-Hofmann, C. 1999. Niche expansion and exploratory behaviour on island: are they linked? In Adams, N.J. & Slotow, R.H. eds. *Proceedings of the 22nd International Ornithological Congress, Durban, South Africa, 1998.* Johannesburg: BirdLife South Africa, pp. 878-885.

Meyer, J-Y. & Florence, J. 1996. Tahiti´s native flora endangered by the invasion of *Miconia calvescens DC.* (Melastomataceae). *J. Biogeogr.* **23**: 775-781.

Milder, S.L. & Schreiber, R.W. 1989. The vocalizations of the Christmas Island Warbler (*Acrocephalus aequinoctialis*) an island endemic. *Ibis* **131**: 99-111.

Moksnes, A. & Røskaft, E. 1995. Egg-morphs and host preference in the common cuckoo (*Cuculus canorus*): an analysis of cuckoo eggs from European museum collections. *J. Zool.* **236**: 625-648.

Moksnes, A., Røskaft, E., Hagen, L.G., Honza, M., Mørk, C. & Olsen, P.H. 2000. Common cuckoo *Cuculus canorus* and host behaviour at reed warbler *Acrocephalus scirpaceus* nests. *Ibis* **142**: 247–258.

Møller, A.P. 1987. Intruders and defenders on avian breeding territories: the effect of sperm competition. *Oikos* **48**: 47-54.

Møller, A.P. 1990. Changes in the size of avian breeding territories in relation to the nesting cycle. *Anim. Behav.* **40**: 1070-1079.

Møller, A.P. 1994. Phenotype-dependent arrival time and its consequences in a migratory bird. *Behav. Ecol. Sociobiol.* **35**: 115-122.

Møller, A.P. 2000. Male parental care, female reproductive success, and extrapair paternity. *Behav. Ecol.* **11**: 161-168.

Møller, A.P., Saino, N., Adamik, P., Ambrosini, R., Antonov, A., Campobello, D., Stokke, B.G., Fossøy, F., Lehikoinen, E., Martin-Vivaldi, M., Moksnes, A., Moskát, C., Røskaft, E., Rubolini, D., Schulze-Hagen, K., Soler, M. & Shykoff, J.A. 2011. Rapid change in host use of the common cuckoo *Cuculus canorus* linked to climate change. *Proc. Roy. Soc. B-Biol. Sci.* **278**: 733-738.

Molnar, B. 1944. The cuckoo in the Hungarian plain. *Aquila* **51**: 100-112.

Mönkkönen, M. 1995. Do migrant birds have more pointed wings? A comparative study. *Evol. Ecol.* **9**: 520-528.

Monnet, C., Thibault, J.C. & Varney, A. 1993. Stability and changes during the twentieth century in the breeding landbirds of Tahiti (Polynesia). *Bird Conserv. Int.* **3**: 261-280.

Moore, I.T., Bonier, F. & Wingfield, J.C. 2005a. Reproductive asynchrony and population divergence between two tropical bird populations. *Behav. Ecol.* **16**: 755-762.

Moore, F.R., Smith, R.J. & Sandberg, R. 2005b. Stopover ecology of intercontinental migrants: en route problems and consequences for reproductive performance. In: Greenberg, R. & Marra, P.P. eds. *Birds of two worlds: the ecology and evolution of migration.* Baltimore: Johns Hopkins University Press, pp. 251-261.

Moreau, R.E. 1972. *The Palaearctic-African bird migration systems.* London: Academic Press.

Morin, M.P., Conant, S. & Conant, P. 1997. Laysan and Nihoa Millerbird (*Acrocepalus familiaris*). In: Poole, A. & Gill, F. eds. *The Birds of North America.* No. 302. Philadelphia: The Academy of Natural Sciences.

Morse, D.H. 1989. *American warblers. An ecological and behavioural perspective.* Cambridge, Mass: Harvard University Press.

Morton, E.S. 1975. Ecological sources of selection on avian sounds. *Am. Nat.* **108**: 17-34.

Mosher, S.M. & Fancy, S.G. 2002. Description of nests, eggs and nestlings of the endangered Nightingale Reed-Warbler on Saipan, Micronesia. *Willson Bull.* **114**: 38991.

Moskát, C. & Honza, M. 2000. Effect of nest and nest site characteristics on the risk of cuckoo *Cuculus canorus* parasitism in the great reed warbler *Acrocephalus arundinaceus*. *Ecography* **23**: 335-341.

Moskát, C. & Honza, M. 2002. European Cuckoo *Cuculus canorus* parasitism and host's rejection behaviour in a heavily parasitized Great Reed Warbler *Acrocephalus arundinaceus* population. *Ibis* **144**: 614-622.

Moskát, C., Szentpéteri, J. & Barta, Z. 2002. Adaptations by great reed warblers to brood parasitism: a comparison of populations in sympatry and allopatry with the common cuckoo. *Behaviour* **139**: 1313-1329.

Moskát , C., Barta, Z., Hauber, M.E. & Honza, M. 2006. High synchrony of egg laying in common cuckoos (*Cuculus canorus*) and their great reed warbler (*Acrocephalus arundinaceus*) hosts. *Ethol. Ecol. Evol.* **18**: 159-167.

Moskát, C., Hansson, B., Barabás, L., Bartol, I. & Karcza, Z. 2008. Common cuckoo *Cuculus canorus* parasitism, antiparasite defence and gene flow in closely located populations of great reed warblers *Acrocephalus arundinaceus*. *J. Avian Biol.* **39**: 663-671.

Moskát , C., Hauber, M.E., Aviles, J.M., Ban, M., Hargitai, R. & Honza, M. 2009. Increased host tolerance of multiple cuckoo eggs leads to higher fledging success of the brood parasite. *Anim. Behav.* **77**: 1281-1290.

Moskát , C., Rosendaal, E.C., Boers, M., Zoelei, A., Ban, M. & Komdeur, J. 2011. Post-ejection nest-desertion of common cuckoo hosts: a second defense mechanism or avoiding reduced reproductive success? *Behav. Ecol. Sociobiol.* **65**: 1045-1053.

Mountjoy, J. & Leger, D.W. 2001. Vireo song repertoires and migratory distance: three sexual selection hypotheses fail to explain the correlation. *Behav. Ecol.* **12** :98-102.

Mueller-Dombois, D. & Fosberg, R.E. 1998. *Vegetation of the tropical Pacific Islands.* Berlin: Springer.

Mukhin, A. 1999. Nocturnal restlessness in caged juvenile reed warblers *Acrocephalus scirpaceus*. *Avian Ecol. Behav.* **3**: 91-97.

Mukhin, A. 2004. Night movements of young Reed Warblers (*Acrocephalus scirpaceus*) in summer: is it postfledging dispersal? *Auk* **121**: 203-209.

Mukhin, A. & Grinkevich, V. 2007. Hidden movements of Reed Warblers - nocturnal life of diurnal birds. In: Hengsberger, R., Hofmann, H., Wagner, R. & Winkler, H. eds. *Abstracts of the 6th European Ornithologists' Union Conference, 24-29 August 2007, Vienna, Austria.* Vienna: Austrian Academy of Sciences, p. 11.

Mukhin, A., Kosarev, V. & Ktitorov, P. 2005 Nocturnal life of young songbirds well before migration. *Proc. Roy. Soc. B-Biol. Sci.* **272**: 1535-1539.

Mukhin, A., Chernetsov, N. & Kishkinev, D. 2008. Acoustic information as a distant cue for habitat recognition by nocturnally migrating passerines during landfall. *Behav. Ecol.* **19**: 716-723.

Mukhin, A., Grinkevich, V. & Helm, B. 2009. Under cover of darkness: nocturnal life of diurnal birds. *J. Biol. Rhythm.* **24**: 225-231.

Murphy, R.C. & Mathews, G.M. 1928. Birds collected during the Whitney South Sea Expedition, V. *Am. Mus. Novit.* **337**: 9-18.

Murphy, R.C. & Mathews, G.M. 1929. Birds collected during the Whitney South Sea Expedition, VI. *Am. Mus. Novit.* **350**: 1-20.

Murray, B.G. 1981. The origins of adaptive interspecific territorialism. *Biol. Rev.* **50**: 1-22.

Muthuri, F.M., Jones, M.B. & Ibanmba, S.K. 1989. Primary productivity of papyrus (*Cyperus papyrus*) in a tropical swamp, Lake Naivasha, Kenya. *Biomass* **18**: 1-14.

Nagy, L.R. & Holmes, R.T. 2004. Factors influencing fecundity in migratory songbirds: is nest predation the most important? *J. Avian Biol.* **35**: 487-491.

Nakamura, H. 1990. Brood parasitism by the Cuckoo *Cuculus canorus* in Japan and the start of new parasitism on the Azure-Winged Magpie *Cyanopica cyana*. *Jpn. J. Ornithol.* **39**: 1-18.

Nakamura, H. & Miyazawa, Y. 1997. Movements, space use and social organization of radio tracked common cuckoos during the breeding season in Japan. *Jpn. J. Ornithol.* **46**: 23-54.

Nemeth, E., Dabelsteen, T., Pedersen, S.B. & Winkler, H. 2006. Rainforests as concert halls for birds: are reverberations improving sound transmission of long song elements? *J. Acoust. Soc. Am.* **119**: 620-626.

Neudorf, D.L.H. 2004. Extrapair paternity in birds: understanding variation among species. *Auk* **121**: 302-307.

Nicholson, J.S., Buchanan, K.L., Marshall, R.C. & Catchpole, C.K. 2007. Song sharing and repertoire size in the sedge warbler, *Acrocephalus schoenobaenus*: changes within and between years. *Anim.Behav.* **74**: 1585-1592.

Niemi, G.J. 1985. Patterns of morphological evolution in bird genera of new world and old world peatlands. *Ecology* **66**: 1215-1228.

Niethammer, G. 1937. *Handbuch der deutschen Vogelkunde.* Bd. 1. Leipzig: Akademische Verlagsgesellschaft.

Nikolaus, G. 1983. A bird cemetery in the Nubian desert, Sudan. *Scopus* **7**: 48.

Nikolaus, G. 1990. Shrikes, Laniidae, feeding on Marsh Warblers (*Acrocephalus palustris*) during migration. *Scopus* **14**: 26-28.

Nisbet, I.C.T. & Medway, L. 1972. Dispersion population ecology and migration of eastern great reed warblers (*Acrocephalus orientalis*) wintering in Malaysia. *Ibis* **114**: 451-494.

Nishiumi, I. 1998. Brood sex ratio is dependent on female mating status in polygynous great reed warblers. *Behav. Ecol. Sociobiol.* **44**: 9-14.

Nishiumi, I., Yamagishi, S., Maekawa, H. & Shimoda, C. 1996. Paternal expenditure is related to brood sex ratio in polygynous great reed warblers. *Behav. Ecol. Sociobiol.* **39**: 211-217.

Norman, S.C. 1997. Juvenile wing shape, wing moult and weight in the family sylviidae. *Ibis* **139**: 617-630.

Noss, R.F. 1996: The naturalists are dying off. *Conserv. Biol.* **10**: 1-3.

Nowakowski, J.J. 1994. Long term variability of wing shape in *Acrocephalus scirpaceus* population. *J. Ornithol.* **135**: 55.

Nowakowski, J.J. 2000. Long-term variability of wing length in a population of the Reed Warbler *Acrocephalus scirpaceus*. *Acta Ornithol.* **35**: 173-181.

Nowakowski, J.J. & Wojciechowski, Z. 2002. What determines long-term variability of wing length in a population of the Swallow *Hirundo rustica* and the Reed Warbler *Acrocephalus scirpaceus*? *Ecol. Quest.* **2**: 79-87.

Nowicki, S. & Searcy, W.A. 2005. Song and mate choice in birds: how the development of behavior helps us understand function. *Auk* **122**. 1-14.

Nowicki, S., Hasselquist, D., Bensch, S. & Peters, S. 2000. Nestling growth and song repertoire size in great reed warblers: evidence for song learning as an indicator mechanism in mate choice. *Proc. Roy. Soc. B-Biol. Sci.* **267**: 2419-2424.

Ohlson, J., Fjeldså, J. & Ericson, P.G.P. 2008. Tyrant flycatchers coming out in the open: phylogeny and ecological radiation of Tyrannidae (Aves, Passeriformes). *Zool. Scr.* **37**: 315-335.

Øien, I.J., Honza, M., Moksnes, A. & Røskaft, E. 1996. The risk of parasitism in relation to the distance from reed warbler nests to cuckoo perches. *J. Anim. Ecol.* **65**: 147-153.

Øien, I.J., Moksnes, A., Røskaft, E. & Honza, M. 1998. Costs of Cuckoo *Cuculus canorus* parasitism to Reed Warblers *Acrocephalus scirpaceus*. *J. Avian Biol.* **29**: 209-215.

Olesen, J.M. & Valido, A. 2003. Lizards as pollinators and seed dispersers: an island phenomenon. *Trends Ecol. Evol.* **18**: 177-181.

Olson, S.L. 1989. Extinction on islands: man as a catastrophe. In: Western, D. & Pearl, M.C. eds. *Conservation biology for the next century*. New York, Oxford: Oxford University Press, pp. 50-55.

Olson, S.L. & James, H.F. 1982. Fossil birds from the Hawaiian Islands: evidence for wholesale extinction by man before western contact. *Science* **217**: 633-635.

Olson, S.L., Irestedt, M., Ericson, P.G.P. & Fjeldså, J. 2005. Independent evolution of two Darwinian marsh-dwelling Ovenbirds (Furnariidae:Limnornis, Limnoctites). *Ornitol. Neotrop.* **16**: 347-359.

Opaev, A.S., Marova, I.M. & Ivanitskii, V.V. 2005. The comparative study of vocalizations of the western *Acrocephalus arundinaceus* and eastern *A. orientalis* great reed warblers. *Abstracts of the 5th Conference of the European Ornithologists' Union, 19-25 August 2005, Strasbourg, France*, p. 314.

Opaev, A.S., Marova, I.M. & Ivanitskii, V.V. 2009. Morphological differentiation and geographic variation in Great (*Acrocephalus arundinaceus*), Oriental (*Acrocephalus orientalis*), and Clamorous (*Acrocephalus stentoreus*) Reed-Wablers (Sylviidae, Passeriformes). *Zool. Zh.* **88**: 871-882.

Orians, G.H. 1961. The ecology of Blackbird (*Agelaius*) social systems. *Ecol. Monogr.* **31**: 285-312.

Orians, G.H. 1969. On the evolution of mating systems in birds and mammals. *Am. Nat.* **103**: 589-603.

Orians, G.H. ed. 1980. *Some adaptations of marsh-nesting blackbirds*. Monographs in Population Biology, 14. Princeton, NJ: Princeton University Press.

Orians, G.H. 1985. *Blackbirds of the Americas*. Seattle and London: University of Washington Press.

Orians, G.H. 2000. Biodiversity and ecosystem processes in tropical ecosystems. *Rev. Biol.Trop.* **48**: 297-303.

Orians, G.H. & Willson, M.F. 1964. Interspecific territories of birds. *Ecology* **45**: 736-745.

Ostendorp, W. 1993. Schilf als Lebensraum. *Beih.Veröff. Natursch. Landschaftspfl. Bad.-Württ.* **68**: 173-280.

Ottosson, U., Bensch, S., Svensson, L. & Waldenström, J. 2005. Differentiation and phylogeny of the olivaceous warbler *Hippolais pallida* species complex. *J. Ornithol.* **146**: 127-136.

Pain, D.J., Green, R.E., Giebetaing, B., Kozulin, A., Poluda, A., Ottosson, U., Flade, M. & Hilton, G.M. 2004. Using stable isotopes to investigate migratory connectivity of the globally threatened aquatic warbler (*Acrocephalus paludicola*). *Oecologia* **138**: 168-174.

Pambour, B. 1990. Vertical and horizontal distribution of five wetland passerine birds during the postbreeding migration period in a reed-bed of the Camargue, France. *Ring. Migr.* **11**: 52-56.

Panov, E.N., Nepomnyashchikh, V.A. & Rubtsov, A.S. 2004. Organization of song in the sedge warbler, *Acrocephalus schoenobaenus* (Aves, Sylviidae). *Zool. Zh.* **83**: 464-479.

Paradis, E., Baillie, S.R., Sutherland, W.J. & Gregory, R.D. 1998. Patterns of natal and breeding dispersal in birds. *J. Anim. Ecol.* **67**: 518-536.

Pärt, T., Arlt, D.,Doligez, B.,Low, M. & Qvarnström, A. 2011. Prospectors combine social and environmental information to improve habitat selection in the subsequent year. *Abstracts of the 8th Conference of the European Ornithologists' Union, Riga, Latvia, 27-30 August*, p.295.

Patterson, C.B., Erckmann, W.J. & Orians, G.H. 1980. An experimental study of parental investment and polygyny in male blackbirds. *Am. Nat.* **116**: 757-769.

Peach, W., Baillie, S. & Underhill, L. 1991. Surival of British Sedge Warbler (*Acrocephalus schoenobaenus*) in relation to west African rainfall. *Ibis* **133**: 300-305.

Peach, W.J., Hanmer, D.B. & Oatley, T.B. 2001. Do southern African songbirds live longer than their European counterparts? *Oikos* **93**: 235-249.

Pearson, D.J. 1973. Moult of some Palaearctic Warblers wintering in Uganda. *Bird Study* **20**: 24-36.

Pearson, D.J. 1990. Palearctic passerine migrants in Kenya and Uganda: temporal and special pattens of their movements. In: Gwinner, E. ed. *Bird migration, physiology and ecophysiology*. Berlin, Heidelberg: Springer, pp. 44-58.

Pearson, D.J. & Backhurst, G.C. 1988. Characters and taxonomie position of Basra Reed Warbler. *Brit. Birds* **81**: 171-178.

Pearson, D.J. & Lack, P.C. 1992. Migration patterns and habitat use by passerine and near-passerine migrant birds in eastern Africa. *Ibis* **134**: 89-98.

Pearson, D.J., Ash, J.S. & Bensch, S. 2004. The identity of some *Hippolais* specimens from Eritrea and the United Arab Emirates examined by mtDNA analysis: a record of Sykes`s Warbler *H.rama* in Africa. *Ibis* **146**: 683-684.

Peintinger, M. & Schuster, S. 2005. Changes in first arrival dates of common migratory bird species in southwestern Germany. *Vogelwarte* **43**: 161-169.

Peiro, I.G. 2003. Intraspecific variation in the wing shape of the long-distance migrant Reed Warbler (*Acroephalus scirpaceus*): effects of age and distance of migration. *Ardeola* **50**: 31-37.

Peron, G., Henry, P.Y., Provost, P., Dehorter, O. & Julliard, R. 2007. Climate changes and post-nuptial migration strategy by two reedbed passerine. *Clim. Res.* **35**: 147-157.

Perrins, C.B. & Birkhead, T.R. 1983. *Avian ecology*. Glasgow, London: Blackie.

Persson, K-G. 1983. Rörsångare och Sävsångare. *Fåglar i Södra Vätterbygden* 22: 55-56.

Petrie, M. & Kempenaers, B. 1998. Extra-pair paternity in birds: explaining variation between species and populations. *Trends Ecol. Evol.* 13: 52-58.

Picman, J. 1988. Behavioral interactions between North American marsh-nesting blackbirds and marsh wrens and their influence on reproductive strategies of these passerines. In: Ouellet, H. ed. *Acta XIX Congressus Internationalis Ornithologici*. Ottawa, Ont: For National Museum of Natural Sciences by University of Ottawa Press, pp. 2624-2634.

Picman, J., Leonard, M. & Horn, A. 1988. Antipredation role of clumped nesting by marsh-nesting red-winged blackbirds. *Behav. Ecol. Sociobiol.* 22: 9-15.

Picman, J., Milks, M.L. & Leptich, M. 1993. Patterns of predation on passerine nests in marshes: effects of water depth and distance from edge. *Auk* 110: 89-94.

Pike, T.W. 2005. Sex ratio manipulation in response to maternal condition in pigeons: evidence for pre-ovulatory follicle selection. *Behav. Ecol. Sociobiol.* 58: 407-413.

Pike, T.W. & Petrie, M. 2006. Experimental evidence that corticosterone affects offspring sex ratios in quail. *Proc. Roy. Soc. B-Biol. Sci.* 273: 1093-1098.

Podos, J. & Nowicki, S. 2004. Beaks, adaptation, and vocal evolution in Darwin`s Finches. *Bioscience* 54: 501-510.

Polo, V. & Carrascal, L.M. 1999. Shaping the body mass distribution of passeriformes: habitat use and body mass are evolutionarily and ecologically related. *J. Anim. Ecol.* 68: 324-337.

Poot, M., Engelen, F. & van der Winden, J. 1999. A mixed breeding pair of Blyth's Reed Warbler *Acrocephalus dumetorum* and Marsh Warbler *A. palustris* near Utrecht in spring 1998. *Limosa* 72: 151-157.

Post, W. 1988. Reproductive success in New World marsh-nesting passerines. In: Ouellet, H. ed. *Acta XIX Congressus Internationalis Ornithologici*. Ottawa, Ont: For National Museum of Natural Sciences by University of Ottawa Press, pp. 2645-2667.

Post, W. & Greenlaw, J.S. 1982. Comparative costs of promiscuity and monogamy: a test of reproductive effort theory. *Behav. Ecol. Sociobiol.* 10: 101-107.

Poulin, B. & Lefebvre, G. 2002. Effect of winter cutting on the passerine breeding assemblage in French Mediterranean reedbeds. *Biodivers. Conserv.* 11: 1567-1581.

Poulin, B., Lefebvre, G. & Metref, S. 2000. Spatial distribution of nesting and foraging sites of two *Acrocephalus* warblers in a Mediterranean reedbed. *Acta Ornithol.* 35: 117-121.

Poulin, B., Lefebvre, G. & Mauchamp, A. 2002. Habitat requirements of passerines and reedbed management in southern France. *Biol. Conserv.* 107: 315-325.

Pozgayová, M., Procházka, P. & Honza, M. 2009. Sex roles in antiparasitic nest defence in two Common Cuckoo hosts. In: Keller, V. & O'Halloran, J. eds. *Abstracts of the 7th Conference of the European Ornithologists' Union Zürich, Switzerland, 21-26 August 2009*, pp. 135-136.

Pratt, H.D., Bruner, P.L. & Berrett, D.G. 1987. *A field guide to the birds of Hawaii and the Tropical Pacific*. Princeton, NJ: Princeton University Press.

Pravosudov, V.V., Sanford, K. & Hahn, T.P. 2007. On the evolution of brain size in relation to migratory behaviour in birds. *Anim. Behav.* 73: 535-539.

Preiszner, B. & Csörgö, T. 2008. Habitat preference of Sylviidae warblers in a fragmented wetland. *Acta Zool. Acad. Sci. Hung.* 54: 111-122.

Price, T. 2008. *Speciation in birds*. Greenwood Village, CO: Roberts and Company.

Price, T., Kirkpatrick, M. & Arnold, S.J. 1988. Directional selection and the evolution of breeding date in birds. *Science* 240: 798-799.

Procházka, P. 2000. Nest site selection and breeding biology in the reed warbler *Acrocephalus scirpaceus* in the littoral stands of the fish ponds in south Bohemia, Czech Republic. *Acta Ornithol.* 35: 123-128.

Procházka, P. & Reif, J. 2002. Movements and settling patterns of sedge warblers (*A.schoenobaenus*) in the Czech Republic and Slovakia: an analysis of ringing recoveries. *Ring* 24: 3-13.

Procházka, P., Hobson, K.A., Karcza, Z. & Kralj, J. 2008. Birds of a feather winter together: migratory connectivity in the Reed Warbler *Acrocephalus scirpaceus*. *J. Ornithol.* 149: 141-150.

Procházka, P., Stokke, B.G., Jensen, H., Fainova, D., Bellinvia, E., Fossøy, F., Vikan, J.R., Bryja, J. & Soler, M. 2011. Low genetic differentiation among reed warbler *Acrocephalus scirpaceus* populations across Europe. *J. Avian Biol.* 42: 103-113.

Prokesova, J. & Kocian, L. 2004. Habitat selection of two *Acrocephalus* warblers breeding in reed beds near Malacky (Western Slovakia). *Biologia* 59: 637-644.

Prum, R.O. 1998. Phylogeny and social behaviour evolution of the cotingas (Cotingidae). *Ostrich* 69: 411.

Pulliam, H.R. & Danielson, B.J. 1991. Sources, sinks, and habitat selection: a landscape perspective on population dynamics. *Am. Nat.* 137 (Suppl.): 50-66.

Raach, A. & Leisler, B. 1989. Auswirkung der Jugenderfahrung auf die Wahl von Habitatstrukturen und auf das Erkundungsverhalten des Mariskensängers (*Acrocephalus melanopogon*). *J. Ornithol.* 130: 256-259.

Raess, M. 2008. Continental efforts: migration speed in spring and autumn in an inner-Asian migrant. *J. Avian Biol.* 39: 13-18.

Read, A.F. & Weary, D.M. 1992. The evolution of bird song: comparative analyses. *Phil. Trans. Soc. B-Biol. Sci*.338: 165-187.

Redfern, C.P.T. 1978. Survival in relation to sex in Reed Warbler populations. *Wicken Fen Group Rep.* 10: 28-33.

Redfern, C.P.F. & Alker, P.J. 1996. Plumage development and post-juvenile moult in the Sedge Warbler (*Acrocephalus schoenobaenus*). *J. Avian Biol.* 27: 157-163.

Reed, T.M. 1982. Interspecific territoriality in the chaffinch and great tit on islands and the mainland of Scotland: playback and removal experiments. *Anim. Behav.* 30: 171-181.

Reichel, J.D., Wiles, G.J. & Glass, P.O. 1992. Island extinctions: the case of the endangered nightingale reed-warbler. *Wilson Bull.* 104: 44-54.

Remes, V. 2007. Avian growth and development rates and age-specific mortality: the roles of nest predation and adult mortality. *J. Evol. Biol.* 20: 320-325.

Remsen, J.V. 2003. Family Furnariidae (Ovenbirds). In: del Hoyo, J., Elliott, A. & Christie, D.A. eds. *Handbook of the birds of the world*. Vol. 8: *Broadbills to Tapaculos*. Barcelona: Lynx Edicions, pp. 162-239.

Remsen, J.V. & Robinson, S. 1990. A classification scheme for foraging behavior of birds in terrestrial habitats. *Stud. Avian Biol.* 13: 144-160.

Reullier, J., Perez-Tris, J., Bensch, S. & Secondi, J. 2006. Diversity, distribution and exchange of blood parasites meeting at an avian moving contact zone. *Mol. Ecol.* 15: 753-763.

Rguibi-Idrissi, H., Julliard, R. & Bairlein, F. 2003. Variation in the stopover duration of Reed Warbler (*Acrocephalus scirpaceus*) in Morocco: effects of season, age and site. *Ibis* 145: 650-656.

Rice, J. 1978. Ecological relationships of two interspecifically territorial vireos. *Ecology* 59: 526-538.

Richardson, D.S., Jury, F.L., Blaakmeer, K., Komdeur, J. & Burke, T. 2001. Parentage assignment and extra-group paternity in a cooperative breeder: the Seychelles warbler (*Acrocephalus sechellensis*). *Mol. Ecol.* 10: 2263-2273.

Richardson, D.S., Burke, T. & Komdeur, J. 2002. Direct benefits and the evolution of female-biased cooperative breeding in Seychelles warblers. *Evolution* 56: 2313-2321.

Richardson, D.S., Komdeur, J. & Burke, T. 2003. Altruism and infidelity among warblers. *Nature* 422: 580.

Richardson, D.S., Komdeur, J. & Burke, T. 2004. Inbreeding in the Seychelles warbler: environment-dependent maternal effects. *Evolution* 58: 2037-2048.

Richardson, D.S., Komdeur, J., Burke, T. & von Schantz, T. 2005. MHC-based patterns of social and extra-pair mate choice in the Seychelles warbler. *Proc. Roy. Soc. B-Biol. Sci.* 272: 759-767.

Richardson, D.S., Burke, T. & Komdeur, J. 2007. Grandparent helpers: the adaptive significance of older, postdominant helpers in the Seychelles warbler. *Evolution* 61: 2790-2800.

Richman, A.D. & Price, T. 1992. Evolution of ecological differences in the Old World leaf warblers. *Nature* 355: 817-821.

Ricklefs, R.E. 1969. An analysis of nesting mortality in birds. *Smithson. Contrib. Zool.* 9: 1-48.

Ricklefs, R.E. 1980. Geographical variation in clutch size among passerine birds: Ashmole's hypothesis. *Auk* 97: 38-49.

Ricklefs, R.E. 2010. Evolutionary diversification, coevolution between populations and their antagonists, and the filling of niche space. *Proc. Natl Acad. Sci. USA* 107: 1265-1272.

Ricklefs, R.E. & Miles, D.B. 1994. Ecological and evolutionary inferences from morphology: an ecological perspective. In: Wainwright, P.C. & Reilly, S.M. eds. *Integrative organismal biology*. Chicago: University of Chicago Press, pp. 13-41.

Ricklefs, R.E. & Wikelski, M. 2002. The physiology/life history nexus. *Trends Ecol. Evol.* 17: 462-468.

Riegner, M.F. 2008. Parallel evolution of plumage pattern and coloration in birds: implications for defining avian morphospace. *Condor* 110: 599-614.

Rjabitzev, W.K. 2001. *Die Vögel des Ural, des Ural-Vorlandes und Westsibiriens*. Jekaterinburg: Verlag der Ural-Universität (in Russian).

Robinson, S.K. & Terborgh, J. 1995. Interspecific aggression and habitat selection by Amazonian birds. *J. Anim. Ecol.* 64: 1-11.

Robinson, W.D., Hau, M., Klasing, K.C., Wikelski, M., Brawn, J.D., Austin, S.H., Tarwater, C.E. & Ricklefs, R.E. 2010. Diversification of life histories in New World birds. *Auk* 127: 253-262.

Rodenhouse, N.L., Sherry, T.W. & Holmes, R.T. 1997. Site-dependent regulation of population size: a new synthesis. *Ecology* 78: 2025-2042.

Rodewald-Rudescu, L. 1974. *Das Schilfrohr (Phragmites communis trinius)*. Die Binnengewässer Bd 27. Stuttgart: E. Schweizerbart.

Rogge, D. 1959. Beobachtungen und Untersuchungen zur Brutbiologie des Schilfrohrsängers (*Acrocephalus schoenobaenus* L.). Potsdam: Päd. Hochschule. [Typescript]

Rolando, A. & Palestrini, C. 1989. Habitat selection and interspecific territoriality in sympatric warblers at two Italian marshland areas. *Ethol. Ecol. Evol.* 1: 169-183.

Root, R.B. 1967. The niche exploitation pattern of the Blue-Gray Gnatcatcher. *Ecol. Monogr.* 37: 317-347.

Roselaar, C.S. 1995. *Taxonomy, morphology, and distribution of the songbirds of Turkey: an atlas of biodiversity of Turkish passerine birds*. London: Pica Press.

Røskaft, E., Moksnes, A., Meilvang, D., Bicík, V., Jemelíková, J. & Honza, M. 2002a. No evidence for recognition errors in *Acrocephalus* warblers. *J. Avian Biol.* 33: 31-38.

Røskaft, E., Moksnes, A., Stokke, B.G., Bicík, V. & Moskát, C. 2002b. Aggression to dummy cuckoos by potential European cuckoo hosts. *Behaviour* 139: 613-628.

Røskaft, E., Takasu, F., Moksnes, A. & Stokke, B.G. 2006. Importance of spatial habitat structure on establishment of host defenses against brood parasitism. *Behav. Ecol.* 17: 700-708.

Round, P.D. & Rumsey, S.J. 2003. Habitat use, moult and biometrics in the Manchurian Reed Warbler (*Acrocephalus tangorum*) wintering in Thailand. *Ring. Migr.* 21: 215-221.

Round, P.D., Hansson, B., Pearson, D.J., Kennerley, P.R. & Bensch, S. 2007. Lost and found: the enigmatic large-billed reed warbler *Acrocephalus orinus* rediscovered after 139 years. *J. Avian Biol.* 38: 133-138.

Rowan, W. 1926. On photoperiodism, reproductive periodicity and the annual migrations of birds and certain fishes. *Proc. Boston Soc. Nat. Hist.* 38: 147-189.

Russell, E.M. 2000. Avian life histories: is extended parental care the southern secret? *Emu* 100: 377-399.

Russell, E.M., Yom-Tov, Y. & Geffen, E. 2004. Extended parental care and delayed dispersal: northern, tropical and southern passerines compared. *Behav. Ecol.* 15: 831-838.

Saether, B.E., Engen, S., Lande, R., Møller, A.P., Bensch, S., Hasselquist, D., Beier, J. & Leisler, B. 2004. Time to extinction in relation to mating system and type of density regulation in populations with two sexes. *J. Anim. Ecol.* 73: 925-934.

Saino, N. 1989. Breeding microhabitats of three sympatric Acrocephalinae species (Aves) in northwestern Italy. *Boll. Zool.* 56: 47-53.

Salewski, V., Almasi, B. & Schlageter, A. 2006. Nectarivory of Palearctic migrants at a stopover site in the Sahara. *Brit. Birds* 99: 299-305.

Salewski, V., Almasi, B., Heuman, A., Thoma, M. & Schlageter, A. 2007.

Agonistic behaviour of Palaearctic passerine migrants at a stopover site suggests interference competition. *Ostrich* 78: 349-355.

Samways, M.J., Hitchins, P.M., Bourquin, O. & Henwood, J. 2010. Restoration of a tropical island: Cousine Island, Seychelles. *Biodivers. Conserv.* 19: 425-434.

Sanderson, F.J., Donald, P.F., Pain, D.J., Burfield, I.J. & van Bommel, F.P.J. 2006. Long-term population declines in Afro-Palaearctic migrant birds. *Biol. Conserv.* 131: 93-105.

Schaefer, H.M., Naef-Daenzer, B., Leisler, B., Schmidt, V., Müller, J.K. & Schulze-Hagen, K. 2000. Spatial behaviour in the Aquatic Warbler (*Acrocephalus paludicola*) during mating and breeding. *J. Ornithol.* 141: 418-424.

Schaefer, T., Ledebur, G., Beier, J. & Leisler, B. 2006. Reproductive responses of two related coexisting songbird species to environmental changes: global warming, competition, and population sizes. *J. Ornithol.* 147: 47-56.

Schaub, M. & Jenni, L. 2000a. Body mass of six long-distance migrant passerine species along the autumn migration route. *J. Ornithol.* 141: 441-460.

Schaub, M. & Jenni, L. 2000b. Fuel deposition of three passerine bird species along the migration route. *Oecologia* 122: 306-317.

Schaub, M. & Jenni, L. 2001a. Stopover durations of three warbler species along their autumn migration route. *Oecologia* 128: 217-227.

Schaub, M. & Jenni, L. 2001b. Variation of fuelling rates among sites, days and individuals in migrating passerine birds. *Funct. Ecol.* 15: 584-594.

Schaub, M., Pradel, R., Jenni, L. & Lebreton, J.D. 2001. Migrating birds stop over longer than usually thought: an improved capture-recapture analysis. *Ecology* 82: 852-859.

Schaub, M., Jenni, L. & Bairlein, F. 2008. Fuel stores, fuel accumulation, and the decision to depart from a migration stopover site. *Behav. Ecol.* 19: 657-666.

Schlenker, R. 1988. Zum Zug der Neusiedlersee (Österreich)-Population des Teichrohrsängers (*Acrocephalus scirpaceus*) nach Ringfunden. *Vogelwarte* 34: 337-343.

Schluter, D. 1986. Tests for similarity and convergence of finch communities. *Ecology* 67: 1073-1085.

Schluter, D., Price, T., Mooers, A. & Ludwig, D. 1997. Likelihood of ancestor states in adaptive radiation. *Evolution* 51: 1699-1711.

Schmaljohann, H., Liechti, F. & Bruderer, B. 2007. Songbird migration across the Sahara: the non-stop hypothesis rejected! *Proc. Roy. Soc. B-Biol. Sci.* 274: 735-739.

Schmaljohann, H., Becker, P.J.J., Karaardic, H., Liechti, F., Naef-Daenzer, B. & Grande, C. 2011. Nocturnal exploratory flights, departure time, and direction in a migratory songbird. *J. Ornithol.* 152: 439-452.

Schmidly, D.J. 2005. What it means to be a naturalist and the future of natural history at American universities. *J. Mammol.* 86: 449-456.

Schmidt, E. & Hüppop, K. 2007. First observation and start of birdsong of 97 bird species in a community in the county of Parchim (Mecklenburg-Vorpommern) in the years 1963 to 2006. *Vogelwarte* 45: 27-58.

Schmidt, V., Schaefer, H.M. & Leisler, B. 1999. Song behaviour and range use in the polygamous Aquatic Warbler *Acrocephalus paludicola*. *Acta Ornithol.* 34: 209-213.

Schmidt, M.H., Lefebvre, G., Poulin, B. & Tscharntke, T. 2005. Reed

cutting affects arthropod communities, potentially reducing food for passerine birds. *Biol. Conserv.* 121: 157-166.

Schottler, B. 1995. Songs of blue tits *Parus caeruleus palmensis* from La Palma (Canary Islands): a test of hypotheses. *Bioacoustics* 6: 135-152.

Schreiber, R.W. 1979. The egg and nest of the Bokikokiko *Acrocephalus aequinoctialis*. *Bull. Brit. Ornithol. Club* 99:120-124.

Schulze-Hagen, K. 1975. Habitat und Bruterfolg beim Sumpfrohrsänger (*Acrocephalus palustris*) im Rheinland. [Diplom] Bonn: Zoologischen Institut der Rheinischen Friedrich-Wilhelms Universität.

Schulze-Hagen, K. 1984. Bruterfolg des Sumpfrohrsängers (*Acrocephalus palustris*) in Abhängigkeit von der Nistplatzwahl. *J. Ornithol.* 125: 201-208.

Schulze-Hagen, K. 1991a. *Acrocephalus paludicola* – Seggenrohrsänger. In: Glutz von Blotzheim, U.N. & Bauer, K.M. eds. *Handbuch der Vögel Mitteleuropas*. Wiesbaden: Aula, pp. 252-291.

Schulze-Hagen, K. 1991b. *Acrocephalus palustris* – Sumpfrohrsänger. In: Glutz von Blotzheim, U.N. & Bauer, K.M. eds. *Handbuch der Vögel Mitteleuropas*. Wiesbaden: Aula, pp. 377-433.

Schulze-Hagen, K. 1991c. *Acrocephalus scirpaceus* – Teichrohrsänger. In: Glutz von Blotzheim, U.N. & Bauer, K.M. eds. *Handbuch der Vögel Mitteleuropas*. Wiesbaden: Aula, pp. 433-486.

Schulze-Hagen, K. 1992. Parasitierung und Brutverluste durch den Kuckuck (*Cuculus canorus*) bei Teich- und Sumpfrohrsänger (*Acrocephalus scirpaceus, A. palustris*) in Mittel- und Westeuropa. *J. Ornithol.* 133: 237-249.

Schulze-Hagen, K. 2004. Allmenden und ihr Vogelreichtum - Wandel von Landschaft, Landwirtschaft und Avifaunen in den letzten 250 Jahren. *Charadrius* 40: 97-121.

Schulze-Hagen, K. & Flinks, H. 1989. Nestlingsnahrung von Sumpfrohrsängern *Acrocephalus palustris*. *Vogelwelt* 110: 112-125.

Schulze-Hagen, K. & Leisler, B. (in press). Von der Einehe zur Keinehe – Reproduktionsbiologische Historie(n) eines Unscheinbaren, des Seggenrohrsängers. *Ökol. Vögel* 34.

Schulze-Hagen, K. & Sennert, G. 1990. Nestverteidigung bei Teich- und Sumpfrohrsänger (*Acrocephalus scirpaceus, A. palustris*): ein Vergleich. *Ökol. Vögel* 12: 1-11.

Schulze-Hagen, K., Flinks, H. & Dyrcz, A. 1989. Brutzeitliche Beutewahl beim *Acrocephalus paludicola*. *J. Ornithol.* 130: 251-255.

Schulze-Hagen, K., Leisler, B. & Winkler, H. 1992. Teich- und Sumpfrohrsänger in gemeinsamem Habitat: Brutzeit und Bruterfolg. *Vogelwelt* 113: 89-98.

Schulze-Hagen, K., Swatschek, I., Dyrcz, A. & Wink, M. 1993. Multiple Vaterschaften in Bruten des Seggenrohrsängers *Acrocephalus paludicola*: erste Ergebnisse des DNA-Fingerprintings. *J. Ornithol.* 134: 145-154.

Schulze-Hagen, K., Leisler, B., Birkhead, T.R. & Dyrcz, A. 1995. Prolonged copulation, sperm reserves and sperm competition in the aquatic warbler *Acrocephalus paludicola*. *Ibis* 137: 85-91.

Schulze-Hagen, K., Pleines, S. & Sennert, G. 1996a. Rasche Zunahme der Kuckucksparasitierung in einer lokalen Teichrohrsänger-Population. *Vogelwelt* 117: 83-86.

Schulze-Hagen, K., Leisler, B. & Winkler, H. 1996b. Breeding success and reproductive strategies of two Acrocephalus warblers. *J. Ornithol.* 137: 181-192.

Schulze-Hagen, K., Leisler, B., Schäfer, H.M. & Schmidt, V. 1999.

The breeding system of the Aquatic Warbler *Acrocephalus paludicola*: a review of new results. *Vogelwelt* 120: 87-96.

Schulze-Hagen, K., Stokke, B.G. & Birkhead, T.R. 2009. Reproductive biology of the European Cuckoo *Cuculus canorus*: early insights, persistent errors and the acquisition of knowledge. *J. Ornithol.* 150: 1-16.

Scott, S.N., Clegg, S.M., Blomberg, S.P., Kikkawa, J. & Owens, I.P.F. 2003. Morphological shifts in island-dwelling birds: the roles of generalist foraging and niche expansion. *Evolution* 57: 2147-2156.

Secondi, J., Bretagnolle, V., Compagnon, C. & Faivre, B. 2003. Species-specific song convergence in a moving hybrid zone between two passerines. *Biol. J. Linn. Soc.* 80: 507-517.

Secondi, J., Faivre, B. & Bensch, S. 2006. Spreading introgression in the wake of a moving contact zone. *Mol. Ecol.* 15: 2463-2475.

Sejberg, D., Bensch, S. & Hasselquist, D. 2000. Nestling provisioning in polygynous great reed warblers (*Acrocephalus arundinaceus*): do males bring larger prey to compensate for fewer nest visits? *Behav. Ecol. Sociobiol.* 47: 213-219.

Shirihai, H. 1996. *The birds of Israel*. London: Academic Press.

Shirihai, H., Roselaar, C.S., Helbig, A.J., Barthel, P.H. & van Loon, A.J. 1995. Identification and taxonomy of large *Acrocephalus* warblers. *Dutch Birding* 17: 229-239.

Shriver, W.G., Vickery, P.D., Hodgeman, T.P. & Gibbs, J.P. 2007. Flood tides affect breeding ecology of two sympatric sharp-tailed sparrows. *Auk* 124: 552-560.

Shurulinkov, P. & Chakarov, N. 2006. Prevalence of blood parasites in different local populations of reed warbler (*Acrocephalus scirpaceus*) and great reed warbler (*Acrocephalus arundinaceus*). *Parasitol. Res.* 99: 588-592.

Sibley, C.G. & Ahlquist, J.E. 1990. *Phylogeny and classification of birds*. New Haven & London: Yale University Press.

Sibley, C.G. & Monroe, B.L.J. 1990. *Distribution and taxonomy of birds of the world*. New Haven & London: Yale University Press.

Slabbekoorn, H., Ellers, J. & Smith, T.B. 2002. Birdsong and sound transmission: the benefits of reverberations. *Condor* 104: 564-573.

Slikas, B., Jones, I.B., Derrickson, S.R. & Fleischer, R.C. 2000. Phylogenetic relationships of Micronesian white-eyes based on mitochondrial sequence data. *Auk* 117: 355-365.

Smith, S.M. 1988. Extra-pair copulations in Black-capped Chickadees: the role of the female. *Behaviour* 107: 15-23.

Sol, D., Lefebvre, L. & Rodriguez-Teijeiro, J.D. 2005. Brain size, innovative propensity and migratory behaviour in temperate Palaearctic birds. *Proc. Roy. Soc. B-Biol. Sci.* 272: 1433-1441.

Soler, J.J., Vivaldi, M.M. & Møller, A.P. 2009. Geographic distribution of suitable hosts explains the evolution of specialized gentes in the European cuckoo *Cuculus canorus*. *BMC Evol. Biol.* 9: 88.

Sparks, T.H., Huber, K., Bland, R.L., Crick, H.Q.P., Croxton, P.J., Flood, J., Loxton, R.G., Mason, C.F., Newnham, J.A. & Tryjanowski, P. 2007. How consistent are trends in arrival (and departure) dates of migrant birds in the UK? *J. Ornithol.* 148: 503-511.

Spina, F. 1986. Ecological aspects of Reed Warbler migration in northern Italy. . In: Ouellet, H. ed. *Acta XIX Congressus Internationalis Ornithologici: Posters*. Ottawa, Ont: For National Museum of Natural Sciences by University of Ottawa Press.

Spottiswoode, C.N., Tottrup, A.P. & Coppack, T. 2006. Sexual selection

predicts advancement of avian spring migration in response to climate change. *Proc. Roy. Soc. B-Biol. Sci.* 273: 3023-3029.

Springer, H. 1960. Studien an Rohrsängern. *Anz. Ornithol .Ges. Bay.* 5:389-433.

Springer, M.S., Meredith, R.W., Janecka, J.E. & Murphy, W.J. (in press). The historical biogeography of mammalia. *Phil. Trans. Soc. B-Biol. Sci.*

Stapleton, M.K., Kleven, O., Lifjeld, J.T. & Robertson, R.J. 2007. Female tree swallows (*Tachycineta bicolor*) increase offspring heterozygosity through extrapair mating. *Behav. Ecol. Sociobio.* 61: 1725-1733.

Steadman, D.W. 2006. *Extinction and biogeography of tropical Pacific birds*. Chicago: University of Chicago Press.

Stoddard, M.C. & Stephens, M. 2010. Pattern mimicry of host eggs by the common cuckoo as seen through a bird's eye. *Proc. Roy. Soc. B-Biol. Sci.* 277: 1387–1393.

Stokke, B., Moksnes, A., Røskaft, E., Rudolfsen, G. & Honza, M. 1999. Rejection of artificial cuckoo (*Cuculus canorus*) eggs in relation to variation in egg appearance among reed warblers (*Acrocephalus scirpaceus*). *Proc. Roy. Soc. B-Biol. Sci.* 266: 1483-1488.

Stokke, B.G., Honza, M., Moksnes, A., Røskaft, E. & Rudolfsen, G. 2002. Costs associated with recognition and rejection of parasitic eggs in two European passerines. *Behaviour* 139: 629-644.

Stokke, B.G., Moksnes, A. & Røskaft, E. 2005. The enigma of imperfect adaptations in hosts of avian brood parasites. *Ornithol. Sci.* 4: 17-29.

Stokke, B.G., Hafstad, I., Rudolfsen, G., Bargain, B., Beier, J., Bigas Campàs, D., Dyrcz, A., Honza, M., Leisler, B., Pap, P.L., Patapavičius, R., Procházka , P., Schulze-Hagen, K., Thomas, R., Moksnes, A., Møller, A.P., Røskaft, E. & Soler, M. 2007a. Host density predicts presence of cuckoo parasitism in reed warblers. *Oikos* 116: 913-922.

Stokke, B.G., Takasu, F., Moksnes, A. & Røskaft, E. 2007b. The importance of clutch characteristics and learning for antiparasite adaptions in hosts of avian brood parasites. *Evolution* 61: 2212-2228.

Stokke, B.G., Hafstad, I., Rudolfsen, G., Moksnes, A., Møller, A.P., Røskaft, E. & Soler, M. 2008. Predictors of resistance to brood parasitism within and among reed warbler populations. *Behav. Ecol.* 19: 612-620.

Stolt, B.O. 1999. The Swedish reed warbler *Acrocephalus scirpaceus* population estimated by a capture-recapture technique. *Ornis Svecica* 9: 35-46.

Stresemann, E. & Arnold, J. 1949. Speciation in the group of great reed-warblers. *J. Bombay Nat. Hist. Soc.* 48: 428-443.

Stutchbury, B.J.M. & Morton, E.S. 2001. *Behavioral ecology of tropical birds*. San Diego: AcademicPress.

Sullivan, M.S. 1994. Mate choice as an information gathering process under time constraint: implications for behavior and signal-design. *Anim. Behav.* 47: 141-151.

Svensson, S.E. 1978. Territorial exclusion of *Acrocephalus schoenobaenus* by *A.scirpaceus* in reedbeds. *Oikos* 30: 467-474.

Svensson, L. 1992. *Identification guide to European passerines*. 4th ed. Stockholm: Svensson.

Svensson, L. 2001. Identification of Western and Eastern Olivaceous, Booted and Sykes´s Warblers. *Birding World* 14: 192-219.

Svensson, E. & Hedenström, A. 1999. A phylogenetic analysis of the evolution of moult strategies in Western Palearctic warblers (Aves: Sylviidae). *Biol. J. Linn. Soc.* 67: 263-276.

Svensson, L., Prys-Jones, R., Rasmussen, P.C. & Olsson, U. 2008.

Discovery of ten new specimens of large-billed reed warbler *Acrocephalus orinus*, and new insights into its distributional range. *J. Avian Biol.* **39**: 605-610.

Swilch, R., Mantovani, R., Spina, F. & Jenni, L. 2001. Nectar consumption of warblers after long-distance flights during spring migration. *Ibis* **143**: 24-32.

Takasu, F. & Moskát, C. 2011 Modeling the consequence of increased host tolerance toward avian brood parasitism. *Popul. Ecol.* **53**: 187-193.

Tanneberger, F. & Wichtmann, W. (eds) 2011. *Carbon credits from peatland rewetting: climate - biodiversity - land use.* (Science, policy, implementation and recommendations of a pilot project in Belarus). Stuttgart: E. Schweizerbart.

Tanneberger, F., Bellebaum, J., Fartmann, T., Haferland, H-J., Helmecke, A., Jehle, P., Just, P. & Sadlik, J. 2008. Rapid deterioration of Aquatic Warbler *Acrocephalus paludicola* habitats at the western margin of the breeding range. *J. Ornithol.* **149**: 105-115.

Tanneberger, F., Tegetmeyer, C., Dylawerski, M., Flade, M. & Joosten, H. 2009. Commercially cut reed as a new and sustainable habitat for the globally threatened Aquatic Warbler. *Biodivers. Conserv.* **18**: 1475-1489.

Tanneberger, F., Flade, M., Preiksa, Z. & Schroeder, B. 2010. Habitat selection of the globally threatened Aquatic Warbler *Acrocephalus paludicola* at the western margin of its breeding range and implications for management. *Ibis* **152**: 347-358.

Taylor, R. 1993. Habitat and feeding ecology of *Acrocephalus melanopogon* and the impact of recent fires and management practices at S'Albufera de Mallorca. [M.Sc. dissertation] University College London.

Tello, J.G., Moyle, R.G., Marchese, D.J. & Cracraft, J. 2009. Phylogeny and phylogenetic classification of the tyrant flycatchers, cotingas, manakins, and their allies (Aves: Tyrannides). *Cladistics* **25**: 429-467.

Teuschl, Y., Taborsky, B. & Taborsky, M. 1998. How do cuckoos find their hosts? The role of habitat imprinting. *Anim. Behav.* **56**: 1425-1433.

Thaxter, C.B., Redfern, C.P.F. & Bevan, R.M. 2006. Survival rates of adult Reed Warblers (*Acrocephalus scirpaceus*) at a northern and southern site in England. *Ring. Migr.* **23**: 65-79.

Thévenot, M., Vernon, R. & Bergier, P. 2003. *The birds of Morocco.* BOU Checklist No. 20. Tring: Natural History Museum.

Thibault, J.C. & Cibois, A. 2006. The natural history and conservation of *Acrocephalus rimitarae*, the endemic reed-warbler of Rimatara Island, Oceania. *Bull. Brit.Ornithol.Club.* **126**: 201-207.

Thibault, J.C. & Guyot, I. 1987. Recent changes in the avifauna of Makatea Island (Tuamotus, Central Pacific). *Atoll Res. Bull.* **300**: 1-14.

Thibault, J.C., Martin, J.L., Penloup, A. & Meyer, J.Y. 2002. Understanding the decline and extinction of monarchs (Aves) in Polynesian Islands. *Biol. Conserv.* **108**: 161-174.

Thielcke, G. 1973. On the origin of divergence of learned signals (songs) in isolated populations. *Ibis* **115**: 511-516.

Thorne, C.J.R. 1975. Wing length of Reed Warbelers. *Wicken Fen Group Rep.* **7**: 10-13.

Thorup, K. 2006. Does the migration programme constrain dispersal and range sizes of migratory birds? *J. Biogeogr.* **33**: 1166-1171.

Timmins, R.J., Mostafawafi, N., Rajabi, A.M., Noori, H., Ostrowski, S., Olsson, U., Svensson, L. & Poole, C.M. 2009. The discovery of large-billed reed warblers *Acrocephalus orinus* in north-eastern Afghanistan. *Birding Asia* **12**: 42-45.

Timmins, R.J., Ostrowski, S., Mostafawi, N., Noori, H., Rajabi, A.M., Svensson, L., Olsson, U. & Poole, C.M. 2010. New information on the Large-billed Reed Warbler *Acrocephalus orinus*, including its song and breeding habitat in north-eastern Afghanistan. *Forktail* **26**: 9-23.

Tinbergen, N. 1963. On aims and methods in ethology. *Z. Tierpsychol.* **20**: 410-433.

Tökölyi, J. & Barta, Z. 2011. Breeding phenology determines evolutionary transitions in migratory behaviour in finches and allies. *Oikos* **120**: 184-193.

Tokue, K. & Ueda, K. 2010. Mangrove Geryones *Geryone laevigaster* eject Little Bronze-cuckoo *Chalcites minutillus* hatchlings from parasitized nests. *Ibis* **152**: 835-839.

Trivers, R.L. 1972. Parental investment and sexual selection. In: Campbell, B. ed. *Sexual selection and the descent of man: 1871–1971.* Chicago: Aldine, pp. 136-179.

Trivers, R.L. & Willard, D.E. 1973. Natural selection of parental ability to vary the sex ratio of offspring. *Science* **179**: 90-92.

Trnka, A. & Prokop, P. 2010. Does social mating system influence nest defence behaviour in Great Reed Warbler (*Acrocephalus arundinaceus*) males? *Ethology* **116**: 1075-1083.

Trnka, A. & Prokop, P. 2011. The use and function of snake skins in the nests of Great Reed Warblers *Acrocephalus arundinaceus*. *Ibis* **153**: 627-630.

Trnka, A., Szinai, P. & Hosek, V. 2006. Daytime activity of reed passerine birds based on mist-netting. *Acta Zool. Acad. Sci. Hung.* **52**: 417-425.

Trnka, A., Prokop, P. & Batáry, P. 2010. Infanticide or interference: does the great reed warbler selectively destroy eggs? *Ann. Zool. Fenn.* **47**: 272-277.

Urano, E. 1990. Factors affecting the cost of polygynous breeding for female Great Reed Warblers *Acrocephalus arundinaceus*. *Ibis* **132**: 584-594.

Urano, E. 1992. Early settling the following spring: a long-term benefit of mate desertion by male Great Reed Warblers (*Acrocephalus arundinaceus*). *Ibis* **134**: 83-86.

Urano, E., Ezaki, Y. & Yamagishi, S. 1995. The ecology and mating system of the Great Reed Warbler: an inhabitant of reed marshes. *Jpn. J. Ornithol.* **44**: not paginated.

Urban, E.K., Fry, C.H. & Keith, S. (eds) 1997. *The birds of Africa.* Vol. V: *Thrushes to Puffback Flycatchers.* London: Academic Press.

Vadász, C., Németh, A., Králl, A., Biró, C. & Csörgő, T. 2007. The seasonal movements of the Moustached Warbler population breeding in the Carpathian Basin. In: Hengsberger, R., Hofmann, H., Wagner, R. & Winkler, H. eds. *Abstracts of the 6th European Ornithologists' Union Conference, 24-29 August 2007, Vienna, Austria*. Vienna: Austrian Academy of Sciences, p. 47.

van de Crommenacker, J., Komdeur, J., Burke, T. & Richardson, D.S. 2011. Spatio-temporal variation in territory quality and oxidative status: a natural experiment in the Seychelles warbler (*Acrocephalus sechellensis*). *J. Anim. Ecol.* 80: 668-680.

van der Hut, R.M.G. 1986. Habitat choice and temporal differentiation in reed passerines of a Dutch marsh. *Ardea* 74: 159-176.

van der Putten, W.H. 1997. Die-back of *Phragmites australis* in European wetlands: an overview of the European Research Programme on Reed Die-Back and Progression (1993-1994). *Aquat. Bot.* 59: 263-275.

van Dobben, W.H. 1949. Nestbuilding technique of Icterine Warbler and Chaffinch. *Ardea* 37: 89-97.

van Eerde, K.A.1999. Hybrid Sedge X European Reed Warbler at Makkum in August 1997. *Dutch Birding* 21: 35-37.

Vergeichik, L. & Kozulin, A. 2006a. Changing nesting dates and nest placement as adaptations of Aquatic Warbler *Acrocephalus paludicola* to unstable nesting conditions on fen mires in Belarus. *Vogelwelt* 127: 145-155.

Vergeichik, L. & Kozulin, A. 2006b. Breeding ecology of Aquatic Warblers *Acrocephalus paludicola* in their key habitats in SW Belarus. *Acta Ornithol.* 41: 153-161.

Verner, J. 1971. Survival and dispersal of male long-billed marsh wrens. *Bird Banding* 42: 92-98.

Verner, J. & Willson, M.F. 1966. The influence of habitats on mating systems of North American passerine birds. *Ecology* 47: 143-147.

Voelker, G. 1999. Dispersal, vicariance, and clocks: Historical biogeography and speciation in a cosmopolitan passerine genus (*Anthus*: Motacillidae). *Evolution* 53: 1536-1552.

Voous, K.H. 1977. *List of recent Holarctic bird species*. London: British Ornithologists' Union/Academic Press.

Wagner, R.H. 1998. Hidden leks: mechanisms of sexual selection and colony formation . In Adams, N.J. & Slotow, R.H. eds. *Abstracts of the 22nd International Ornithological Congress, Durban, South Africa, 1998*. Johannesburg: BirdLife South Africa, p. 67.

Wallschläger, D. 1980. Correlation of song frequency and body weight in passerine birds. *Experientia* 36: 412.

Wallschläger, D. 1985. Der Einfluß struktureller und abiotischer ökologischer Faktoren auf den Reviergesang von Passeriformes. *Mitt. Zool. Mus. Berl.* 61: 39-69.

Walther, B.A., Wisz, M.S. & Rahbek, C. 2004. Known and predicted African winter distributions and habitat use of the endangered Basra reed warbler (*Acrocephalus griseldis*) and the near-threatened cinereous bunting (*Emberiza cineracea*). *J. Ornithol.* 145: 287-299.

Ward, S., Lampe, H.M. & Slater, P.J.B. 2004. Singing is not energetically demanding for pied flycatchers, *Ficedula hypoleuca*. *Behav. Ecol.* 15: 477-484.

Wawrzyniak, H. & Sohns, G. 1977. *Der Seggenrohrsänger, Acrocephalus paludicola*. Wittenberg Lutherstadt: A. Ziemsen.

Weatherhead, P.J. & Forbes, M.R.L. 1994. Natal philopatry and the costs of dispersal in birds. *Behav. Ecol.* 5: 426-433.

Weber, T.P., Borgudd, J., Hedenström, A., Persson, K. & Sandberg, G. 2005. Resistance of flight feathers to mechanical fatigue covaries with moult strategy in two warbler species. *Biol. Lett.* 1: 27-30.

Webster, M.S., Marra, P.P., Haig, S.M., Bensch, S. & Holmes, R.T. 2002. Links between worlds: unraveling migratory connectivity. *Trends Ecol. Evol.* 17: 76-83.

Wegrzyn, E. & Leniowski, K. 2010. Syllable sharing and changes in syllable repertoire size and composition within and between years in the great reed warbler, *Acrocephalus arundinaceus*. *J. Ornithol.* 151: 255-267.

Wegrzyn, E., Leniowski, K. & Osiejuk, T.S. 2010. Whistle duration and consistency reflect philopatry and harem size in great reed warblers. *Anim.Behav.* 79: 1363-1372.

Weir, J.T. & Schluter, D. 2008. Calibrating the avian molecular clock. *Mol. Ecol.* 17: 2321-2328.

Welbergen, J.A. & Davies, N.B. 2009. Strategic variation in mobbing as a front line of defense against brood parasitism. *Curr. Biol.* 19: 235-240.

Welbergen, J.A. & Davies, N.B. 2011. A parasite in wolf's clothing: hawk mimicry reduces mobbing of cuckoos by hosts. *Behav. Ecol.* 22: 574-579.

Westerdahl, H. 2004. No evidence of an MHC-based female mating preference in great reed warblers. *Mol. Ecol.* 13: 2465-2470.

Westerdahl, H., Bensch, S., Hansson, B., Hasselquist, D. & von Schantz, T. 2000. Brood sex ratios, female harem status and resources for nestling provisioning in the great reed warbler (*Acrocephalus arundinaceus*). *Behav. Ecol. Sociobiol.* 47: 312-318.

Westerdahl, H., Hansson, B., Bensch, S. & Hasselquist, D. 2004. Between-year variation of MCH allele frequencies in great reed warblers: selection or drift? *J. Evol. Biol.* 17: 485-492.

Westerdahl, H., Waldenström, J., Hansson, B., Hasselquist, D., von Schantz, T. & Bensch, S. 2005. Associations between malaria and MHC genes in a migratory songbird. *Proc. Roy. Soc. B-Biol. Sci.* 272: 1511-1518.

Westneat, D.F. & Stewart, R.K. 2003. Extra-pair paternity in birds: causes, correlates, and conflict. *Annu. Rev. Ecol. Evol. Syst.* 34: 365-396.

Wetmore, A. 1924. A warbler from Nihoa. *Condor* 26: 177-78

Wetmore, A. 1926: Observations on the birds of Argentina, Paraguay, Uruguay and Chile. *Bull. U.S. Nat. Mus.* 133: 1-448.

Wikelski, M., Tarlow, E.M., Raim, A., Diehl, R.H., Larkin, R.P. & Visser, G.H. 2003. Costs of migration in free-flying songbirds. *Nature* 423: 704-704.

Wiley, R.H. 1991. Associations of song properties with habitats for territorial Oscine birds of eastern North America. *Am. Nat.* 138: 973-993.

Wiley, R.H. 2000. A new sense of the complexities of bird song. *Auk* 117: 861-868.

Wiley, R.H. & Richards, D.G. 1982. Adaptations for acoustic communication in birds: sound transmission and signal detection. In: Kroodsma, D. & Miller, K. eds. *Acoustic communication in birds*. New York: Academic Press, pp. 131-181.

Williamson, K. 1968. *Identification for ringers: the genera Cettia, Locustella, Acrocephalus and Hippolais*. BTO field guide no.7; BTO identification guide no. 1. Tring: British Trust for Ornithology.

Williamson, M. 1981. *Island populations.* Oxford: Oxford University Press.

Willoughby, E.J. 1991. Molt of the genus Spizella (Passeriformes Emberizidae) in relation to ecological factors affecting plumage wear. *Proc. West. Found. Vertebr.Zool.* **4** :247-286.

Willson, M.F. & Armesto, J.J. 2006. Is natural history really dead? Toward the rebirth of natural history. *Rev. Chil. Hist. Nat.* **79**: 279-283.

Wilson, J.M. & Cresswell, W. 2010. Densities of Palearctic warblers and Afrotropical species within the same guild in Sahelian West Africa. *Ostrich* **81**: 225-232.

Winkler, H. 1988. An examination of concepts and methods in ecomorphology. In: Ouellet, H. ed. *Acta XIX Congressus Internationalis Ornithologici.* Ottawa, Ont: For National Museum of Natural Sciences by University of Ottawa Press, pp. 2246-2253.

Winkler, H. 2001. The ecology of avian acoustical signals. In: Barth, F.G. & Schmid, A. eds. *Ecology of sensing*: Berlin; NY: Springer, pp. 79-104.

Winkler, H. 2010: The comparative study of bird brains. *Abstracts of the 25th International Ornithological Congress, 22-28 August 2010, Campos do Jordão, SP, Brazil*, p. 320.

Winkler, H. & Leisler, B. 1992. On the ecomorphology of migrants. *Ibis* **134** (suppl.): 21-28.

Winkler, H. & Leisler, B. 1999. Exploration and curiosity in birds: functions and mechanisms. In: Adams, N.J. & Slotow, R.H. eds. *Proceedings of the 22nd International Ornithological Congress, Durban, South Africa, 1998.* Johannesburg: BirdLife South Africa, pp. 915-932.

Winkler, H. & Leisler, B. 2005. To be a migrant: ecomorphological burdens and chances. In: Greenberg, R. & Marra, P.P. eds. *Birds of two worlds: the ecology and evolution of migration.* Baltimore and London: Johns Hopkins University Press, pp. 79-86.

Winkler, H. & Leisler, B. 2006. Evolution of morphology and behavior in island passerines. Poster presented at the 24th International Ornithological Congress, 13-19 August 2006, Hamburg, Germany. *J. Ornithol.* **147** (Suppl.1): 273.

Winkler, H., Leisler, B. & Bernroider, G. 2004. Ecological constraints on the evolution of avian brains. *J. Ornithol.* **145**: 238-244.

Wirtz, P. 1999. Mother species-father species: unidirectional hybridization in animals with female choice. *Anim. Behav.* **58**: 1-12.

Wolfson, A. 1954. Sperm storage at lower-than-body temperature outside the body cavity in some passerine birds. *Science* **120**: 68-71.

Wolters, H.E. 1982. *Die Vogelarten der Erde: eine systematische Liste mit Verbreitungsangaben sowie deutschen und englischen Namen.* Hamburg; Berlin: Parey.

Wüst-Graf, R. 1992. Auswirkungen von Biotoppflegemassnahmen auf den Brutbestand des Teichrohrsangers *Acrocephalus scirpaceus* am Mauensee. *Ornithol. Beobacht.* **89**: 267-271.

Yohannes, E., Hobson, K.A. & Pearson, D.J. 2007. Feather stable-isotope profiles reveal stopover habitat selection and site fidelity in nine migratory species moving through sub-Saharan Africa. *J. Avian Biol.* **38**: 347-355.

Yohannes, E., Bensch, S. & Lee, R. 2008a. Philopatry of winter moult area in migratory Great Reed Warblers *Acrocephalus arundinaceus* demonstrated by stable isotope profiles. *J. Ornithol.* **149**: 261-265.

Yohannes, E., Hansson, B., Lee, R.W., Waldenström, J., Westerdahl, H., Akesson, M., Hasselquist, D. & Bensch, S. 2008b. Isotope signatures in winter moulted feathers predict malaria prevalence in a breeding avian host. *Oecologia* **158**: 299-306.

Yohannes, E., Biebach, H., Nikolaus, G. & Pearson, D.J. 2009a. Migration speeds among eleven species of long-distance migrating passerines across Europe, the desert and eastern Africa. *J. Avian Biol.* **40**: 126-134.

Yohannes, E., Biebach, H., Nikolaus, G. & Pearson, D.J. 2009b. Passerine migration strategies and body mass variation along geographic sectors across East Africa, the Middle East and the Arabian Peninsula. *J. Ornithol.* **150**: 369-381.

Yosef, R. & Chernetsov, N. 2005. Longer is fatter: body mass changes of migrant Reed Warblers (*Acrocephalus scirpaceus*) staging at Eilat, Israel. *Ostrich* **76**: 142-147.

Zajac, T. & Solarz, W. 2004. Low incidence of polygyny revealed in a long term study of the Sedge Warbler *Acrocephalus schoenobaenus* in natural wetlands of the S Poland. *Acta Ornithol.* **39**: 83-86.

Zajac, T., Solarz, W. & Bielanski, W. 2006. Adaptive settlement in sedge warblers *Acrocephalus schoenobaenus*: focus on the scale of individuals. *Acta Oecol.* **29**: 123-134.

Zajac, T., Bielanski, W. & Solarz, W. 2008. On the song resumption, polyterritorial behaviour and their population context in the Sedge Warbler *Acrocephalus schoenobaenus*. *J.Ornithol.* **149**: 49-57.

Zehtindjiev, P., Ilieva, M. & Akesson, S. 2010. Autumn orientation behaviour of paddyfield warblers, *Acrocephalus agricola*, from a recently expanded breeding range on the western Black Sea coast. *Behav. Process.* **85**: 167-171.

Zehtindjiev, P., Ilieva, M., Hansson, B., Oparina, O., Oparin, M. & Bensch, S. 2011. Population genetic structure in the paddyfield warbler (*Acrocephalus agricola* Jerd.). *Curr. Zool.* **57**: 63-71.

Zwarts, L., Bijlsma, R.G., van der Kamp, J. & Wymenga, E. 2009. *Living on the edge: wetlands and birds in a changing Sahel.* Zeist, The Netherlands: KNNV Publishing.

Zwicker, E. & Grüll, A. 1984. Zu den räumlich-zeitlichen Beziehungen zwischen Schilfvögeln und ihrem Lebensraum. *Wiss. Arb. Burgenland* **72**: 413-445.

Index

territoriality 105-6, 109, **114**
testis length 199
winter site fidelity 219
evolutionary adaptations 90-4, 99-102
    clutch size 150
    convergence 268, 271, 274-6
    dispersal 217
    egg morphs 167
    host/parasite 'arms race' 162, 167, 176
    intermediate 47, 278
    island species 234, 237
    mating systems **199-200**, 279
    migration 89, 94, **96-9**, 204
    moult 223-6
    nest building 144
    song 136
    systematics 29, 42-3
    see also phylogeny; sexual selection
exploration 93, **217-10**, 235, 239
extinction 230-3, 236, 241-3, 256
extra-pair paternity 107, 111, 131-2, 184-97

F farming 242, 248-63
fat accumulation **205-10**, 215-17, 220-1, 226
feathers
    isotope analysis 215-16, 220
    see also flight; moult; plumage
feeding techniques 73-4, 81-4
    Cuckoo chicks 169
    island birds 237-40, 244
    migration 205
    see also diet and foraging
fitness factors 12
    climate change 260
    competition 109-10, 115-16
    cooperative breeding 187-93
    migration 220
    polygyny 194
    reproduction 106, 148, **156-8**, 183
    territoriality 109-10, 115-16
flexibility 246-65
    brood parasitism 173-6
    foraging skills 84, 87
    generalism 248
    habitat 52
    mating systems 279
    migration 136
    mobility 248
    moult 221, 225-7
    see also climate change
flight
    apparatus 90, 95-9, 237-8, 271
    exploratory 217-18
    feathers 96, 203
    feeding techniques 83
    island species 235, 238, 244
    migratory 96-100, 204-5, 209-10, 220
    moult 221, 225
    nocturnal 217-18

song 133, 185
    see also tail/s; wing/s
food supply 75-8
    abundance 77-9, 258-60, 274-7
    adult *vs* nestling 78
    aerial 73, 238
    and breeding, *see* breeding and food
    climate change 257-8
    fruit 74, 207, 238
    and habitat 63-4, **74-8, 84-6**, 260
    islands 234-5, 238, 241
    migration **204-10**, 215-17, 220-1, 226
    moult 223-4, 226
    nectar 74, 87, 238
    plants 74
    prey mobility 55, 205-6
    prey size **79-82**, 195, 237-40
    prey spectrum **78-80**, 205
    regional variation 75
    renewable 63
    selection 78-9
    song 136
    territoriality 106-7, 114
    see also diet and foraging; fat accumulation; arthropods
foot morphology 28, 35, 92-6, 271-2
    see also claws; toes
foraging, *see* diet and foraging
forests and woodlands 17, 63, **66-7**
    climate change 258-9
    islands 233, 243
    threats to 247
frugivory 74, 207, 238
fuelling, *see* fat accumulation

G gender
    battle of the sexes 178-201
    body size 89
    cooperative breeding 189-91
    determination of 192
    dispersal 218-19
    incubation 166, 181-2
    leucism 241
    signals 278
    see also mate guarding; maternal care; pair bonds; paternal care
genetic factors 12, 28
    climbing ability 92
    convergence 270
    cooperative breeding 187, 191
    dispersal 218-19
    diversity loss 263-5
    gene flow 44-7, 167, **173**, 218
    genetic distances 43-6
    heterozygosity 158, 185, 191
    island colonization 230
    mating systems **183-5**, 197, 200
    MHC complex 185, 191, 197, 220
    migration 215, 220
    moult 227
    population/s 46, 217, 255, **263-5**

Colophon

Authors: Bernd Leisler, Karl Schulze-Hagen

Translation: Brian Hillcoat

Edited by: Marie-Anne Martin

Artwork: David Quinn

Graphic lay out and design: x-hoogte - Tilburg

Printer: DZS Grafik

The appendix of the Reedwarblers is available on www.knnvuitgeverij.nl/EN/appendix-the-reedwarblers

Publication of KNNV Publishing

© KNNV Publishing – Zeist – the Netherlands 2011

ISBN: 978 90 5011 3915

NUR: 942

www.knnvpublishing.nl

This book is published in cooperation with the Max Planck Institute for Ornithology.

KNNV Publishing

Discovering nature and getting close to it

KNNV Publishing specialises in unique works on nature and landscape: easily accessible field guides, conservation manuals, distribution atlases and much more. KNNV Publishing also produces chronicles on nature conservation, beautifully illustrated books on natural, cultural and landscape history, travel guides, children's books and, last but not least, the journal 'Natura'. All these works help to make valuable knowledge gathered by scientists and amateur researchers available to a broad public. By producing these works, KNNV Publishing contributes to nature conservation and to the enjoyment of nature in the Netherlands.